A Field Theory Approach
to Photonics

Online at: https://doi.org/10.1088/978-0-7503-5789-0

A Field Theory Approach to Photonics

Marco Ornigotti

Faculty of Engineering and Natural Sciences, Physics Unit, Photonics Laboratory, Tampere University, Korkeakoulunkatu 7, 33720 Tampere, Finland

IOP Publishing, Bristol, UK

ISBN   978-0-7503-5789-0 (ebook)
ISBN   978-0-7503-5787-6 (print)
ISBN   978-0-7503-5790-6 (myPrint)
ISBN   978-0-7503-5788-3 (mobi)

DOI   10.1088/978-0-7503-5789-0

Version: 20250701

IOP ebooks

British Library Cataloguing-in-Publication Data: A catalogue record for this book is available from the British Library.

Published by IOP Publishing, wholly owned by The Institute of Physics, London

IOP Publishing, No.2 The Distillery, Glassfields, Avon Street, Bristol, BS2 0GR, UK

US Office: IOP Publishing, Inc., 190 North Independence Mall West, Suite 601, Philadelphia, PA 19106, USA

To Stephanie, whose constant support and encouragement are what made this book become reality.

Contents

Preface

Photonics is a very rich and thriving field of research, with virtually unlimited potential for disruptive applications in several technological sectors, as well as a fascinating connection of fundamental aspects and effects, linking this discipline to all other fields of physics.

The invention of lasers in 1960 by Theodore Maiman marked an essential turning point for the use of light-based technologies for 'practical' applications. Nowadays, in fact, it is difficult to think about a specific technology that doesn't involve, directly or indirectly, photonics. Barcode readers, laser pointers, smartphones, smart watches, tablets, e-readers, LED and OLED TVs, biometric sensors in fitness devices, optical fibres, CD- and DVD-ROMs, laser surgery, microscopy and diagnostics, LIDAR, Wi-Fi, smart light bulbs. These are only a few of the examples of how pervasive light-based technologies are in our present day, and how easy is to enter into contact with them in our everyday life, without even realising it.

Because of this tremendous success, one could say that photonics is to the 21st century what electronics was for the 20th century, and define photonics as the next step in the development of our technology, with nearly infinite potential for new and disruptive applications, including quantum information and quantum computing, which promise to revolutionise the way we look at and think about computers.

A more general definition of photonics can be found on Wikipedia, and reads 'a branch of optics that involves the application of generation, detection, and manipulation of light in the form of photons through emission, transmission, modulation, signal processing, switching, amplification and sensing'[1].

All these definitions of photonics highlight the tremendous impact this discipline has had in our society, and highlight too the bright future it will unlock, if we invest in the promise of its technological advance and progress. They do not, however, fully highlight the wonders the theoretical foundations of photonics are capable of generating, not only in terms of disruptive applications, but, more importantly, in terms of mathematical elegance. At its very foundation, photonics is deeply connected with all the other fields of physics, sometimes in a visible and explicit manner, some other times in a more cryptic and esoteric fashion, drawing, to quote H P Lovecraft, 'lines and curves that could be made to point out directions leading through the walls of space to other spaces beyond'[2].

In the last two to three decades, many different areas of physics have been connected with photonics: Hawking radiation and cosmology with light propagating through nonlinear media, thermodynamics and black holes with optical fibres, curved spacetime and gravity with transformation optics and invisibility cloaks, waveguide optics and quantum field theory, supersymmetry and optical design, quantum weak measurements and optical beam shifts, differential geometry and

---

[1] https://en.wikipedia.org/wiki/Photonics.
[2] H P Lovecraft, *Dream in the Witch House*, in *Necronomicon: the Best Weird Tales of H P Lovecraft*, (Gollancz, London, 2008).

light's polarisation, phase transitions and non-Hermitian photonics. These are only few of the many examples of the extent to which the roots of photonics are touching other areas of physics.

This deep interconnection between photonics and the rest of physics is not only elegant and astonishing mathematical trivia. It is also an opportunity to use photonics as a platform to investigate in a simple and reliable manner the validity and extent of different physical models, enabling cross-fertilisation between disciplines, which ultimately could lead to new ways to think about photonics and its applications and, ultimately, to new innovations. This, to a certain extent, is the conceptual foundation of this book, and its general mindset.

This book, however, does not pretend to be a comprehensive guide to navigate all the possible connections between photonics and the rest of science. This is simply a task too big (and daunting) for a single person to accomplish, and, ultimately, of little use, since reading such a book would ultimately require a deep knowledge in all the aspects of physics.

Instead, this book focuses on a specific connection, one that is more naturally embedded into the theory of electromagnetism, at the foundations of photonics. A connection that gives the possibility to reinterpret the general framework of photonics from a more general perspective, unlocking methods and tools from different areas of physics to tackle complicated problems in an easier manner, with vewpoint of regard for those tools that allows analytical or semi-analytical treatment, rather than a full brute-force numerical approach. This framework is that of quantum field theory (QFT) and Feynman path integrals.

By exploring how several problems in photonics can be framed within the language of QFT and path integrals, this book aims at providing the reader with a *new set of old tools*, which are traditionally not part of the standard toolbox for researchers working in the field of theoretical optics and photonics. This will hopefully provide them with a fresh perspective on the theory of electromagnetism and its interaction with matter, both at the linear and nonlinear level.

This book is essentially divided into three parts: part I comprises chapters 2–4, and deals with the classical theory of electromagnetism. Chapter 2 briefly reviews familiar concepts of electromagnetism from the standard Maxwell's equation perspective, derives the wave and paraxial equation, and introduces the basic concepts of light–matter interaction. This chapter also introduces the concept of Green's function (providing a simple calculation for both the Helmholtz and paraxial equation), which will be an important player in the rest of the book. Chapter 3 gives instead a brief summary of the main results of field theory, including the Lagrangian and Hamiltonian description of a physical system, the principle of least action, and the relation between symmetries and conserved quantities of a physical system through Noether's theorem. This will serve as background for the rest of the book. Chapter 4 then applies the concepts and methods of chapter 3 to the electromagnetic field and presents its standard field theory formulation. In particular, this chapter makes use of Noether's theorem to classify the conserved quantities of the field, with particular emphasis on spin and orbital angular momentum of light, as well as optical helicity. Chapter 4 then concludes with a brief mention on how to understand transformation optics in terms of electromagnetic field in a curved spacetime background.

Part II includes chapters 5 and 6, and delves into the quantum theory of light. Chapter 5 presents the essential tools to describe, both from a Hamiltonian (i.e., canonical) and Lagrangian (i.e., path integral) perspective, the properties of a quantum field, using the simple example of a real, scalar field. In particular, both the case of a free and interacting field are discussed, introducing important concepts like Fock space, Wick's theorem and Feynman diagrams. Chapter 6 discusses the quantisation of the electromagnetic field, and it is divided into three parts: in part I, the canonical quantisation of the free electromagnetic field (in the usual box quantisation framework, familiar to quantum optics) is presented and used to derive some of the properties of the basic quantum states of the electromagnetic field, i.e., number and coherent states. Then, the quantisation framework is extended to optical beam and the quantisation in terms of optical modes is briefly discussed. Part II applies the results of chapter 5 to an electromagnetic field interacting in a nonlinear manner with matter, and presents second-order nonlinear processes as a case study to explore the formalism of interacting quantum fields applied to nonlinear optics. Finally, part III presents the path integral quantisation of the electromagnetic field through the Fadeev–Popov quantisation scheme and derives the form of the electromagnetic partition function.

Part III consists of chapters 7 and 8, and presents an application of the path integrals framework to solve different classes of problems in photonics. Chapter 7 presents three examples of this. First, nonrelativistic path integrals, á la Feynman–Hibbs, are used to describe the propagation of optical rays in an inhomogeneous medium. Then, the same framework is applied to nonlinear optics, and the dynamics of an electromagnetic field in a parametric amplifier is taken as case study to showcase how path integrals in phase space can be a useful tool for solving nonlinear optics problems. The third example, covering the whole second part of this chapter, presents the case of a dispersive, lossy medium and shows how to calculate the linear and nonlinear optical properties of such a medium starting from a path integral formulation of the light–matter interaction problem. Moreover, the connection between path integral and canonical quantisation is discussed, taking the case of a dispersive, lossy material as an example to derive the usual Vogel–Welsch quantisation scheme. Finally, chapter 8 presents a rather recent framework, that of pseudo quantum electrodynamics, used to solve light–matter interaction problems involving 2D materials. Here, graphene is taken as a case study and, as an example of application of this framework, the calculation of the DC conductivity of graphene is presented in detail.

This book assumes that the reader is familiar with standard electrodynamics and basic quantum mechanics. Complex analysis would also be beneficial, as it is implicitly used throughout the book. Chapters 3 and 5 provide the necessary background in QFT and path integrals as tools that are then applied to the electromagnetic field to make the reader familiar with them. For this reason, none of the formal aspects of these theories are discussed in detail here, and we refer the interested reader to more specific books on the topic, in case they are interested.

All the nontrivial integral formulas and relations are presented and derived either in the main text, or in the appendices. Additionally, most of the calculations presented in this book are derived explicitly step by step, or, if not so, the method to arrive at the final result is thoroughly described. Finally, at the beginning of each chapter, further literature on the subject of the chapter is given, to allow the reader to delve deeper into the topic of the chapter, if they so wish.

I hope you, dear reader, will enjoy reading this book as much as I did writing it, and will appreciate the beauty and elegance of QTF and path integrals applied to photonics as much as I do.

Marco Ornigotti
Tampere, February 2025.

Author biography

Marco Ornigotti

Marco Ornigotti is an Associate Professor for optics and photonics at Tampere University in Tampere, Finland. He received his MSc degree in photonics from Polytechnic Institute of Milan (Milan, Italy) in 2006, and a PhD in physics from the same institute in 2010, with a thesis on quantum optical analogies in waveguide systems. In 2017, he received the habilitation in theoretical physics from Friedrich Schiller University (Jena, Germany). He has spent several years in Germany as a postdoc, first at the Max Planck Institute for the Science of Light in Erlangen, then at Friedrich Schiller University in Jena and the University of Rostock. Since 2019 he has joined the Photonics Laboratory of Tampere University, where he now leads the Theoretical Optics and Photonics group.

He enjoys theoretical physics in all its aspects, with a soft spot for field theory and topology, and he is particularly interested in unravelling the hidden connections between photonics and other disciplines of physics. His research activity includes investigation of the fundamental aspects of light–matter interaction, both in the classical and quantum regime, and at the mesoscopic and nanoscale, structured light and its application, optical nonlinearities in 2D and epsilon-near-zero materials.

Notation and conventions

Throughout this book, if not specified otherwise, the following notations and conventions are implicitly adopted.

- **Natural units**: $c = \hbar = m_e = 1$.
- **Einstein summation convention**: we adopt the Einstein summation convention over repeated indices, i.e., $a_i b_i = \sum_i a_i b_i$. Greek indices will be used to indicate spacetime coordinates, i.e., $\mu, \nu, \sigma, \ldots = \{t, x, y, z\} \equiv \{0, 1, 2, 3\}$, while Latin indices only run over spatial indices, i.e., $i, j, k, \ldots = \{x, y, z\} \equiv \{1, 2, 3\}$.
- **Covariant and contravariant indices**: Covariant (x^μ) and contravariant (x_μ) indices are raised and lowered with the Minkowski metric $\eta_{\mu\nu}$, i.e., $x^\mu = \eta^{\mu\alpha} x_\alpha$ and $x_\mu = \eta_{\mu\alpha} x^\alpha$.
- **Minkowski metric**: For the Minkowski metric, we adopt the mostly positive signature convention, i.e.,

$$
\eta_{\mu\nu} = \begin{pmatrix} -1 & 0 & 0 & 0 \\ 0 & 1 & 0 & 0 \\ 0 & 0 & 1 & 0 \\ 0 & 0 & 0 & 1 \end{pmatrix};
$$

- **Derivatives**: we use the shorthand ∂_μ to indicate the derivative with respect to x^μ, i.e., $\partial_\mu = \partial/\partial x^\mu$.
- **Functionals**: when dealing with functionals, we drop their dependence from the fields, i.e., we write, for example, $\mathscr{L}$ instead of $\mathscr{L}[\psi, \phi]$. We will restore the full notation whenever the shorthand form is ambiguous and cannot be easily determined from context. Moreover, functional derivation is indicated with normal partial derivation instead of the usual symbol δ, i.e., $\partial\mathscr{L}/\partial\psi$ is to be understood as $\delta\mathscr{L}/\delta\psi$.
- **Multiple integration**: when dealing with multiple integrals, if not specified otherwise we adopt the following notation for the product of integration measures: $[dx]_n = dx_1 \, dx_2 \, \ldots \, dx_n$.
- **Functions of multiple variables**: when dealing with functions of multiple variables, if not specified otherwise we adopt the following notation to denote all the variables (or fields, in case of functionals) the function depends on: $f([x]_n) = f(x_1, x_2, \ldots, x_n)$.

IOP Publishing

A Field Theory Approach to Photonics

Marco Ornigotti

Chapter 1

Introduction

The word *photonics* is a crasis between 'photos' (the genitive case of the Greek word 'phos', meaning *light*) and 'electronics', and nicely reflects its original meaning of a research field aimed at using light to perform tasks that up to that moment in history were in the exclusive domain of electronics, such as communication, or information processing. Photonics has nowadays become an umbrella term for any kind of research that deals with investigating the propagation, interaction, and manipulation of light, and, by extension, the electromagnetic field.

Even though the term itself was introduced in the 1960s, as a consequence of the invention of lasers and the prominent rise of optical technology as the *new electronics*, with high potential for life-changing applications, the roots of photonics are to be found much deeper in the history of physics.

Although the history of optics can be tracked back as far as the Egyptians and the Mesopotamians (the first known civilisations to know how to make lenses), passing through the Greek ideas of emission and intromission theories tackled, amongst others, by Euclid (who basically laid the foundation for ray optics), and their resurgence during the middle ages thanks to muslim scientists, all the way up to the 17th century and Kepler, who expanded the old concepts from Greek optics including the study of curved mirrors, the principle of pinhole cameras, and the inverse square-law for the intensity of light [1], it is Sir Isaac Newton who is frequently considered the father of optics.

In his treaty *Opticks, or a treatise on Reflections, Refractions, Inflections, and Colours of Light*, published in 1704, Sir Isaac Newton in fact lays the modern foundation of modern optics, reporting and explaining several phenomena related to light, such as reflection, refraction, dispersion, and diffraction (which he calls *inflection*). His studies on diffraction were built upon the work of Francesco Maria Grimaldi, who was the first to observe the phenomenon of diffraction and name it so in 1660. In his work, in fact, we read

'It has illuminated for us another, fourth way, which we now make known and call 'diffraction' [i.e., shattering], because we sometimes observe light break up; that is, that parts of the compound [i.e., the beam of light], separated by division, advance farther through the medium but in different [directions], as we will soon show.' [2]

Newton's work on optics also comprised a corpuscular theory of light (predating quantum electrodynamics by about two centuries!), for which he famously introduced the *luminiferous aether* (literally, the light-bearing ether) as a means to describe wave effects as vibrations of a medium pervading everything

'Doth not this aethereal medium in passing out of water, glass, crystal, and other compact and dense bodies in empty spaces, grow denser and denser by degrees, and by that means refract the rays of light not in a point, but by bending them gradually in curve lines? ... Is not this medium much rarer within the dense bodies of the Sun, stars, planets and comets, than in the empty celestial space between them? And in passing from them to great distances, doth it not grow denser and denser perpetually, and thereby cause the gravity of those great bodies towards one another, and of their parts towards the bodies; every body endeavouring to go from the denser parts of the medium towards the rarer?' [3]

This inspired Maxwell to create a theory of electric and magnetic waves propagating through the luminiferous aether, which eventually led to what we nowadays know as Maxwell's equations, unifying the concepts of electricity, magnetism and light in a single phenomenon, i.e., the electromagnetic field [4]. The work of Maxwell has been so paramount for the development of modern physics and modern society, that Richard Feynman, in his *Feynman Lectures on Physics*, recognises the contribution of Maxwell to the history of humanity with the following words:

'From a long view of the history of mankind—seen from, say, ten thousand years from now—there can be little doubt that the most significant event of the 19th century will be judged as Maxwell's discovery of the laws of electrodynamics. The American Civil War will pale into provincial insignificance in comparison with this important scientific event of the same decade.' [5]

The 19th century is easily the century of electromagnetism, with the early studies of Ørsted and Ampère on magnetic field generated by electric currents in the early 1820s, and the unification of electricity and magnetism carried out by Maxwell and Faraday, later on refined at the mathematical level by Heaviside and Hertz (to which we owe the modern formulation of Maxwell's equations, for example), all the way up to the early days of special relativity, from the work of Heaviside in 1888, showing that the electric field generated by a charged sphere loses its spherical symmetry if the sphere is put in motion [6], passing through the work of Michelson and Morley, (which eventually contributed to disprove the concept of aether) [7], and Lorentz and Larmor trying to figure out under which transformations Maxwell's equation would remain invariant when transformed from the luminiferous aether to a moving frame (leading to the Lorentz transformations, appearing for the first time in 1892 [8]), and finally culminating in the special theory of relativity in 1905 by Einstein [9–11].

Contextually, the works of Young [12], Poisson [13], Arago [14], and Fresnel [15] (partially revamping the work of Huygens [16]) contributed to cement the wave nature of light, thus firmly rebutting Newton's idea of the corpuscular nature of light once and for all. But the 20th century, and the birth of quantum mechanics, had other plans for Newton's idea. Thanks to the work of the founding fathers of quantum theory [17], the wave–particle duality became a paradigm of the newly developed Quantum Physics, and quickly made its way to the electromagnetic field, where it reconciled Newton's particle and Maxwell's wave views on the subject, up until 1926, when Lewis created the word *photon* as a mean of uniquely addressing the new indivisible entity responsible of mediating energy exchange between atoms. In Lewis' own words:

> '*It would seem inappropriate to speak of one of these hypothetical entities as a particle of light, a corpuscle of light, a light quantum, or a light quanta, if we are to assume that it spends only a minute fraction of its existence as a carrier of radiant energy, while the rest of the time it remains as an important structural element within the atom. [...] I therefore take the liberty of proposing for this hypothetical new atom, which is not light but plays an essential part in every process of radiation, the name photon.*' [18]

Not many years later, the theory of the electromagnetic field was ready to be yet again the vessel for a new idea that would revolutionise physics. This time it was thanks to Hermann Weyl and his attempt, in 1918, at adapting the ideas of Einstein about the geometry of spacetime to the electromagnetic field, which pointed out how the electromagnetic field possesses an intrinsic *Eichinvarianz* (scale, or *gauge* invariance), and that this should appear as a local symmetry of general relativity [19]. In a second paper in 1929, Weyl then lays the foundations for modern gauge theory, using the electromagnetic field as prime example for this new framework [20]. Interestingly enough, this innate scale invariance of electromagnetism was also suggested by Maxwell as a curiosity deriving from the fact, that if Λ is a scalar function, and $\nabla \Lambda$ its gradient, so that $\nabla \times \nabla \Lambda = 0$, then the magnetic field **B** is not affected by the transformation $\mathbf{A} \rightarrow \mathbf{A} + \nabla \Lambda$ [21]. Adding gauge invariance to the picture allowed modern theoretical physics to rephrase the whole electromagnetic theory in terms of differential geometry, thus allowing us to understand the electromagnetic field as a connection of a $U(1)$-bundle over some manifold space [22], which is very far away, in terms of formalism, from the works of Newton and Maxwell, but it is still very faithful to them, conceptually.

This brief historical excursus, which by no means pretends to be complete, highlights how electromagnetism represented an important starting point for the development of different aspects of theoretical physics, from quantum mechanics to geometry, acting as a useful playground for cross-fertilisation of ideas between different areas of physics and mathematics. This is only corroborated if we look at more recent developments of the fields, where photonics has unravelled an astonishing connection with many other disciplines of physics, opening new intriguing channels, that can be exploited to find new ways to control and modify the properties of light for new applications and technologies, or create a reliable and easily accessible platform to test theories and conjectures from various fields of mathematics and physics, that would be difficult to observe directly in their environment of origin. Examples of these can be found in photonic crystals, borrowing from condensed matter physics [23], waveguides and their similarities with quantum mechanics [24], black hole physics and its connection with nonlinear optics [25–27], optical cloaking and invisibility and its connection with general relativity [28], supersymmetry as a guiding principle to design optical waveguides and fibres [29], topological photonics bringing topological physics into optics [30], non-Hermitian photonics [31], thermodynamics to explain light propagation in optical fibres [32], quantum gravity [33], and many more.

If photonics has been so successful and impactful in modern society and science, part of the reason is also because of the endless effort the scientific community, from the founding fathers to the present day, has put into creating a comprehensive, solid, and ever-giving framework, that not only resulted in transforming the ideas and concepts behind electromagnetism in life-changing applications, such as the invention of the electric bulb, the optical fibre, or any other modern photonics-based technology, but also contributed to create a thriving and multifaceted field of research, driven by technological progress on one side, and by the unique opportunity to explore all aspects of physics using the intuitive and, to certain extents, easily accessible theory of light to improve and push forward our understanding of the world.

The following chapters try to shed some light on a small corner of this vast landscape, by providing the reader with a set of tools and techniques (mainly those of quantum field theory and path integrals), that are traditionally part of the toolset of modern theoretical physics, but do not find a place in basic theoretical optics and photonics programmes. Rekindling the connection between photonics and field theory might give a new perspective on the field, allowing for the flow of new ideas and methods, that could provide a different way to tackle modern problems in photonics, which are becoming more and more complex and computationally demanding.

If the framework of field theory, the centrepiece of this book, provides on one side an elegant and simple environment to investigate the properties of light, both in vacuum and within the premises of light–matter interaction, it also creates a natural environment for more cross-fertilisation to happen, as it provides a unique insight into more modern mathematical physics topics, such as geometry, knot theory, and topology, some of which have already made their appearance in photonics in recent years (see, e.g., reference [34] for a primer on optical knots). Although these topics are not covered here, understanding how to rephrase problems in photonics in the jargon of field theory and path integrals is the first step towards a general theoretical framework for implementing and efficiently exploiting the geometrical and topo-logical properties of photonic systems.

References

[1] Ilardi V 2007 *Renaissance Vision from Spectacles to Telescopes* (Philadelphia, PA: American Philosophical Society)

[2] Grimaldi F M 1665 *Physico Mathesis de Lumine, Coloribus, et Iride, Aliisque Annexis Libri Duo* (Bologna: Vittorio Bonati)

[3] Newton I 2024 *Newton's Opticks: A Treatise on Reflections, Refractions, Inflections, and Colours of Light* (Bristol: Read and Co. Books)

[4] Maxwell J K 1954 *A Treatise on the Theory of Electricity and Magnetism* (New York: Dover)

[5] Feynman R P, Leighton R B and Sands M 2011 *The Feynman Lectures on Physics* (New York: Basic Books)

[6] Heaviside O 1888 The electromagnetic effects of a moving charge *Electrician* **22** 147

[7] Michelson A A and Morley E W 1887 On the relative motion of the Earth and the luminiferous aether *Am. J. Sci.* **34** 333

[8] Brown H R 2003 Michelson, Fitzgerald and Lorentz: the origins of relativity revisited. Available at https://philsci-archive.pitt.edu/987/

[9] Einstein A 1905 Zur elekrodynamik bewegter körper *Ann. Phys.* **17** 891

[10] Einstein A 1917 *Über die Spezielle und die Allgemeine RelativitÄtstheorie* (Braunschweig: F. Vieweg und Sohn) Available at https://echo-old.mpiwg-berlin.mpg.de/ECHOdocuView?url=/permanent/library/RB68AZVS/pageimg&pn=5&mode=imagepath

[11] Einstein A, (translator) and Lawson R W 1920 *Relativity: The Special and the General Theory: Popular Exposition* (Sheffield: University of Sheffield)

[12] Young T 1804 The Bakerian Lecture: experiments and calculations relative to physical optics *Phil. Trans. R. Soc. Lond.* **94** 1

[13] Born M and Wolf E 2020 *Principles of Optics* (Cambridge: Cambridge University Press)

[14] Brewster D 1831 *A Treatise on Optics* (London) (Longman, Rees, Orme, Brown and Green and John Taylor)

[15] Fresnel A J 1818 Mémoire sur la diffraction de la lumiére *Mém. Acad. R. Sci. Inst. France* **339** 1826

[16] Huygens C 1690 *Traité de la Lumiére* (Leiden: Van der Aa)

[17] Bacciagaluppi G 2013 *Quantum Theory at the Crossroads: Reconsidering The 1927 Solvay Conference* (Cambridge: Cambridge University Press)

[18] Lewis G N 1926 The conservation of photons *Nature* **118** 874

[19] Weyl H 1918 Gravitation un elektrizität *Siz. Kön. Preuss. Akad. Wiss.* 465

[20] O'Raifeartaigh L and Straumann N 2000 Gauge theory: historical origins and some modern developments *Rev. Mod. Phys.* **72** 1

[21] Maxwell J K 1865 A dynamical theory of the electromagnetic field *Phil. Trans. R. Soc. Lond.* **155** 459

[22] Nakahara M 2003 *Geometry, Topology, and Physics* (Boca Raton, FL: CRC Press)

[23] Ilardi V 2007 *Renaissance Vision from Spectacles to Telescopes* (Philadelphia, PA: American Philosophical Society)

[24] Longhi S 2009 Quantum optical analogies using photonic structures *Laser Photon. Rev.* **3** 243

[25] Philbin T G, Kuklewicz C, Robertson S, Hill S, Konig F and Leonhardt U 2007 Fiber-optical analogue of the event horizon *Science* **319** 1367

[26] Belgiorno F, Cacciatori S L and Dalla Piazza F 2015 Hawking effect in dielectric media and the Hopfield model *Phys. Rev. D* **91** 124063

[27] Belgiorno F, Cacciatori S L, Clerici M, Gorini V, Ortenzi G, Rizzi L, Rubino E, Sala V G and Faccio D 2010 Hawking radiation from ultrashort laser pulse filaments *Phys. Rev. Lett.* **105** 203901

[28] Leonhardt U 2012 *Geometry and Light: The Science of Invisibility* (New York: Dover)

[29] Miri M-A, Heinrich M, El-Ganainy R and Christodoulides D N 2013 Supersymmetric optical structures *Phys. Rev. Lett.* **110** 233902

[30] Ozawa H M *et al* 2019 Topological photonics *Rev. Mod. Phys.* **91** 015006

[31] El-Ganainy R, Khajavikhan M, Christodoulides D N and Ozdemir S K 2019 The dawn of non-Hermitian optics *Commun. Phys.* **2** 37

[32] Zhong Q, Wu F O, Hassan A U, El-Ganainy R and Christodoulides D N 2023 Universality of light thermalization in multimoded nonlinear optical systems *Nat. Commun.* **14** 370

[33] Conti C 2014 Quantum gravity simulation by nonparaxial nonlinear optics *Phys. Rev. A* **89** (061801) (R)

[34] Ricca R L and Liu X 2024 *Knotted Fields* (Berlin: Springer)

Part I

Classical theory of light

IOP Publishing

A Field Theory Approach to Photonics

Marco Ornigotti

Chapter 2

Electromagnetic field and light–matter interaction

A fundamental building block of any photonic system is the electromagnetic field. Understanding how light propagates, both in free space and inside dielectric and non-dielectric media, and being able to describe its interaction with matter and light of other colours, i.e., electromagnetic radiation of different frequency, is a fundamental requirement for everyone studying or working in photonics.

In this chapter, we will briefly review some fundamental concepts about electrodynamics, starting from Maxwell's equations and delving into its paraxial and nonparaxial representation, to introduce the concept of electromagnetic modes. Furthermore, we will look at a general way to describe the interaction of the electromagnetic field with matter within the two most common approximations, i.e., the dipole and the minimal coupling approximation. Then, we will present a way, based on the so-called Power–Wooley–Zienau expansion, to connect these two interaction pictures and understand how they are related to each other.

The topic of this chapter is well-rooted in the fundamental theory of electrodynamics and light–matter interaction, and there are a lot of excellent textbooks out there, for the interested reader to deepen their knowledge of the topics, like the classic books by Stratton [1] and Jackson [2], or the two-series books on electrodnyamics by Cohen-Tannoudji [3, 4]. A detailed discussion of the paraxial modes of the electromagnetic field can be found in most laser physics books, such as the book by Svelto [5] or Siegman [6]. The derivation of the Power–Wooley–Zienau formula presented here, instead, is mostly taken from the book on molecular quantum electrodynamycs by Thirumamachandra [7].

2.1 Maxwell's equations, wave equation, and the Helmholtz equation

The starting point of every problem in electromagnetism is Maxwell's equations, a set of four, first-order coupled partial differential equations governing the propagation of the electric and magnetic field through space and time, that read

doi:10.1088/978-0-7503-5789-0ch2   2-1

$$\nabla \times \mathbf{E}(\mathbf{r},\, t) = -\frac{\partial \mathbf{B}(\mathbf{r},\, t)}{\partial t}, \tag{2.1a}$$

$$\nabla \cdot \mathbf{D}(\mathbf{r},\, t) = \rho(\mathbf{r},\, t), \tag{2.1b}$$

$$\nabla \cdot \mathbf{B}(\mathbf{r},\, t) = 0, \tag{2.1c}$$

$$\nabla \times \mathbf{H}(\mathbf{r},\, t) = \mathbf{J}(\mathbf{r},\, t) + \frac{\partial \mathbf{D}(\mathbf{r},\, t)}{\partial t}, \tag{2.1d}$$

where $\nabla = \partial_x \hat{\mathbf{x}} + \partial_y \hat{\mathbf{y}} + \partial_z \hat{\mathbf{z}}$, ρ is the charge distribution, source of the electric field $\mathbf{E}(\mathbf{r},\, t)$, $\mathbf{J}(\mathbf{r},\, t)$ is the free current density, source of the magnetic field $\mathbf{B}(\mathbf{r},\, t)$, $\mathbf{D}(\mathbf{r},\, t) = \varepsilon_0\, \mathbf{E}(\mathbf{r},\, t) + \mathbf{P}(\mathbf{r},\, t)$ is the displacement vector, with $\mathbf{P}(\mathbf{r},\, t)$ being the medium polarisation, accounting for the optical properties of the medium, and $\mathbf{H}(\mathbf{r},\, t) = \mathbf{B}/\mu_0 + \mathbf{M}(\mathbf{r},\, t)$ is the magnetic induction[1], with $\mathbf{M}(\mathbf{r},\, t)$ being the magnetisation vector, that accounts for magnetic properties of the medium, in which the electromagnetic field is propagating.

In free space, and in the absence of electric and magnetic sources, Maxwell's equations can only be described in terms of $\mathbf{E}(\mathbf{r},\, t)$ and $\mathbf{B}(\mathbf{r},\, t)$ as

$$\nabla \times \mathbf{E}(\mathbf{r},\, t) = -\frac{\partial \mathbf{B}(\mathbf{r},\, t)}{\partial t}, \tag{2.2a}$$

$$\nabla \cdot \mathbf{E}(\mathbf{r},\, t) = 0, \tag{2.2b}$$

$$\nabla \cdot \mathbf{B}(\mathbf{r},\, t) = 0, \tag{2.2c}$$

$$\nabla \times \mathbf{B}(\mathbf{r},\, t) = \frac{1}{c^2} \frac{\partial \mathbf{E}(\mathbf{r},\, t)}{\partial t}, \tag{2.2d}$$

where we have defined the speed of light in vacuum as $c^2 = 1/\varepsilon_0 \mu_0$ and $\nabla^2 = \partial_x^2 + \partial_y^2 + \partial_z^2$ as the Laplacian operator. If we now apply $\nabla \times$ to both sides of the first equation and use the vector identity $\nabla \times \nabla \times \mathbf{V} = \nabla\,(\nabla \cdot \mathbf{V}) = \nabla^2\, \mathbf{V}$ [8], we can get to a set of two, uncoupled, second order differential equations for $\mathbf{E}$ and $\mathbf{B}$ separately, i.e., the so-called wave equation, which is of the form

$$\nabla^2 \mathbf{E}(\mathbf{r},\, t) - \frac{1}{c^2} \frac{\partial^2 \mathbf{E}(\mathbf{r},\, t)}{\partial t^2} = 0. \tag{2.3}$$

A similar wave equation also holds for the magnetic field, but here, without loss of generality, we limit our attention to that for the electric field solely. The equation above is the starting point for the analysis of any electromagnetic wave phenomenon

---

[1] The general literature on electrodynamics is a bit confusing on the nomenclature of **H** versus **B**. Some textbooks, in fact, call **B** the magnetic field and **H** the magnetic induction, or magnetic H-field, while others use the opposite convention of naming **H** the magnetic field, and **B** the magnetic induction. Here, I have decided to adopt the former convention, with **B** being the magnetic field and **H** the magnetic induction, purely based on the fact that this is the convention, with which I have learned electromagnetism at university.

in free space[2]. Another version of the wave equation, obtained by substituting $\mathbf{D} = \varepsilon_0 \mathbf{E} + \mathbf{P}$ instead of $\mathbf{E}$ in the last of equation (2.2), is the following

$$\nabla^2 \mathbf{E}(\mathbf{r},\, t) - \frac{1}{c^2} \frac{\partial^2 \mathbf{E}(\mathbf{r},\, t)}{\partial t^2} = \mu_0 \frac{\partial^2 \mathbf{P}}{\partial t^2}, \tag{2.4}$$

which allows us to describe the linear and nonlinear optical response of a material, whose properties are encoded in the material polarisation vector $\mathbf{P}$ in the presence of an electromagnetic field.

Although the wave equations above can handle time-varying fields, a common approximation made in photonics, especially for low power applications, is the monochromatic approximation, where the electric field (and, eventually the polarisation vector) is assumed to be oscillating in time at a single frequency ω_0 and written as $\mathbf{E}(\mathbf{r},\, t) = \mathbf{E}(\mathbf{r}) \exp(-i\omega_0 t)$ (and similarly for $\mathbf{P}$). Substituting this Ansatz in the wave equation above then leads to the Helmholtz equation, i.e.,

$$\nabla^2 \mathbf{E}(\mathbf{r}) + k_0^2 \, \mathbf{E}(\mathbf{r}) = 0, \tag{2.5}$$

where $k_0^2 = \omega_0^2/c^2$ is the dispersion relation of the vacuum [2]. This equation will be the subject of the rest of this chapter. Note, how in natural units, $c = 1$ and $k_0 = \omega_0$, so the wave equation above is insensitive to the unit convention used.

2.2 The propagator for the Helmholtz equation

In deriving the Helmholtz equation above, we have assumed that there are no sources of electric or magnetic field in the domain where the equation is defined. However, if there were sources, equation (2.5) would get modified as follows

$$\nabla^2 \mathbf{E}(\mathbf{r}) + k_0^2 \mathbf{E}(\mathbf{r}) = \mathbf{F}(\mathbf{r}), \tag{2.6}$$

where the explicit form of the source term $\mathbf{F}(\mathbf{r})$ depends on the particular problem at hand. For example, if we assume to have a domain region without free charges ($\rho = 0$), the source term assumes the usual matter-polarisation form $\mathbf{F}(\mathbf{r}) = \mu_0 \partial_t \mathbf{J}(\mathbf{r}) = \mu_0 \partial_t^2 \mathbf{P}(\mathbf{r})$ where we used the constitutive relation for bound currents $\mathbf{J} = \nabla \times \mathbf{M} + \partial_t \mathbf{P}$ for the case of non-magnetic interactions (i.e., zero magnetisation material) [2]. Here, $\mathbf{P}(\mathbf{r}) = \varepsilon_0 \chi \mathbf{E}(\mathbf{r})$ is the matter polarisation, and χ is the material susceptibility, here assumed to be constant and with no nonlinear components [9].

In this case, the general solution of equation (2.6) is easily written in terms of the Green's function[3] $\mathbf{G}(\mathbf{r},\, \mathbf{r}')$ as

---

[2] The extension to a homogeneous, linear or nonlinear medium, is rather trivial and can be found in standard textbooks, such as Jackson [2] or Stratton [1]. For the extension to a nonlinear medium, instead, a good reference is the book on nonlinear optics by Boyd [9].

[3] In field theory, the Green's function is also referred to as the propagator. These two names are equivalent in terms of physical meaning, and will be used as synonyms throughout this book, preferring one over the other according to context.

$$\mathbf{E}(\mathbf{r}) = \int d^3r' \; \mathbf{G}(\mathbf{r} - \mathbf{r}') \cdot \mathbf{F}(\mathbf{r}'). \tag{2.7}$$

Notice that in the general case, $\mathbf{G}(\mathbf{r} - \mathbf{r}')$ is a tensor quantity (frequently referred to as the dyadic Green's function), since both the source term and the electric field are vector quantities.

There are many methods to find the explicit expression for the Green's function of a given differential problem, and the interested reader is referred to the book by Tai [10], as an extensive reference on dyadic Green's functions and how to compute them, or to the book by Byron and Fueller [8] for a primer on Green's functions. Here, we want to showcase a general method to find an explicit expression for the Green's function that will serve as guidance for future similar calculations in the following chapters. For the sake of simplicity, and without any loss of generality, we then limit ourselves to the scalar version of the Helmholtz equation, since it is easier to handle analytically. The same procedure then applies to the dyadic case, and a detailed derivation of such case can be found in literature, for example in reference [10].

To start with, let us rewrite equation (2.6) for a scalar electric field[4]

$$\nabla^2 E(\mathbf{r}) + k_0^2 E(\mathbf{r}) = F(\mathbf{r}), \tag{2.8}$$

which admits a general solution in terms of the (scalar) propagator $G(\mathbf{r}, \mathbf{r}')$ as

$$E(\mathbf{r}) = \int d^3r' \; G(\mathbf{r} - \mathbf{r}')F(\mathbf{r}'). \tag{2.9}$$

If we substitute this solution back into equation (2.8), it is not difficult to show, that $G(\mathbf{r}, \mathbf{r}')$ must obey

$$\nabla^2 G(\mathbf{r}, \mathbf{r}') + k_0^2 G(\mathbf{r}, \mathbf{r}') = \delta(\mathbf{r} - \mathbf{r}'), \tag{2.10}$$

where $\delta(\mathbf{r})$ is the Dirac delta function in three dimensions, defined as

$$\delta(\mathbf{r} - \mathbf{r}') = \begin{cases} 0 & \mathbf{r} \neq \mathbf{r}', \\ \infty & \mathbf{r} = \mathbf{r}', \end{cases} \tag{2.11}$$

with the property that $\int d^3r \; \delta(\mathbf{r} - \mathbf{r}') = 1$ [8].

Equation (2.10) is a partial differential equation involving a distribution, i.e., the Dirac delta function, and its solution contains some pitfalls, especially if one is not familiar with the theory of distributions. However, we can get around this problem by solving equation (2.10) in Fourier space, where the Dirac delta becomes a constant, and the differential equation transforms into an algebraic one. If we introduce the (spatial) Fourier transform of the propagator as

$$G(\mathbf{r}) = \left(\frac{1}{2\pi}\right)^3 \int d^3k \; D(\mathbf{k}) \exp(i\mathbf{k} \cdot \mathbf{r}), \tag{2.12}$$

---

[4] From a physical perspective, we can assume, as will be done in the following sections, that the polarisation of the field is homogeneous throughout the field distribution, i.e., it is spatially independent, and it can then be neglected when computing the field propagation, since it doesn't change.

and use the Fourier transform of the Dirac delta function

$$\delta(\mathbf{r} - \mathbf{r}') = \left(\frac{1}{2\pi}\right)^3 \int d^3k \, \exp[i\mathbf{k} \cdot (\mathbf{r} - \mathbf{r}')], \qquad (2.13)$$

we get, upon substitution into equation (2.10) the following algebraic equation

$$\left(-k^2 + k_0^2\right)D(\mathbf{k}) = \exp(-i\mathbf{k} \cdot \mathbf{r}') \qquad (2.14)$$

whose solution can be simply calculated to be $D(\mathbf{k}) = \exp(-i\mathbf{k} \cdot \mathbf{r}')/(k_0^2 - k^2)$. Using this result, we can then write the explicit expression of the propagator using equation (2.12) as

$$G(\mathbf{r} - \mathbf{r}') = \left(\frac{1}{2\pi}\right)^3 \int d^3k \, \frac{\exp[i\mathbf{k} \cdot (\mathbf{r} - \mathbf{r}')]}{k_0^2 - k^2}. \qquad (2.15)$$

We can solve this integral by introducing spherical coordinates in both r- and k-space, such that $d^3k = k^2 \sin\theta \, dkd\theta d\phi$ and $\mathbf{k} \cdot (\mathbf{r} - \mathbf{r}') = k\rho \cos(\theta - \vartheta)$, where $\rho = |\mathbf{r} - \mathbf{r}'|$, and ϑ is the polar angle in r-space. The integral in ϕ is then immediate, since the integrand in equation (2.15) does not depend on ϕ, and gives 2π as result. We can also perform the integral over θ using

$$\int_0^\pi d\theta \, \sin\theta \, \exp[ik\rho \cos(\theta - \vartheta)] = \frac{1}{ik\rho}[\exp(ik\rho) - \exp(-ik\rho)], \qquad (2.16)$$

so that equation (2.15) now becomes

$$G(\mathbf{r} - \mathbf{r}') = \frac{1}{i\rho}\left(\frac{1}{2\pi}\right)^2 \int_0^\infty dk \, \frac{k}{(k_0^2 - k^2)}[\exp(ik\rho) - \exp(-ik\rho)]. \qquad (2.17)$$

Notice that the integrand above is an even function of k; if we replace k with $-k$, in fact, the integral remains invariant, apart from a change of integration boundaries, from $(0, \infty)$ to $(-\infty, 0)$. Since then $\int_0^\infty dk = \int_{-\infty}^0 dk$, this allows us replace the integral $\int_0^\infty dk$ with $(1/2)\int_{-\infty}^\infty dk$, which will allow us to calculate the k-integral in an easier manner. The integral we now have to solve is then given by

$$G(\mathbf{r} - \mathbf{r}') = \frac{1}{2i\rho}\left(\frac{1}{2\pi}\right)^2 \int_{-\infty}^\infty dk \, \frac{k}{(k_0^2 - k^2)}[\exp(ik\rho) - \exp(-ik\rho)]. \qquad (2.18)$$

This integral contains two poles at $k = \pm k_0$. This means, that when $k = \pm k_0$, the integrand diverges and the propagator is infinite. The presence of poles in an integral is a hint at the fact that analytic continuation and the residue theorem might be a helpful strategy to find a closed-form expression for the integral. To solve it, therefore, we first complexify the integral variable k by adding a small imaginary part, i.e., $k \rightarrow k + i\varepsilon$ (with $\varepsilon \ll 1$), so that the poles are shifted into the complex plane. This allows us to use the residue theorem for a holomorphic function $f(z)$

integrated over a simply oriented closed contour[5] Γ containing some of (or all) the poles of $f(z)$ [8], i.e.,

$$\oint_\Gamma dz\, f(z) = 2\pi i \sum_{n=1}^{N} \mathrm{Res}(f, z_k), \tag{2.19}$$

where $\mathrm{Res}(f, z_k)$ stands for the residue of the function $f(z)$ at the pole z_k and, for the case of simple poles as those of equation (2.15) it is calculated as $\mathrm{Res}(f, z_k) = \lim_{z \to z_k}(z - z_k)f(z)$ [8].

Using the residue theorem requires the definition of a closed contour Γ in the complex plane, along which we need to perform the integral in (2.18) using equation (2.19). To choose such an integration contour properly, it is convenient to split the integral in equation (2.18) into two terms, one containing $\exp(ik\rho)$ and the other containing $\exp(-ik\rho)$. In this way, we can define two different contours, Γ_1 and Γ_2 (shown in red and blue, respectively, in figure 2.1), so that Γ_1 only contains the term $\exp(ik\rho)$ and can close at the infinity in the upper-half plane, while Γ_2 only contains the term $\exp(-ik\rho)$ and can therefore close at infinity in the lower-half plane. By doing so, we ensure that Jordan's lemma holds [8] and we can then employ the residue theorem to calculate the integral in equation (2.18), which we can then write as follows

$$G(\mathbf{r} - \mathbf{r}') = \frac{1}{2i\rho}\left(\frac{1}{2\pi}\right)^2 \left[\int_{\Gamma_1} dk\, \frac{k\exp(ik\rho)}{(k_0^2 - k^2)} - \int_{\Gamma_2} dk\, \frac{k\exp(-ik\rho)}{(k_0^2 - k^2)}\right]. \tag{2.20}$$

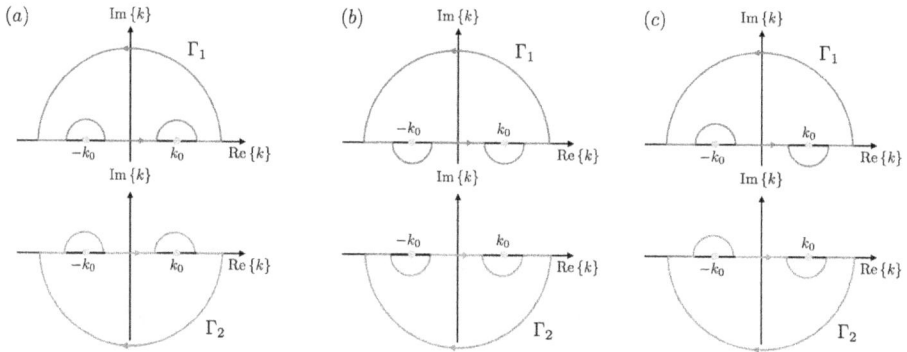

Figure 2.1. Possible integration contours for the complex integrals in equation (2.20). The path Γ_1 is depicted in red, and runs through the upper-half plane, while Γ_2 is depicted in blue, and runs through the lower-half plane. (a) Retarded propagator: Γ_1 (top) doesn't encircle any pole, while Γ_2 (bottom) encircles both poles. (b) Advanced propagator: Γ_1 (top) encircles both poles counterclockwise, while Γ_2 (bottom) encircles no poles. (c) Feynman propagator: Γ_1 (top) encircles the pole at $k = k_0$ counterclockwise, while Γ_2 (bottom) encircles the pole at $k = -k_0$ clockwise.

---

[5] A simple closed curve is one that creates a loop without self-intersecting. Moreover, it is positively oriented if the interior of the curve is always on the left while travelling on the curve, i.e., if the curve is travelling in a counterclockwise sense.

The possible physically meaningful choices are depicted schematically in figure 2.1 and are essentially three: Γ_1 doesn't encircle any pole, while Γ_2 encircles both poles clockwise (figure 2.1(a)), Γ_1 encircles both poles counterclockwise, while Γ_2 doesn't encircle any pole (figure 2.1(b)), or Γ_1 encircles the pole at $k = k_0$ counterclockwise, while Γ_2 encircles the pole at $k = -k_0$ clockwise (figure 2.1(c)). These three choices lead to three different expressions for the propagator called, respectively, the *retarded*, *anticipated*, and *Feynman* propagator. The meaning of the first two is immediate, as the retarded propagator only describes events for which, starting at the point **r** at time t, the point **r**′ is visited at a time $t' > t$, while the advanced propagator implements the opposite situation, where **r**′ is visited at a time $t' < t$. The Feynman propagator, on the other hand, is a mixture of the two, and it is the one frequently used in field theory.

We solve equation (2.20) using the contour in figure 2.1(c), i.e., we calculate the Feynman propagator associated with the Helmholtz equation. The calculation of the advanced and retarded propagators is done similarly, and it is left to the reader as an exercise. As can be seen from figure 2.1(c), the contour Γ_1 encircles the pole at $k = k_0$ counterclockwise, so when using the residue theorem, the right-hand side of equation (2.19) has a positive sign, because the contour Γ_1 around the pole $k = k_0$ is positively oriented. We then have

$$\int_{\Gamma_1} dk \, \frac{k \exp(ik\rho)}{(k_0^2 - k^2)} = 2\pi i \, \text{Res}\left(\frac{k \exp(ik\rho)}{(k_0^2 - k^2)}, k_0\right)$$

$$= 2\pi i \lim_{k \to k_0}(k - k_0)\frac{k \exp(ik\rho)}{(k_0^2 - k^2)} = \frac{1}{2}\exp(ik\rho). \quad (2.21)$$

The contour Γ_2 encircles instead the pole $k = -k_0$ clockwise, so Γ_2 is negatively oriented in the vicinity of $k = k_0$, which means that the right-hand side of equation (2.19) will have a minus sign to account for the opposite orientation of the contour. We then have

$$\int_{\Gamma_2} dk \, \frac{k \exp(-ik\rho)}{(k_0^2 - k^2)} = -2\pi i \, \text{Res}\left(\frac{k \exp(ik\rho)}{(k_0^2 - k^2)}, -k_0\right)$$

$$= 2\pi i \lim_{k \to k_0}(k - k_0)\frac{k \exp(ik\rho)}{(k_0^2 - k^2)} = -\frac{1}{2}\exp(ik_0\rho). \quad (2.22)$$

Substituting these results into equation (2.20) gives then the following final form of the propagator for the Helmholtz equation

$$G(\mathbf{r}, \mathbf{r}') = -\frac{1}{4\pi}\frac{\exp(ik_0|\mathbf{r} - \mathbf{r}'|)}{|\mathbf{r} - \mathbf{r}'|}. \quad (2.23)$$

2.3 Helmholtz equation in cylindrical coordinates: Bessel beams

We now turn our attention to the homogeneous Helmholtz equation, i.e., we set $\mathbf{F}(\mathbf{r}, t) = 0$ in (2.5) to obtain

$$\nabla^2 \, \mathbf{E}(\mathbf{r}) + k_0^2 \, \mathbf{E}(\mathbf{r}) = 0. \quad (2.24)$$

The expression above can be interpreted as an eigenvalue problem, where $E(\mathbf{r})$ is the eigenmode[6] of the Laplacian operator, and k_0^2 its associated eigenvalue. Eigenmodes of the Helmholtz equation (and its paraxial counterpart, as discussed in the next section) are then electromagnetic fields that maintain their functional shape upon propagation, up to a scale transformation. This means, that upon propagation, an optical beam can at most acquire a phase and its intensity distribution can spread in space, while maintaining the same shape. Physically, this is nothing else than diffraction [11].

In this section, we discuss a particular family of eigenmodes of the Helmholtz equation, namely Bessel beams. These are solutions of the Helmholtz equation in cylindrical coordinates and have several interesting properties, making them very interesting in photonics. The first of these properties is that they carry orbital angular momentum [12], and can be therefore used as the nonparaxial counterparts of Laguerre–Gaussian beams (see next section). Other interesting features of Bessel beams are their ability propagate along a given direction without exhibiting diffraction [13], and to exhibit self-healing, i.e., the ability to reconstruct themselves after passing an obstacle, even if most of the energy they carry has been absorbed by the obstacle itself. To achieve these features, however, Bessel beams need to carry infinite energy, i.e., their transverse intensity distribution is not square-integrable, a feature that makes them, just like plane waves, physically unattainable. Nevertheless, finite-energy Bessel beams, the so-called Bessel–Gauss beams, can be generated pretty easily in optics laboratories nowadays, either in the focus of a conical lens [13] or using a spatial light modulator [14], and they exhibit non-diffractive propagation in a limited propagation length, that can nowadays extend up to several hundreds of meters [15], making them quite attractive for free-space optical communications. In addition to this, Bessel beams are very useful in the analysis of strongly focussed electromagnetic fields and light propagation in wave-guides and optical fibres (see, e.g., reference [16] for the use of Bessel beams in integrated optics).

To start with, let us make again the scalar field approximation, i.e., let us assume that the polarisation of the electric field does not depend on the spatial coordinates, i.e., it is homogeneous throughout the transverse field profile, so that the electric field can be written as $\mathbf{E}(R, \theta, z) = \psi(R, \theta, z)\hat{\mathbf{f}}$, where $\hat{\mathbf{f}} = f_x\hat{\mathbf{x}} + f_y\hat{\mathbf{y}}$ encodes the field's polarisation, and $|f_x|^2 + |f_y|^2 = 1$ is assumed. With this assumption at hand, we can write Helmholtz equation for the scalar field $\psi(R, \theta, z)$ by writing the Laplacian in cylindrical coordinates to obtain [8]

$$\frac{1}{r}\frac{\partial}{\partial r}\left[r\frac{\partial}{\partial r} + \frac{1}{r^2}\frac{\partial^2}{\partial \theta^2} + \frac{\partial^2}{\partial z^2} + k_0^2\right]\psi(R, \theta, z) = 0. \qquad (2.25)$$

Cylindrical coordinates are one of the eleven reference frames in which Helmholtz equation is separable, so we can employ the method of separation of variables to

---

[6] In photonics, eigenmodes are frequently called optical beams, or optical modes.

solve the equation above. By assuming $\psi(R, \theta, z) = \xi(R)\Phi(\theta)Z(z)$, we can then rewrite equation (2.25) as

$$\frac{1}{\xi}\left[\frac{1}{r}\frac{\partial}{\partial r}\left(r\frac{\partial \xi}{\partial r}\right) + k_0^2 \xi\right] + \frac{1}{\Phi}\left(\frac{1}{r^2}\frac{\partial^2 \Phi}{\partial \theta^2}\right) + \frac{1}{Z}\left(\frac{\partial^2 Z}{\partial z^2}\right) = 0. \qquad (2.26)$$

This equation can now be split into three independent ordinary differential equations in the variables $\{R, \theta, z\}$ by simply focussing on one variable at a time, and setting the other parts of the Helmholtz equation to a suitable constant while solving that part. The equation for $Z(z)$, for example, can be easily brought to that of a harmonic oscillator by requiring that the first two terms of equation (2.26) are equal to a constant, i.e., k_z^2. By doing so, we get, for the z-dependent part of the solution $Z(z) = \exp(ik_z z)$, i.e., the solution along the z-direction has the same form as that of a plane wave propagating along that direction, with wave vector $k_z \in \mathbb{R}$. Similarly, the θ-equation can be also reduced to

$$\frac{\partial^2 \Phi}{\partial \theta^2} - m^2 \Phi = 0, \qquad (2.27)$$

whose solution, provided that $m \in \mathbb{Z}$, is given by $\Phi(\theta) = \exp(im\theta)$. Finally, using these two results, the equation for the radial function $\xi(R)$ becomes

$$\frac{1}{r}\frac{\partial}{\partial r}\left(r\frac{\partial \xi}{\partial r}\right) + \left[(k_0^2 - k_z^2) - \frac{m^2}{r^2}\right]\xi = 0, \qquad (2.28)$$

which is the Bessel equation, whose (nonsingular) solutions are given in the form of Bessel function of the first kind, i.e., $\xi(R) = J_m(k_\perp R)$, where $k_\perp = \sqrt{k_0^2 - k_z^2}$ is the transverse wave vector. Putting everything together, the electric field solution of the Helmholtz equation in cylindrical coordinates can be written as

$$\mathbf{E}(R, \theta, z, t) = J_m(k_\perp R)\exp[i(m\theta + k_z z - \omega t)]\hat{\mathbf{f}}. \qquad (2.29)$$

An example of the transverse amplitude and phase distribution of Bessel beams for different values of the parameter m is shown in figure 2.2. From it, and from equation (2.29), we can immediately observe that the intensity distribution of a Bessel beam is z-independent, i.e., that $\partial_z |\mathbf{E}|^2 = 0$. This corresponds to the beam being diffractionless, i.e., the beam maintains its shape unaltered during propagation.

Another interesting feature that we can read out from equation (2.29) is that Bessel beams are eigenstates of the z-component of the angular momentum operator $\hat{L}_z = i\hbar\partial_\theta$ [17], so that

$$\hat{L}_z\mathbf{E}(R, \theta, z) = i\hbar m\,\mathbf{E}(R, \theta, z). \qquad (2.30)$$

For this reason, Bessel beams are said to carry orbital angular momentum (OAM), specifically $\hbar m$ units per photon [12]. As a consequence of this, the phase front of Bessel beams with $m \neq 0$ is characteristically twisted into an m-folded helix, a signature of OAM-carrying beams.

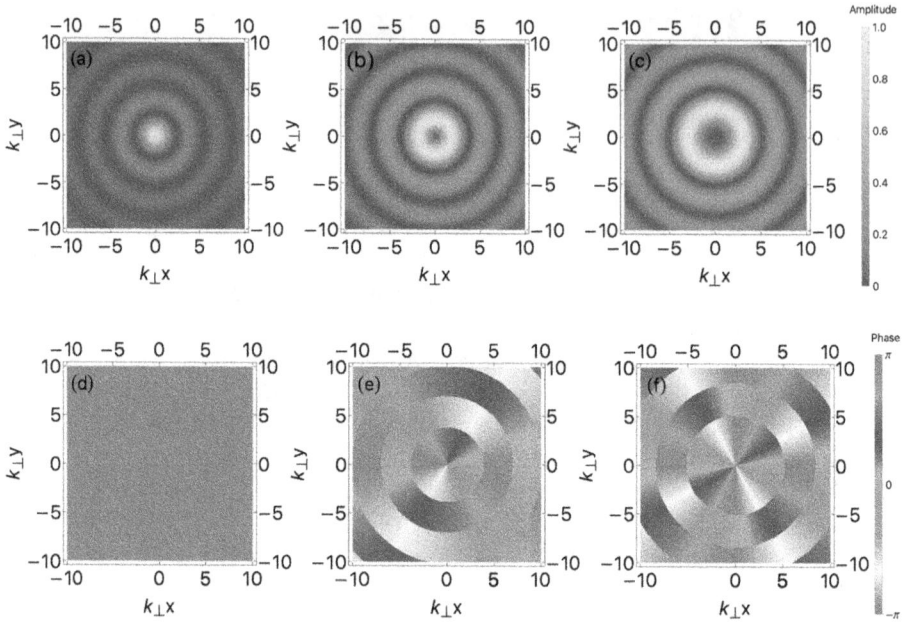

Figure 2.2. Transverse amplitude (top row) and phase (bottom row) distribution of a Bessel beam in $z = 0$, for various values of the OAM parameter: $m = 0$ [panels (a) and (d)], $m = 1$ [panels (b) and (e)], and $m = 2$ [panels (c) and (f)]. As can be seen, while the phase distribution for the $m = 0$ Bessel beam is flat, denoting constant phase overall the whole wave front, when $m \neq 0$ the situation drastically changes, and a phase singularity (optical vortex) appears at the origin. The phase then winds around the origin once for $m - =1$ [panel (e)], accumulating a total phase jump of 2π, and twice for $m = 2$ [panel (f)], accumulating two phase jumps of 2π, for a total of 4π phase shift over one full rotation.

2.3.1 Green's function in cylindrical coordinates

A useful result to keep at hand when dealing with photonics problems involving OAM and nonparaxial beams is the expression of the Green's function in cylindrical coordinates, as this provides a natural framework for including OAM in the Green's function formalism. To do so, we can use the same procedure highlighted in section 2.2 to obtain the Green's function for the Helmholtz equation, with the difference that we are going to use cylindrical, rahter than spherical, coordinates to solve the integral appearing in equation (2.15). In doing so (as we will see below) one obtaines an integral expression for the Green's function in cylindrical coordinates that has no closed-form solution, and must therefore be implemented numerically. For the sake of completeness, we present here the details of such calculation, as an example of application of the Green's function formalism and as an excuse to introduce some useful series expansions and integral representations involving Bessel functions.

We start by rewriting the integral in equation (2.15) in cylindrical coordinates. To this aim, we can use the relation $\mathbf{k} \cdot \mathbf{r} = k_{\perp} R \cos(\theta - \phi) + k_z z$, where $k_{\perp} = \sqrt{k_x^2 + k_y^2}$, $R = \sqrt{x^2 + y^2}$, $\theta = \arctan(y/x)$ and $\phi = \arctan(k_y/k_x)$, to write the exponential function appearing in equation (2.15) as $\mathbf{k} \cdot (\mathbf{r} - \mathbf{r}') = k_\perp R \cos(\theta - \phi)$ $-k_\perp R' \cos(\theta' - \phi) + k_z(z - z')$. Then, we use the Jacobi–Anger expansion [8]

$$\exp(i\,z\cos\theta) = \sum_{n=-\infty}^{\infty} i^n J_n(z)\exp(i\,n\,\theta), \tag{2.31}$$

to write the resulting exponential in terms of Bessel functions as follows

$$\exp\{i[k_\perp R\cos(\theta - \phi) - k_\perp R'\cos(\theta' - \phi) + k_z(z - z')]\}$$
$$= \sum_{\ell=-\infty}^{\infty}\sum_{m=-\infty}^{\infty} i^{\ell+m} J_\ell(k_\perp R)J_m(k_\perp R')\exp\{i[\ell(\theta - \phi) - m(\theta' - \phi)]\}. \tag{2.32}$$

This allows us to decouple $k_\perp$ and ϕ and thus gives us the possibility to perform the ϕ-integral. Recalling that, in cylindrical coordinates, $d^3k = k_\perp\, dk_\perp\, d\phi\, dk_z$, substituting the expansion above into equation (2.15) gives

$$G(\mathbf{r} - \mathbf{r}') = \left(\frac{1}{2\pi}\right)^3 \sum_{\ell,m} i^{\ell+m}\exp[i(\ell\theta - m\theta')]\int_0^\infty dk_\perp\, J_\ell(k_\perp R)J_m(k_\perp R')$$
$$\times \int_{-\infty}^\infty dk_z\, \frac{\exp[ik_z(z - z')]}{k_0^2 - k_\perp^2 - k_z^2}\int_0^{2\pi} d\phi\, \exp[i(m - \ell)\phi]. \tag{2.33}$$

The angular integral can be straightforwardly performed, giving

$$\int_0^{2\pi} d\phi\, e^{i(m-\ell)\phi} = 2\pi\, \delta_{m,\ell}. \tag{2.34}$$

The integral in k_z can also be readily calculated using the residue theorem and choosing a suitable contour of integration in the complex plane. Analogously to what we have done in section 2.2, we chose again the Feynman propagator, i.e., the contour in figure 2.1(c). Notice, that depending on whether $z - z'$ is positive or negative, when chosing the contour, we need to ensure that the exponential term $\exp[ik_z(z - z')]$ converges. This means that for $(z - z') > 0$ we will choose the contour Γ_1 from figure 2.1(c), while for $(z - z') < 0$ we will choose instead the contour Γ_2. To proceed with our calculations, and without loss of generality, we can assume that $(z - z') > 0$ and then choose the contour Γ_1, to get the following result

$$\int_{-\infty}^\infty dk_z\, \frac{e^{ik_z(z-z')}}{k_0^2 - k_\perp^2 - k_z^2} = 2\pi\, i\, \lim_{k_z \to K_z}(k_z - K_z)\frac{e^{ik_z(z-z')}}{(k_z - K_z)(k_z + K_z)}\Big|_{k_z = K_z}$$
$$= \pi\, i\,\frac{e^{i\sqrt{k_0^2 - k_\perp^2}(z-z')}}{\sqrt{k_0^2 - k_\perp^2}}, \tag{2.35}$$

where $K_z = \sqrt{k_0^2 - k_\perp^2}$ is the (positive) pole encircled by Γ_1. Substituting the result for the angular integral and the one for the integral in k_z obtained above into equation (2.33) gives the following final form for the Green's function in cylindrical coordinates

$$G(\mathbf{r}, \mathbf{r}') = \sum_\ell \frac{i\,(-1)^\ell}{4\pi}\int_0^\infty dk_\perp\, \frac{k_\perp e^{i\sqrt{k_0^2 - k_\perp^2}(z-z')}}{\sqrt{k_0^2 - k_\perp^2}}J_\ell(k_\perp R)J_\ell(k_\perp R'). \tag{2.36}$$

This expression is exact but, unfortunately, does not admit a closed-form solution and therefore it can only be handled numerically. Notice, however, that for $k_0 \to 0$, the Green's function above reduces to that of the Poisson equation, which admits a closed-form solution [2]. Although it is possible to numerically implement this expression, care must be used, as the integrand above is an oscillatory and singular function over the integration domain. Efficient numerical implementations using absolutely convergent series and hypergeometric functions have been proposed to generate more implementation-friendly expressions of equation (2.36), but the discussion of these details is beyond the scope of this book. The interested reader can learn more about this from reference [18], for example.

2.4 Paraxial approximation and Gaussian beams

A very useful and widely used approximation to the Helmholtz equation (2.5), which, for example, describes the propagation of a laser beam, is the so-called paraxial approximation. Within this framework, a choice is made to promote one particular direction to propagation direction for the electromagnetic field, and the assmumption is made that, along that direction, the field does not change appreciably (i.e., it changes slowly). For convention's sake, this direction is frequently identified with the z-direction. To derive the so-called paraxial equation, i.e., the Helmholtz equation satisfying the above assumptions, we can again write the electric field as scalar and monochromatic, but emphasise that z now has a 'special role', within the spatial variables, as it indicates the direction along which the field is mainly propagating. This allows us to write, for an electric field polarised along the x-direction, $\mathbf{E}(\mathbf{r}, t) = \mathscr{E}(\mathbf{R}, z)\exp[i(k_0 z - \omega_0 t)]\hat{\mathbf{x}}$, where $\mathbf{R} = \{x, y\}$. From this expression we see that, essentially, the electric field behaves like a plane wave in the z-direction, with a z-dependent envelope that, by assumption, varies slowly along the propagation direction. We can translate this into a mathematical condition by taking the Laplacian in equation (2.5) and separate it into its longitudinal and transverse coordinates, i.e., $\nabla^2 = \partial_z^2 + \nabla_\perp^2$, where $\nabla_\perp^2 = \partial_x^2 + \partial_y^2$, so that equation (2.5) becomes

$$\frac{\partial^2 \mathscr{E}(\mathbf{R}, z)}{\partial z^2} + 2ik_0 \frac{\partial \mathscr{E}(\mathbf{R}, z)}{\partial z} + \nabla_\perp^2 \, \mathscr{E}(\mathbf{R}, z) = 0. \tag{2.37}$$

Requiring that the envelope field $\mathscr{E}(\mathbf{R}, z)$ varies slowly along z is then equivalent to enforce the following condition

$$\left| \frac{\partial^2 \mathscr{E}}{\partial z^2} \right| \ll \frac{1}{k_0} \left| \frac{\partial \mathscr{E}}{\partial z} \right|, \tag{2.38}$$

which results in neglecting the second derivative term in z above, leading to the so-called paraxial equation

$$i\frac{\partial \mathscr{E}(\mathbf{R}, z)}{\partial z} = -\frac{1}{2k_0} \, \nabla_\perp^2 \, \mathscr{E}(\mathbf{R}, z). \tag{2.39}$$

Solutions of this equation can be easily found in Fourier space, where a simple expression of the paraxial propagator can be given in the so-called angular spectrum representation.

We introduce the Fourier transform of the field amplitude $\mathscr{E}(\mathbf{R}, z)$ with respect to the transverse variables $\{x, y\}$ as

$$\mathscr{E}(\mathbf{R}, z) = \left(\frac{1}{2\pi}\right)^2 \int d^2k \; \mathscr{A}(\mathbf{K}, z) \exp(i\mathbf{K} \cdot \mathbf{R}), \tag{2.40}$$

where $d^2k = dk_x dk_y$, and $\mathbf{K} = k_x \hat{\mathbf{x}} + k_y \hat{\mathbf{y}}$ is the transverse wave vector. Substituting this ansatz into equation (2.39) then results in the following propagation equation for the Fourier amplitude $\mathscr{A}(\mathbf{K}, z)$

$$\frac{\partial \mathscr{A}(\mathbf{K}, z)}{\partial z} = \frac{ik_\perp^2}{2k_0} \mathscr{A}(\mathbf{K}, z), \tag{2.41}$$

where $k_\perp^2 = k_x^2 + k_y^2$. The expression above is an ordinary first-order differential equation in z, whose general solution can be written as

$$\mathscr{A}(\mathbf{K}, z) = \mathscr{A}(\mathbf{K}, 0) \exp\left(\frac{ik_\perp^2 z}{2k_0}\right), \tag{2.42}$$

where $\mathscr{A}(\mathbf{K}, 0)$ is the Fourier spectrum of the electromagnetic field at the plane $z = 0$, conventionally taken as the initial plane for the propagation of the field. Substituting this result into equation (8.53) gives the general form of a paraxial field in terms of its Fourier transform as

$$\mathscr{E}(\mathbf{R}, z) = \left(\frac{1}{2\pi}\right)^2 \int d^2k \; \mathscr{A}(\mathbf{K}, 0) \exp\left(\frac{ik_\perp^2 z}{2k_0}\right) \exp(i\,\mathbf{K} \cdot \mathbf{R}). \tag{2.43}$$

Notice, that for $z = 0$ the integral above gives the expression for the electromagnetic field at the moment it starts propagating, i.e., at the plane $z = 0$. For $z \neq 0$, instead, the propagation is accounted for, in Fourier space, by the term $\exp\left(\frac{ik_\perp^2 z}{2k_0}\right)$, which for this reason takes the name of paraxial propagator.

If we assume a Gaussian spectrum for $\mathscr{A}(\mathbf{K}, 0)$, i.e.,

$$\mathscr{A}(\mathbf{K}, 0) = \mathscr{E}_0 \exp-\left(\frac{w_0^2 k_\perp^2}{4}\right), \tag{2.44}$$

with w_0 being the beam waist and $\mathscr{E}_0$ the beam amplitude, equation (8.56) can be solved analytically in terms of Gaussian integrals, leading to the usual expression for a Gaussian beam propagating along z, i.e.,

$$\mathscr{E}(\mathbf{R}, z) = \left(\frac{ik_o \mathscr{E}_0}{4\pi}\right)\frac{1}{q(z)} \exp\left[-\frac{k_0 R^2}{2q(z)}\right], \tag{2.45}$$

where $q(z) = z + iz_R$ is the Gaussian complex q-parameter, and $z_R = k_0 w_0^2/2$ is the Rayleigh range [5]. Expanding the factor $1/q(z)$ in the previous equation gives access to both the beam waist and the beam wave front curvature as a function of z, as well as the Gouy phase, i.e.,

$$w^2(z) = w_0^2 \left[1 + \left(\frac{z}{z_R} \right)^2 \right], \qquad (2.46a)$$

$$\rho(z) = z \left[1 + \left(\frac{z_R}{z} \right)^2 \right], \qquad (2.46b)$$

$$\psi(z) = \arctan \left(\frac{z}{z_R} \right). \qquad (2.46c)$$

An example of the longitudinal and transverse characteristics of Gaussian beams are given in figure 2.3, while the waist function $w(z)$ and the Gouy phase $\psi(z)$ are presented in figure 2.4

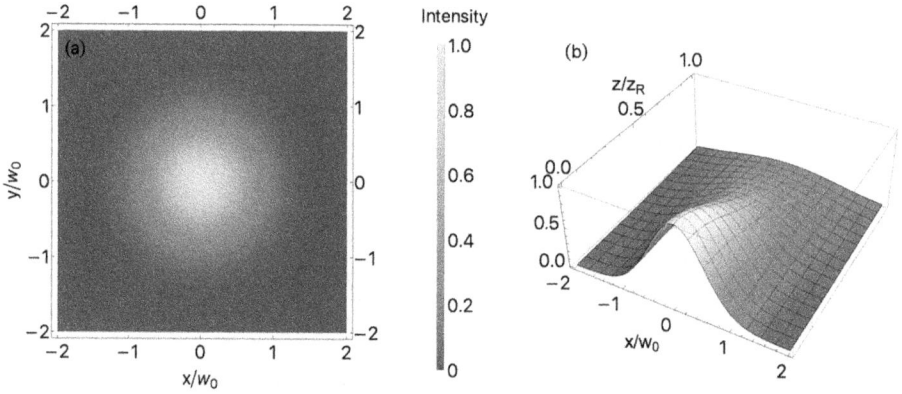

Figure 2.3. (a) Transverse intensity distribution of a Gaussian beam, as a function of the scaled coordinates $(x/w_0, y/w_0)$. (b) Longitudinal intensity distribution (at $y = 0$) of a Gaussian beam, as a function of the scaled transverse coordinate x/w_0 and longitudinal coordinate z/z_R.

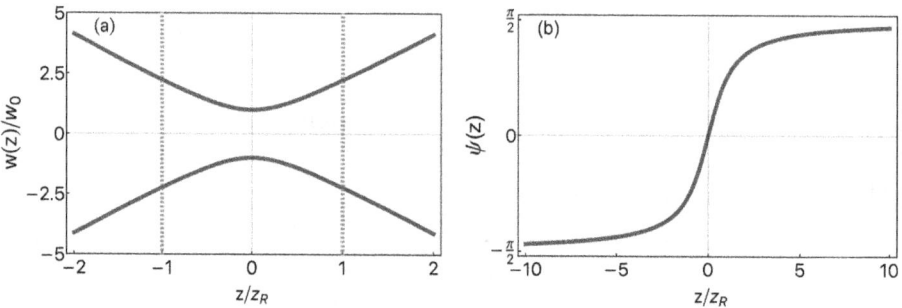

Figure 2.4. (a) Plot of the scaled z-dependent beam waist $w(z)/w_0$, as given in equation (2.46a), as a function of the scaled longitudinal coordinate z/z_R. The red, dashed lines at $z = \pm 1$ (i.e., when the beam has propagated for a Rayleigh range, $z = z_R$) indicate the position where $w(z) = \sqrt{2} \, w_0$, which is usually taken as the definition of Rayleigh range [5]. (b) Plot of the Gouy phase $\psi(z)$ as a function of the scaled longitudinal coordinate z/z_R, as given in equation (2.46c).

2.4.1 Hermite and Laguerre–Gaussian beams

The Gaussian beam solution in equation (2.45) is only the fundamental solution of several different families of higher order paraxial beams, whose shape and expression depend on the particular symmetry considered for the beam itself. Amongst all the possible symmetries possessed by the beam (and, most of the times, inherited by the laser cavity generating the beam), Cartesian and cylindrical symmetry are the most common. The corresponding set of orthogonal paraxial modes are then named Hermite–Gaussian (HG) and Laguerre–Gaussian (LG) beams, respectively, and their explicit expression, given below, can be derived from equation (8.56) by assuming a Cartesian- or a cylindrically-symmetric initial angular spectrum, i.e.,

$$\mathscr{A}_{HG}(\mathbf{K},\,0) = \mathscr{E}_0 H_n\left(\frac{w_0 k_x}{\sqrt{2}}\right) H_m\left(\frac{w_0 k_y}{\sqrt{2}}\right) \exp\left(-\frac{w_0^2 k_\perp^2}{4}\right), \tag{2.47a}$$

$$\mathscr{A}_{LG}(\mathbf{K},\,0) = \mathscr{E}_0 \left(\frac{w_0 k_\perp}{\sqrt{2}}\right) L_p^{|\ell|}\left(\frac{w_0^2 k_\perp^2}{2}\right) \exp\left(-\frac{w_0^2 k_\perp^2}{2}\right), \tag{2.47b}$$

where $H_n(x)$ are the Hermite polynomials, and $L_p^{|\ell|}(x)$ are the associate Laguerre polynomials [19]. Substituting the above ansatz into equation (8.56) gives, after consulting reference [20] for the integrals containing Hermite and Laguerre polynomials, the following form expression

$$\mathscr{E}_{HG}(\mathbf{R},\,z) = \mathscr{E}_0 \frac{w_0}{w(z)} H_n\left(\frac{R\sqrt{2}}{w_0}\right) H_m\left(\frac{R\sqrt{2}}{w_0}\right) \exp\left(-\frac{R^2}{w^2(z)}\right)$$

$$\times \exp\left(-i\frac{k_o R^2}{2\rho(z)}\right) \exp[i(n+m+1)\psi(z)]. \tag{2.48}$$

for HG beams propagating along the z-direction, and

$$\mathscr{E}_{LG}(\mathbf{R},\,z) = \mathscr{E}_0 \frac{w_0}{w(z)} \left(\frac{R\sqrt{2}}{w(z)}\right)^{|\ell|} L_p^{|\ell|}\left(\frac{2R^2}{w^2(z)}\right) \exp\left(-\frac{R^2}{w^2(z)}\right) \exp(i\ell\theta)$$

$$\times \exp\left(-i\frac{k_o R^2}{2\rho(z)}\right) \exp[i(2p+|\ell|+1)\psi(z)]. \tag{2.49}$$

for LG beams propagating along the z-direction. Notice that in the expressions above, all the normalisation and integration constants have been absorbed into the field amplitude $\mathscr{E}_0$ for simplicity of notation. An example of how the transverse intensity distribution for HG and LG modes look is given in figures 2.5–2.7, respectively. The phase profile of LG beams, typicaly of OAM-carrying beams, is instead shown in figure 2.8 for the same choice of indices as figure 2.7. Notice, moreover, that the presence of the azimuthal term $\exp(i\ell\theta)$ in the expression of LG beams, i.e., equation (2.49) is a signature of the fact, that LG beams carry OAM and their wave front is therefore twisted, like that of Bessel beams [12].

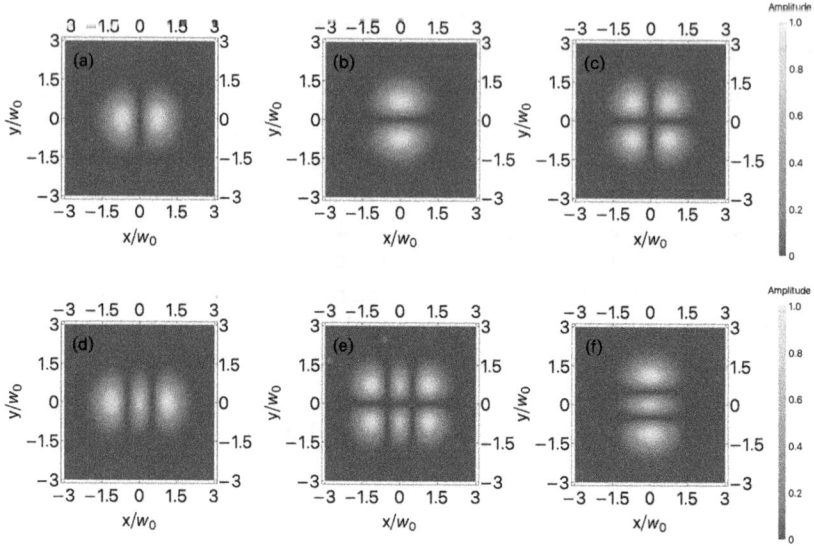

Figure 2.5. Transverse intensity profile of HG beams as a function of the scaled coordinates x/w_0 and y/w_0, with different values of the indices (n, m). (a) $(n, m) = (1, 0)$. (b) $(n, m) = (0, 1)$. (c) $(n, m) = (1, 1)$. (d) $(n, m) = (2, 0)$. (e) $(n, m) = (2, 1)$. (f) $(n, m) = (0, 2)$. As can be seen from these plots, the index n counts the number of zeros of the beam along the x-direction, while the index m counts the number of zeros along the y-direction. Higher order beams are simply generated by adding n zeros along x and m zeros along y.

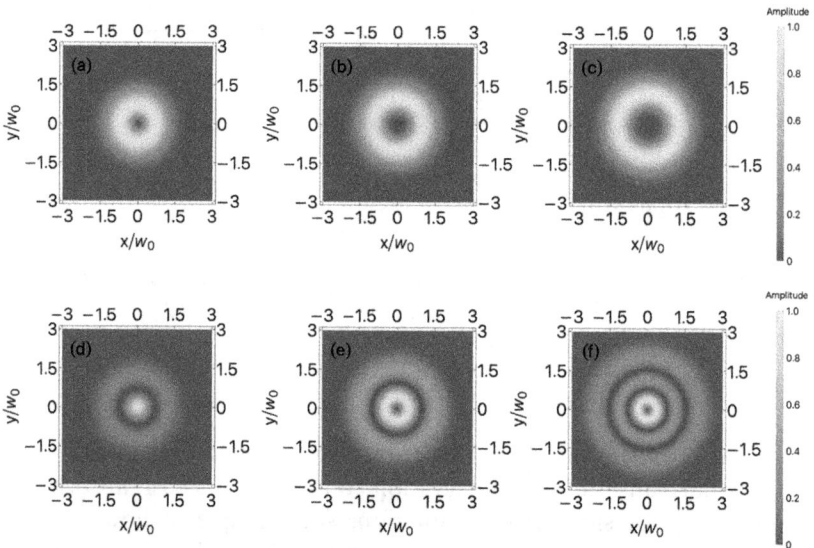

Figure 2.6. Transverse intensity profile of LG beams as a function of the scaled coordinates x/w_0 and y/w_0, with different values of the indices (ℓ, p). (a) $(\ell, p) = (1, 0)$. (b) $(\ell, p) = (2, 0)$. (c) $(\ell, p) = (3, 0)$. (d) $(\ell, p) = (0, 1)$. (e) $(\ell, p) = (1, 1)$. (f) $(\ell, p) = (1, 2)$. Notice, how in the upper row ℓ increasing from $\ell = 1$ to $\ell = 3$ results in the hole at the centre of the beam becoming larger. In fact, the size of the hole scales with $\sqrt{\ell}$ [5]. This is happening because the phase singularity at the centre of the LG beam grows in charge with ℓ and the beam needs to put zero intensity in a larger area to compensate the high-order divergence in the phase.

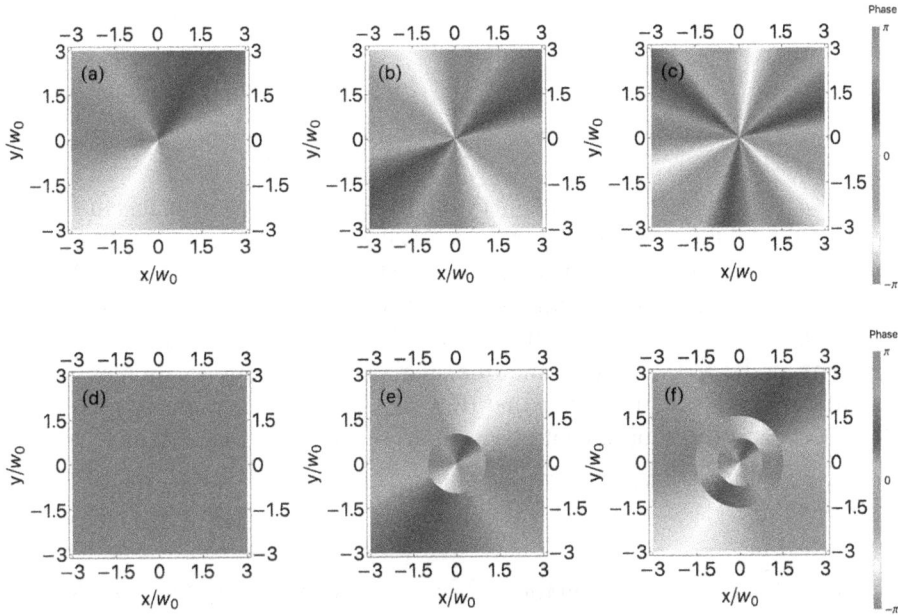

Figure 2.7. Phase profile of LG beams as a function of the scaled coordinates x/w_0 and y/w_0, with different values of the indices (ℓ, p). (a) $(\ell, p) = (1, 0)$. (b) $(\ell, p) = (2, 0)$. (c) $(\ell, p) = (3, 0)$. (d) $(\ell, p) = (0, 1)$. (e) $(\ell, p) = (1, 1)$. (f) $(\ell, p) = (1, 2)$. Notice, how in the upper row ℓ increasing from $\ell = 1$ to $\ell = 3$ results in the charge of the phase singularity (i.e., the OAM carried by the beam) increasing linearly with ℓ, as we observe one, two, and three 2π windings around the phase singularity at the centre, as we go from panel (a) to panel (c). In the lower panels, instead, we see how the radial index p is affecting the phase profile of an LG beam. Essentially, for $p \neq 0$, an extra ring is created in the phase profile (corresponding to the π phase jump associated with the p-index), in which the underlying phase structure dictated by ℓ is replicated with a π phase shift.

2.5 The paraxial propagator

In the previous sections, we have derived the paraxial equation, i.e., the slowly varying limit of the Helmholtz equation (2.6). An interesting question to answer is how does the Green's function for the paraxial equation look, and how it is connected to the Green's function for the Helmholtz equation derived in section 2.2? The rigorous derivation of the Green's function for the paraxial equation can be done in several different ways (for example, by starting from the diffraction integral and imposing the Fresnel approximation [11], or from first principles using path integrals, by exploiting a direct connection between the paraxial equation and the Schrödinger equation [21]). Here, however, to keep the derivation short, we present a direct way to start from the Green's function for the Helmholtz equation in 3D, i.e., equation (2.23), and construct directly the one for the paraxial case by imposing the paraxial assumptions on the Green's function itself. In a way, this is equivalent to imposing the Fresnel and/or Fraunhofer conditions to the diffraction integral and then solve it, with the extra shortcut, that we start from the solution already and manipulate it to fit it into the paraxial case.

To derive a paraxial version of equation (2.23) we first rewrite the term $|\mathbf{r} - \mathbf{r}'|$ as follows

$$|\mathbf{r} - \mathbf{r}'| = \sqrt{z^2 + |\mathbf{R} - \mathbf{R}'|^2} = z\sqrt{1 + \frac{|\mathbf{R} - \mathbf{R}'|^2}{z^2}}, \tag{2.50}$$

where $\mathbf{R}' = x'\hat{\mathbf{x}} + y'\hat{\mathbf{y}}$ and $\mathbf{R} = x\hat{\mathbf{x}} + y\hat{\mathbf{y}}$ are the transverse coordinates in the initial $[(x', y')]$ and final $[(x, y)]$ plane, respectively, and we have assumed, without loss of generality, that the propagation of the wave starts at the plane $z' = 0$. Now, the essence of the paraxial approximation lies essentially in saying, that the wave is propagating along the z-direction far enough, that the characteristic transverse dimensions in both the initial and the final plane are much smaller than the propagation length, i.e, assuming that $|\mathbf{R} - \mathbf{R}'| \ll z$. We can use this assumption to Taylor expand the square root above up to second order to obtain

$$|\mathbf{r} - \mathbf{r}'| = z\sqrt{1 + \frac{|\mathbf{R} - \mathbf{R}'|^2}{z^2}} \simeq z\left(1 + \frac{|\mathbf{R} - \mathbf{R}'|^2}{2z^2} + \mathcal{O}(|\mathbf{R} - \mathbf{R}'|^4)\right). \tag{2.51}$$

Substituting this result into equation (2.23), and keeping only terms $\mathcal{O}(|\mathbf{R} - \mathbf{R}'|)$, leads to the following result

$$G(\mathbf{r} - \mathbf{r}') = -\frac{1}{4\pi} \frac{1}{z\left(1 + \dfrac{|\mathbf{R} - \mathbf{R}'|^2}{2z^2}\right)} \exp\left[ikz\left(1 + \frac{|\mathbf{R} - \mathbf{R}'|^2}{2z^2}\right)\right]. \tag{2.52}$$

The last approximation we need concerns the denominator of the Green's function. In fact, if $|\mathbf{R} - \mathbf{R}'| \ll z$ holds, we can neglect the term proportional to $|\mathbf{R} - \mathbf{R}'|^2$ in the denominator, as the error we will commit in neglecting is of the order $\mathcal{O}(|\mathbf{R} - \mathbf{R}'|^4)$. Employing this further assumption leads to the final expression for the paraxial Green's function

$$G(\mathbf{r} - \mathbf{r}') = \exp(ikz)\left[-\frac{1}{4\pi z}\exp\left(ik\frac{|\mathbf{R} - \mathbf{R}'|^2}{2z}\right)\right] \equiv \exp(ikz)G_p(\mathbf{r} - \mathbf{r}'), \tag{2.53}$$

where $G_p(\mathbf{r} - \mathbf{r}')$ is the paraxial propagator. Notice that we have extracted a term $\exp(ikz)$, since in deriving the paraxial equation we have assumed that the electric field was propagating along the z-direction as $\exp(ikz)$. The presence of this term on the Green's function corroborates the importance of this assumption in deriving the paraxial equation.

2.6 Light–matter interaction

So far we have only considered the dynamics of the electromagnetic field in vacuum. In this section we revert back to the general form of Maxwell's equation with nonzero sources and derive the wave equation in the presence of (free) charges and currents. Using this result, we will then derive the interaction between light and matter in different ways, highlighting the various approaches that lead to different realisations of such interactions and different (but equivalent) models of light–matter interaction. We will derive the wave equation in terms of the electromagnetic potentials, which will lead us to the so-called minimal coupling to describe the interaction with matter. Then, with the help of the Power–Zienau–Wooley transformation, we will revert back to the more familiar electric dipole approximation.

2.6.1 Electromagnetic scalar and vector potentials

Let us start by introducing the electromagnetic scalar ($\phi(\mathbf{r}, t)$) and vector ($\mathbf{A}(\mathbf{r}, t)$) potentials. To do so, let us first recall, that in the electrostatic case, the first of Maxwell's equations, i.e., equation (2.1a), reduces to $\nabla \times \mathbf{E}(\mathbf{r}, t) = 0$, and therefore the electric field can be written in terms of a scalar potential $\phi(\mathbf{r}, t)$ through the relation $\mathbf{E} = -\nabla \phi(\mathbf{r}, t)$ [2]. The scalar field $\phi(\mathbf{r}, t)$ is called the (electric) scalar potential and de-facto, as we will see later in this section, accounts for the electric field generated by the charge distribution. Secondly, we use equation (2.1c) to define the magnetic field in terms of its vector potential $\mathbf{A}(\mathbf{r}, t)$ through the relation $\mathbf{B}(\mathbf{r}, t) = \nabla \times \mathbf{A}(\mathbf{r}, t)$. Using the vector identity $\nabla \cdot (\nabla \times \mathbf{A}) = 0$ it is then immediate to see that this choice of magnetic field still satisfies Maxwell's equation. Moreover, substituting this Ansatz into equation (2.1a) gives us the relation between the electric field and the vector potential as $\mathbf{E}(\mathbf{r}, t) = -\partial \mathbf{A}(\mathbf{r}, t)/\partial t$. The vector potential, then, not only generates the magnetic field, but it is also responsible of the time dependence of the electric field. The relationship between the electric and magnetic field and the field potential can then be summarised as follows

$$\mathbf{E}(\mathbf{r}, t) = -\frac{\partial \mathbf{A}(\mathbf{r}, t)}{\partial t} - \nabla \phi(\mathbf{r}, t), \qquad (2.54a)$$

$$\mathbf{B}(\mathbf{r}, t) = \nabla \times \mathbf{A}(\mathbf{r}, t). \qquad (2.54b)$$

Substituting these relations in the remaining Maxwell's equations, namely equations (2.1b) and (2.1d), gives the following evolution equation for the electromagnetic potentials

$$\nabla[\nabla \cdot \mathbf{A}(\mathbf{r}, t)] - \nabla^2 \mathbf{A}(\mathbf{r}, t) = -\frac{\partial^2 \mathbf{A}(\mathbf{r}, t)}{\partial t^2} - \frac{\partial \nabla \phi(\mathbf{r}, t)}{\partial t}, \qquad (2.55a)$$

$$-\frac{\partial[\nabla \cdot \mathbf{A}(\mathbf{r}, t)]}{\partial t} + \nabla^2 \phi = 0. \qquad (2.55b)$$

These equations carry the same physical meaning as Maxwell's equations, with a significant difference. While Maxwell's equations involve physical quantities such as the electric and magnetic fields, the equations above involve quantities that have no physical sense per se. To illustrate this, let us consider a different set of scalar and vector potentials, that we will indicate with ϕ' and $\mathbf{A}'$, and assume that they are related to the original potentials through the relations

$$\mathbf{A}(\mathbf{r}, t) = \mathbf{A}(\mathbf{r}, t)' - \nabla \Lambda(\mathbf{r}, t), \qquad (2.56a)$$

$$\phi(\mathbf{r}, t) = \phi'(\mathbf{r}, t) + \frac{\partial \, \nabla \, \Lambda(\mathbf{r}, t)}{\partial t}, \tag{2.56b}$$

where $\Lambda(\mathbf{r}, t)$ is an arbitrary (but well-behaved) function. If we now calculate the electric and magnetic fields $\mathbf{E}'$ and $\mathbf{B}'$ generated by the new potentials we get

$$\mathbf{E}'(\mathbf{r}, t) = -\frac{\partial}{\partial t}[\mathbf{A}(\mathbf{r}, t) - \nabla \, \Lambda(\mathbf{r}, t)] - \nabla \left[\phi(\mathbf{r}, t) + \frac{\partial \Lambda(\mathbf{r}, t)}{\partial t} \right], \tag{2.57a}$$

$$\mathbf{B}'(\mathbf{r}, t) = \nabla \times \mathbf{A}(\mathbf{r}, t) - \nabla \times \nabla \, \Lambda(\mathbf{r}, t). \tag{2.57b}$$

It is then not difficult to see, using equations (2.56) that one gets $\mathbf{E}' = \mathbf{E}$ and $\mathbf{B}' = \mathbf{B}$. This means, that different choices of the electromagnetic potentials generate the same electric and magnetic fields. This peculiar property of the electromagnetic field is called *gauge invariance*, and the transformations (2.56) are called *gauge transformations* (and, by consequence, the function $\Lambda(\mathbf{r}, t)$ is called the *gauge function*). The concept of gauge invariance is central for the development of the theory of electromagnetic radiation, and it also constitutes one of the fundamental pillars of modern physics [22–25]. Moreover, the gauge transformations (2.56) have a very important meaning: when dealing with electromagnetic problems, while the electric and magnetic fields are gauge invariant, the potentials are not. A specific gauge, therefore, needs to be chosen when dealing with problems involving the electromagnetic potentials. In the next chapter we will see how, from a field theory perspective, it is possible to automatically incorporate all the possible gauge choices in the Lagrangian of the electromagnetic field. Here, for the moment, we limit ourselves to considering the most popular gauge choice for the free electromagnetic field, namely the Coulomb (ofter also called radiation) gauge, for which $\nabla \cdot \mathbf{A}(\mathbf{r}, t) = 0$. This choice implies that the vector potential is a purely transverse field, and that $\phi(\mathbf{r}, t) = 0$ is a viable choice in free space, which corroborates the interpretation of the scalar potential as the generating function of the electric field due to the charge distribution [2]. Within this gauge, moreover, the vector potential fully describes both the electric and magnetic field, and it obeys the same wave equation as in equation (2.3).

2.6.2 Electromagnetic potentials in the presence of matter

We now consider the electromagnetic field in a region of space where there are field sources, in the form of charge density $\rho(\mathbf{r}, t)$ and current density $\mathbf{J}(\mathbf{r}, t)$. In Coulomb gauge, equations (2.55) modify into

$$\nabla^2 \mathbf{A}(\mathbf{r}, t) - \frac{\partial \, \nabla \, \phi(\mathbf{r}, t)}{\partial t} - \frac{\partial^2 \mathbf{A}(\mathbf{r}, t)}{\partial t^2} = -\mathbf{J}(\mathbf{r}, t), \tag{2.58a}$$

$$\nabla^2 \phi(\mathbf{r}, t) = -\rho(\mathbf{r}, t). \tag{2.58b}$$

From these equations, we see that a charge density acts as a source for the scalar potential, whereas a current density is a source for the vector potential. To make these equations easy to handle, we can use the Helmholtz theorem [2, 26], which states that any vector field $\mathbf{V}$ can be written as the sum of two orthogonal components, one of which has zero divergence (and it is called the transverse, or solenoidal, component), while the other one has zero curl (and it is called the longitudinal, or irrotational, component), i.e., $\mathbf{V} = \mathbf{V}_T + \mathbf{V}_L$. If we employ Helmholtz theorem on the current density and write $\mathbf{J}(\mathbf{r}, t) = \mathbf{J}_L(\mathbf{r}, t) + \mathbf{J}_T(\mathbf{r}, t)$, this will reflect into a decomposition of the vector potential into a longitudinal and transverse part as $\mathbf{A}(\mathbf{r}, t) = \mathbf{A}_T(\mathbf{r}, t) + \mathbf{A}_L(\mathbf{r}, t)$. However, because we have assumed working in the Coulomb gauge, only the transverse part of the vector potential will really contribute in generating the correct electric and magnetic fields, as the essence of the Coulomb gague is precisely to consider the vector potential as a purely transverse vector field. For equation (2.57a) to remain consistent, therefore, the term proportional to $\nabla\phi(\mathbf{r}, t)$ must be purely longitudinal. Using these results, we can then decompose equations (2.57) into the following set of uncoupled equations

$$\nabla^2\mathbf{A}(\mathbf{r}, t) - \frac{\partial^2\mathbf{A}(\mathbf{r}, t)}{\partial t^2} = \mathbf{J}_T(\mathbf{r}, t), \qquad (2.59a)$$

$$\frac{1}{c^2}\frac{\partial\,\nabla\,\phi(\mathbf{r}, t)}{\partial t} = \mathbf{J}_L(\mathbf{r}, t), \qquad (2.59b)$$

$$\nabla^2\phi(\mathbf{r}, t) = -\rho(\mathbf{r}), \qquad (2.59c)$$

where the subscript $_T$ has been dropped from the vector potential, since it is a purely transverse field anyway.

The equations above are fully equivalent to Maxwell's equation, together with the continuity equation linking charge density and current density. This, in fact, can be readily obtained from the equations above by taking the divergence of the second equation and using the third one to eliminate the scalar potential, obtaining

$$\nabla \cdot \mathbf{J}_L(\mathbf{r}, t) = \frac{\partial\rho(\mathbf{r}, t)}{\partial t}. \qquad (2.60)$$

Moreover, from the relations above we can also get the expressions for the longitudinal and transverse electric and magnetic fields, which read as follows

$$\mathbf{E}_T(\mathbf{r}, t) = -\frac{\partial\mathbf{A}(\mathbf{r}, t)}{\partial t}, \qquad (2.61a)$$

$$\mathbf{E}_L(\mathbf{r}, t) = -\nabla\,\phi(\mathbf{r}, t), \qquad (2.61b)$$

$$\mathbf{B}_T(\mathbf{r}, t) = \nabla \times \mathbf{A}(\mathbf{r}, t), \qquad (2.61c)$$

$$\mathbf{B}_L(\mathbf{r}, t) = 0. \qquad (2.61d)$$

The reader might check that these fields are exact solutions of Maxwell's equations, with the longitudinal electric field being related to the charge distribution $\rho(\mathbf{r}, t)$ and the transverse electric and magnetic field are instead associated to the current density $\mathbf{J}_T(\mathbf{r}, t)$.

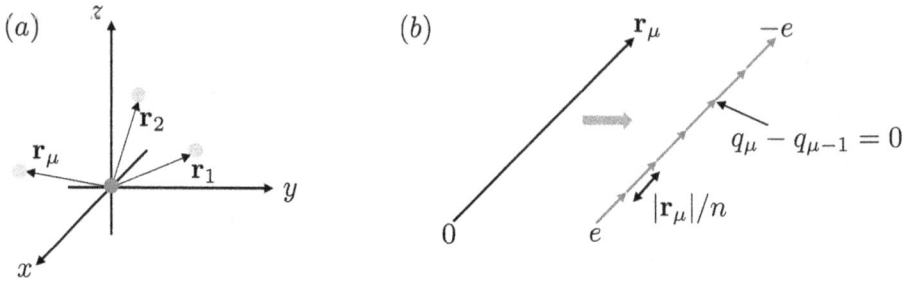

Figure 2.8. (a) Schematic representation of the choice of reference frame to describe atom–field interaction: the nucleus (red) sits in the centre of the reference frame, while the electrons (orange) orbiting around it are labelled by the index μ and located at position r_μ. (b) Schematic representation for the approximation used to calculate $\mathbf{P}(\mathbf{r}, t)$. The actual electric dipole composed by the proton–electron system (left, black arrow) is decomposed into a set of elementary, non-point-like dipoles (red arrows), each constituted by two opposite charges of charge $|q_\mu|$ separated by a distance $|\mathbf{r}_\mu|/n$, and oriented along the direction of $\mathbf{r}_\mu$. Because of the arrangement shown in the figure, the tail of the μth dipole superimposes with the tip of the $(\mu - 1)$th dipole, thus ensuring charge neutrality along the direction $\mathbf{r}_\mu$.

2.6.3 Multipole expansion of the electromagnetic field

Let us now consider a simple, but instructive, case of light–matter interaction, namely that of a single atom interacting with an electromagnetic field. This can be considered as the archetype of light–matter interaction, from which one can easily generalise to obtain the optical properties of, e.g., solid state systems [2, 27]. For the sake of clarity, let us put ourselves into a reference frame solidal with the atom position, i.e., the nucleus of the atom sits at $\mathbf{r} = 0$, and indicate with $\mathbf{r}_\mu$ the position of the the μth electron orbiting around the nucleus, as depicted in figure 2.8(a). Within this framework, we can associate a charge density to the atom by simply summing the charge of all protons in the nucleus, and all electrons orbiting around it, obtaining

$$\rho(\mathbf{r}, t) = -e\sum_\mu \delta(\mathbf{r} - \mathbf{r}_\mu) + Z\, e\delta(\mathbf{r}), \tag{2.62}$$

where e is the electron charge and Z the atomic number (i.e., the number of protons in the atom) and the sum over μ runs over all electrons present in the atom. Because of the motion of the electrons around the nucleus, we can also associate a current density to the atom, namely

$$\mathbf{J}(\mathbf{r}, t) = -e\sum_\mu \dot{\mathbf{r}}_\mu \delta(\mathbf{r} - \mathbf{r}_\mu), \tag{2.63}$$

where the dot indicates derivation with respect to time. The interaction of the atom with the electromagnetic field is then best captured in terms of polarisation and magnetisation, related to the atomic charge and current via the constitutive relations [1, 2]

$$\rho(\mathbf{r}, t) = -\nabla \cdot \mathbf{P}(\mathbf{r}, t), \tag{2.64a}$$

$$\mathbf{J}(\mathbf{r},\ t) = \dot{\mathbf{P}}(\mathbf{r},\ t) + \nabla \times \mathbf{M}(\mathbf{r},\ t). \tag{2.64b}$$

Substituting equations (2.62) and (2.63) into the relations above, we can obtain the following integral representation for the polarisation and the magnetisation

$$\mathbf{P}(\mathbf{r},\ t) = -e\sum_{\mu}\mathbf{r}_{\mu}\ \int_{0}^{1}\ d\xi\ \delta(\mathbf{r} - \xi\ \mathbf{r}_{\mu}), \tag{2.65a}$$

$$\mathbf{M}(\mathbf{r},\ t) = -e\sum_{\mu}\mathbf{r}_{\mu} \times \dot{\mathbf{r}}_{\mu}\ \int_{0}^{1}\ d\xi\ \xi\ \delta(\mathbf{r} - \xi\ \mathbf{r}_{\mu}). \tag{2.65b}$$

To understand how it is possible to go from equations (2.64) to (2.65), let us consider explicitly the calculation for the polarisation. Then, deriving the expression for the magnetisation can be done analogously, and it is left to the reader as an exercise.

To prove equation (2.65a), let us first focus on the case of a single proton at the origin, with charge e, around which a single electron, with charge $-e$ is orbiting at a distance r_{μ} from the proton. This simple proton–electron system behaves like an electric dipole, i.e., two opposite electric particles of charge $|e|$ separated by a distance $|\mathbf{r}_{\mu}|$. Instead of thinking at this system as a single dipole, let us imagine representing it as a collection of non-point-like dipoles, each of charge $-q_{\mu}$ and $+q_{\mu}$, oriented along the direction $\mathbf{r}_{\mu}$ and separated by a distance $\mathbf{r}_{\mu}/n$, disposed in such a way that the charge $-q_{\mu}$ of one dipole is superimposed on the charge $+q_{\mu+1}$ of the preceding dipole, and thus cancels it out (see figure 2.8(b)). With this construction in mind, the polarisation can be then written as simply the sum of the contribution given by each of these elementary dipoles as follows

$$\mathbf{P}(\mathbf{r},\ t) = \lim_{n\to\infty} \sum_{p=0}^{n-1} q_{\mu}\left(\frac{\mathbf{r}_{\mu}}{n}\right)\delta\left(\mathbf{r} - \frac{p+1/2}{n}\mathbf{r}_{\mu}\right) = \int_{0}^{1}\ d\xi\ q_{\mu}\mathbf{r}_{\mu}\ \delta(\mathbf{r} - \xi\mathbf{r}_{\mu}), \tag{2.66}$$

where the sum runs over the various elementary dipoles, and to get to the last equality we have simply replaced the limit of the sum with an integral, introducing the continuous variable ξ, which corresponds to $(p + 1/2)/n$. With the same line of reasoning, we can also get to the integral expression for the magnetisation shown in equation (2.65b).

With the help of equations (2.65) we can now write the electric and magnetic contribution to the potential energy of the atom immersed in an electromagnetic field as

$$V_{E} = -\int\ d^{3}r\ \mathbf{P}(\mathbf{r},\ t)\cdot\mathbf{E}_{T}(\mathbf{r},\ t) = e\sum_{\mu}\ \int_{0}^{1}\ d\xi\ \mathbf{r}_{\mu}\cdot\mathbf{E}_{T}(\xi\mathbf{r}_{\mu},\ t), \tag{2.67a}$$

$$V_{M} = -\int\ d^{3}r\ \mathbf{M}(\mathbf{r},\ t)\cdot\mathbf{B}(\mathbf{r},\ t) = e\sum_{\mu}\ \int_{0}^{1}\ d\xi\ (\mathbf{r}_{\mu} \times \dot{\mathbf{r}}_{\mu})\cdot\mathbf{B}(\xi\mathbf{r}_{\mu},\ t). \tag{2.67b}$$

To proceed further, notice that the characteristic size of an atom is of the order of the Bohr's radius $a_0 = 100$ pm, while the typical size of electromagnetic (visible) radiation is of the order of $\lambda = 500$ nm $\ll a_0$. This size mismatch then justifies the Taylor expansion of the electric and magnetic fields around the position of the nucleus, i.e., $\mathbf{r} = 0$. By doing so, we can calculate the integrals explicitly at any order of the expansion and we obtain the following result for the electric and magnetic potentials

$$V_E = e\sum_\mu \left\{ 1 + \frac{1}{2!}\mathbf{r}_\mu \cdot \nabla + \frac{1}{3!}(\mathbf{r}_\alpha \cdot \nabla)^2 + \cdots \right\} \mathbf{r}_\mu \cdot \mathbf{E}_T(0), \qquad (2.68\text{a})$$

$$V_M = \frac{e}{m}\sum_\mu \left\{ \frac{1}{2!} + \frac{2}{3!}\mathbf{r}_\mu \cdots \nabla + \frac{3}{4!}(\mathbf{r}_\mu \cdot \nabla)^2 + \cdots \right\} \mathbf{I}_\mu \cdot \mathbf{B}(0), \qquad (2.68\text{b})$$

where m is the electron mass and $\mathbf{I}_\mu = m\mathbf{r}_\mu \times \dot{\mathbf{r}}_\mu$ is the orbital angular momentum of electron μ around the nucleus. The expansion in equation (2.68a) is known as the *multipole expansion* of the electromagnetic field, and each term accounts for a multipole moment of the atomic charge distribution. The lower order of this expansion corresponds to the electric and magnetic dipole moments, then the second order terms correspond to the quadrupole moments and so on. The expression for the electric and magnetic dipole moments, as well as for the electric quadrupole moment are given below:

$$-e\mathbf{D}_E = -\sum_\mu e\mathbf{r}_\mu, \qquad (2.69\text{a})$$

$$-e\mathbf{D}_M = -\frac{1}{2}\sum_\mu \frac{e}{m}\mathbf{I}_\mu, \qquad (2.69\text{b})$$

$$\mathbf{Q}_E = -\frac{1}{2}\sum_\mu e\mathbf{r}_\mu\mathbf{r}_\mu. \qquad (2.69\text{c})$$

Notice, that while the dipole moment are vectors, the quadrupole moment is instead a tensor of rank two, and the product $\mathbf{r}_\mu\mathbf{r}_\mu$ in its definition is understood as a dyadic product.

2.6.4 Electric versus magnetic dipoles

Very frequently in electrodynamics or quantum mechanics textbooks and articles dealing with light–matter interaction one reads that the interaction of an atom (or, more in general, with matter) with the electromagnetic field is well described via the electric dipole interaction only (at least in first approximation), and that the magnetic dipole is often neglected. The goal of this section is to provide a reasoning for this statement, by looking at the order of magnitude of the various multipolar interactions, thus giving a quantitative measure of the interaction strength of the electric and magnetic dipoles. To do so, we need to give an estimate of the quantities involved in the definition of the electric and magnetic potentials V_E and V_M for the electric dipole, electric quadrupole, and magnetic dipole. We begin by assuming that $|\mathbf{r}_\mu| \simeq a_0 = 0.1$ nm, i.e., that the electron orbits at about one Bohr radius from the nucleus. Then, we also estimate the angular momentum of the single electron to be $|\mathbf{I}_\mu| \simeq \hbar = 1.054 \times 10^{-34}$ Js, in accordance with quantum mechanics [17]. Then, we assume considering an electromagnetic field in the visible range of the spectrum, oscillating at a frequency $\omega \simeq 3 \times 10^{15}$ rad s^{-1}, corresponding to a wavelength of about $\lambda \simeq 600$ nm. Finally, from classical electrodynamics, we can estimate the electric field gradient as $\nabla \mathbf{E}_T(0) \simeq (\omega/c)\mathbf{E}_T(0)$ [2]. With all these assumptions in place, we can then estimate the contribution to the potential energy of the atom given by the electric dipole, electric quadrupole, and magnetic dipole as follows

$$e\mathbf{D}_E \cdot \mathbf{E}_T(0) \simeq \frac{4\pi\varepsilon_0 \hbar^2}{me}E(0), \tag{2.70a}$$

$$-\nabla \cdot \mathbf{Q} \cdot \mathbf{E}_T(0) \simeq \frac{3e\hbar}{16mc}E(0), \tag{2.70b}$$

$$e\mathbf{D}_M \cdot \mathbf{B} \simeq \frac{e\hbar}{2m}B(0) \simeq \frac{e\hbar}{2mc}E(0), \tag{2.70c}$$

where in the last equality we have used the fact that for monochromatic plane waves $B(0) = E(0)/c$. To get an idea of the relative orders of magnitude, let us first take the ratio of the electric quadrupole with the magnetic dipole, which is given by

$$\frac{-\nabla \cdot \mathbf{Q} \cdot \mathbf{E}_T(0)}{e\mathbf{D}_M \cdot \mathbf{B}(0)} \simeq \frac{3}{8} \simeq 1, \tag{2.71}$$

and then the ratio between the electric quadrupole and the electric dipole

$$\frac{-\nabla \cdot \mathbf{Q} \cdot \mathbf{E}_T(0)}{e\mathbf{D}_E \cdot \mathbf{E}_T(0)} \simeq \frac{e^2}{4\pi\varepsilon_0 \hbar c} \equiv \alpha \simeq \frac{1}{137}, \tag{2.72}$$

where α is the fine structure constant [17]. As can be seen, the electric dipole is roughly α times bigger than the electric quadrupole, and since electric quadrupole and magnetic dipole are of the same order of magnitude as shown by equation (2.71), the electric dipole is also approximately α times, i.e., two orders of magnitude, bigger than the magnetic dipole. One could also show, moreover, that the magnitude of the higher order moments is α^n times smaller that the magnitude of the electric dipole, where n is the order of the multipole considered.

This result constitutes a reasonable enough argument to consider the electric dipole transitions in an atomic system (and, more in general, in any material system) as the only form of interaction with the electromagnetic field, at least in the lowest order approximation of interaction. Whenever the electric dipole approximation is not applicable, e.g., for the case of transitions forbidden to the electric dipole, higher order terms must instead be taken into account.

2.6.5 Interaction Hamiltonian and minimal coupling

We now turn our attention to the actual dynamics of an atom interacting with an electromagnetic field. The aim of this section is to introduce the usual interaction

Hamiltonian in the minimal coupling form, where the electromagnetic vector potential acts as a correction term to make the particle momentum a canonically conjugated variable again or, stated in more modern terms, where the electromagnetic vector potential converts the normal spatial derivative (i.e., the particle momentum) into a covariant derivative [23, 24]. This means, essentially, that in order to make the atom–field interaction appear, we need to make the substitution $\mathbf{p}_\mu \to \mathbf{p}_\mu + e\mathbf{A}(\mathbf{r}_\mu)$ in the expression of the Hamiltonian, i.e., total energy of the atom, where $\mathbf{p}_\mu$ is the momentum of the μth atom. To setup this problem, let us consider again an atom, consisting of a certain number of electrons orbiting around the nucleus, interacting with a monochromatic electromagnetic field. The total Hamiltonian, encompassing the energy stored in the atom, in the electromagnetic field, and the energy required for their interaction, is given by, within the minimal coupling approximation,

$$\mathscr{H} = \frac{1}{2m}\sum_\mu [\mathbf{p}_\mu + e\mathbf{A}(\mathbf{r}_\mu)]^2 + \frac{1}{2}\int d^3r \rho(\mathbf{r})\phi(\mathbf{r})$$
$$+ \frac{1}{2}\int d^3r \left[\varepsilon_0 |\mathbf{E}_T(\mathbf{r})|^2 + \frac{1}{\mu_0}|\mathbf{B}(\mathbf{r})|^2 \right],$$

(2.73)

where the first term is the minimally coupled, free-particle Hamiltnoian, the second term (containing the interaction between the charge distribution of the atom anf the electromagnetic scalar potential) is essentially the Coulomb interaction, and the last term is the energy stored in the free electromagnetic field. Notice, moreover, that the time dependence has been dropped from the electric and magnetic field, since we are implicitly assuming that they are both monochromatic. Within this formalism, the interaction of the atom with the electromagnetic field is fully contained in the first term, i.e., in the canonical momentum $\mathbf{p}_\mu + e\mathbf{A}(\mathbf{r}_\mu)$. Developing the square, we get

$$\frac{1}{2m}\sum_\mu [\mathbf{p}_\mu + e\mathbf{A}(\mathbf{r}_\mu)]^2 = \sum_\mu \frac{|\mathbf{p}_\mu|^2}{2m} + \frac{e}{m}\sum_\mu \mathbf{A}(\mathbf{r}_\mu)\cdot\mathbf{p}_\mu + \left(\frac{e}{2m}\right)^2\sum_\mu |\mathbf{A}(\mathbf{r}_\mu)|^2, \quad (2.74)$$

where the first term is just the sum of the kinetic energy of each electron in the atom, while the second term is the actual light–matter interaction Hamiltonian, in its minimally-coupled form[7], i.e.,

$$\mathscr{H}_I = \frac{e}{m}\sum_\mu \mathbf{A}(\mathbf{r}_\mu)\cdot\mathbf{p}_\mu + \left(\frac{e}{2m}\right)^2\sum_\mu |\mathbf{A}(\mathbf{r}_\mu)|^2. \quad (2.75)$$

Notice, that the interaction Hamiltonian does not contain the Coulomb term, dependent on the scalar potential, since this can be considered an electrostatic term

---

[7] Notice, that in developing the square, I have deliberately neglected the fact, that the electrom momentum $\mathbf{p}_\mu = -i\hbar\nabla_\mu$ is in fact an operator, and that the second term in equation (2.74) should be replaced with the commutator between the momentum and the vector potential. However, since here I am considering the electromagnetic field as classical and not quantised, I can neglect this issue, as long as I keep the discussion in momentum space.

and therefore not a dynamical interaction [2, 17]. Another interesting point to remember is that, in the form above, the interaction Hamiltonian is Lorentz invariant (see next chapter), but gauge dependent, as it is cast in terms of the electromagnetic potentials, and not the electric and magnetic fields directly. Within this framework, one should then be careful in choosing a gauge that is suitable to describe the interaction, before starting evaluating the interaction Hamiltonian.

2.6.6 Dipole interaction Hamiltonian

Most of the times, it is more convenient to work with an interaction Hamiltonian that is manifestly gauge invariant, instead of having a gauge-dependent one such as that in equation (2.75). To do so, one would need to find an interaction Hamiltonian that contains the electric and magnetic fields directly, instead of the vector potential. This is the case, for example, for the usual electric dipole interaction term, whose Hamiltonian reads $\mathcal{H}_{ED} = e\mathbf{D} \cdot \mathbf{E}_T(0)$. The aim of this section is then to show how it is possible to go from the minimally coupled interaction Hamiltonian (2.75) to the electric dipole interaction Hamiltonian, by employing the so-called Power–Zienau–Wooley transformations [4, 28, 29]. This transformation amounts, essentially to introducing the following (gauge) transformation

$$U = \exp\left[i \int d^3r \, \mathbf{P}(\mathbf{r}) \cdot \mathbf{A}(\mathbf{r})\right], \tag{2.76}$$

where $\mathbf{P}(\mathbf{r})$ is the polarisation vector defined in equation (2.65a), and applying it to the atom–field Hamiltonian (2.73) as $\mathcal{H}' = U^{\dagger}\mathcal{H}U$, so that $\mathcal{H}'$ can be then written as $\mathcal{H}' = \mathcal{H}_A + \mathcal{H}_F + \mathcal{H}_{ED}$, where the first term is the Hamiltonian corresponding to the free atom, the second one is the Hamiltonian of the free electromagnetic field, and the third term is the sought after electric dipole interaction Hamiltonian defined as

$$\mathcal{H}_{ED} = e\mathbf{D}_E \cdot \mathbf{E}_T(0) \equiv -\int d^3r \, \mathbf{P}(\mathbf{r}) \cdot \mathbf{E}(\mathbf{r}), \tag{2.77}$$

where to get the second equality we have used the series expansion for V_E truncated at the dipole term, assuming that the higher orders can be neglected. The details of the calculation that allow representing the light–matter interaction in terms of the electric dipole interaction Hamiltonian are given in chapter 4, as they involve transforming the Lagrangian of the field, which will be introduced in the next chapter.

We conclude this chapter by reporting also the explicit form of the magnetic dipole interaction, in its gauge-invariant form, where the magnetic field, rather than the vector potential, appears explicitly. In doing so, let us first notice, that the interaction Hamiltonian (2.75) implicitly contains both electric and magnetic interactions, since it involves the vector potential, that generates both of them, as seen in equations (2.61a) and (2.61c). When dealing with the gauge-invariant interaction Hamiltonian, i.e., after having operated a Power–Zienau–Wooley transformation, the electric and magnetic interactions separate, and the interaction Hamiltonian for the magnetic dipole reads

$$\mathscr{H}_{MD} = e\mathbf{D}_M \cdot \mathbf{B}(0) = -\int d^3r\, \mathbf{M}(\mathbf{r}) \cdot \mathbf{B}(\mathbf{r}), \tag{2.78}$$

where, again, for the second equality we have used the series expansion for V_M truncated at the magnetic dipole order, thus neglecting the contribution of the higher order multipoles. Notice, moreover, that this term only accounts for paramagnetic interactions. The diamagnetic interactions are proportional to the square of the magnetic field and will appear during the detailed calculation presented in chapter 4.

References

[1] Stratton J A 2019 *Electromagnetic Theory* (New York: Dover)
[2] Jackson J D 1998 *Classical Electrodynamics* (New York: Wiley)
[3] Cohen-Tannoudji C 1992 *Atom-Photon Interactions: Basic Processes and Applications* (New York: Wiley)
[4] Cohen-Tannoudji C 1989 *Photons and Atoms: Introduction to Quantum Electrodynamics* (New York: Wiley)
[5] Svelto O 1998 *Principles of Lasers* 4th edn (Dordrecht: Kluwer)
[6] Siegman A E 1990 *Lasers New Edition* (Mill Valley, CA: University Science Books)
[7] Craig D P and Thirunamachandran T 1984 *Molecular Quantum Electrodynamics* (New York: Dover)
[8] Byron F W and Fuller R W 1992 *Mathematics of Classical and Quantum Physics* (Mineola, NY: Dover)
[9] Boyd R W 2008 *Nonlinear Optics* 3rd edn (Amsterdam: Elsevier)
[10] Tai C-T 1971 *Dyadic Greenas Functions in Electromagnetic Theory* (Intext Educational Publisher)
[11] Mandel L and Wolf E 1995 *Optical Coherence and Quantum Optics* (Cambridge: Cambridge University Press)
[12] Andrews D L and Babiker M (ed) 2013 *The Angular Momentum of Light* (Cambridge: Cambridge University Press)
[13] Durnin J, Miceli J J and Eberly J H 1987 Diffraction-free beams *Phys. Rev. Lett.* **58** 1499–501
[14] Bolduc E, Bent N, Santamato E, Karimi E and Boyd R W 2013 Exact solution to simultaneous intensity and phase encryption with a single phase-only hologram *Opt. Lett.* **38** 3546
[15] Vetter C, Steinkopf R, Bergner K, Ornigotti M, Nolte S, Gross H and Szameit A 2019 Realization of free-space long-distance self-healing Bessel beams *Laser Photon. Rev.* **13** 1900103
[16] Snyder A W and Love J 1983 *Optical Waveguide Theory* 3rd edn (Berlin: Springer)
[17] Messiah A 2014 *Quantum Mechanics* (New York: Dover)
[18] Conway J T and Cohl H S 2010 Exact Fourier expansion in cylindrical coordinates for the three-dimensional Helmholtz Green function *Z. Angew. Math. Phys.* **61** 425
[19] Olver F W J, Lozier D W, Boisvert R F and Clarck C W (ed) 2010 *NIST Handbook of Mathematical Functions* (Cambridge: Cambridge University Press)
[20] Gradshteyn I S and Ryzhik I M 2014 *Table of Integrals, Series and Products* 8th edn (New York: Academic)
[21] Feynman R P and Hibbs A R 2010 *Quantum Mechanics and Path Integrals* (New York: Dover)

[22] Coleman S 1988 *Aspects of Symmetry* (Cambridge: Cambridge University Press)

[23] Frampton P H 2000 *Gauge Field Theories* (New York: Wiley)

[24] Baez J and Munian J P 1994 *Gauge Fields, Knots and Gravity* (Singapore: World Scientific)

[25] Zeidler E 2011 *Quantum Field Theory III: Gauge Theory–A Bridge between Mathematicians and Physicists* (Berlin: Springer)

[26] Helmholtz H 1858 über integrale der hydrodynamischen gleichungen, welche den wirbelbewegungen entsprechen *J. Reine Angew. Math.* **55** 25

[27] Kittel C 2004 *Introduction to Solid State Physics* 8th edn (New York: Wiley)

[28] Wooley R G 1980 Gauge invariant wave mechanics and the Power-Zienau-Woolley transformation *J. Phys. A Math. Gen.* **13** 2795

[29] Babiker M and Loudon R 1983 Derivation of the Power-Zienau-Woolley Hamiltonian in quantum electro-dynamics by gauge transformation *Proc. R. Soc. A Math. Phys. Sci.* **385** 439

IOP Publishing

A Field Theory Approach to Photonics

Marco Ornigotti

Chapter 3

Field theory in a nutshell

One of the most elegant formulations of physics, which constitutes the basic building blocks of modern physics, is field theory. This, for example, is the natural language of particle physics and the standard model, but it can also describe a vast category of problems in classical physics, such as mechanics, gravitation, and, through the calculus of variations, which stems from the Lagrangian formalism at the core of field theory, optimisation problems. A detailed discussion of all the aspects of field theory is outside the scope of this book. The interested reader is nevertheless invited to peruse the vast amount of literature available on the subject, should they be interested on specific aspects of this subject. Some nice references to start doing so are [1–7]. Throughout this book, however, we will frequently make use of concepts and methods borrowed from field theory to describe the interaction of light, both at the classical and quantum level, with matter. As a general strategy, the relevant parts of field theory that will be needed for achieving such a task will be introduced contextually with the discussion.

As an invitation to the subject, and to lay some foundations for the rest of the book, this chapter aims at introducing the reader to the basic formalism of field theory, namely the Lagrangian and Hamiltonian approach to the dynamics of a field, how to derive the equations of motion of the field from these quantities, and introduce Noether's theorem [8], which will allow us to show how the various conserved quantities of a field (such as momentum and angular momentum, for example) can be derived from the Lagrangian density.

The approach we will take in this chapter is rather conventional, i.e., we will use the simple case of a real, scalar field to introduce the relevant concepts and methods. In addition of being the standard approach in many textbooks in field theory, looking at the properties of a scalar field also gives us immediate access to the dynamics of scalar electromagnetic fields, i.e., electromagnetic fields whose polarisation doesn't change upon propagation of interaction with matter.

doi:10.1088/978-0-7503-5789-0ch3 3-1 © IOP Publishing Ltd 2025. All rights,

3.1 Lagrangian, Hamiltonian, and Noether's theorem

Without loss of generality, and to keep the mathematics as simple as possible, in this section we consider a single real scalar field $\psi \equiv \psi(\mathbf{x}, t)$. Einstein summation convention and the relativistic notation is also implicitly assumed.

3.1.1 Lagrangian dynamics

We construct the Lagrangian density associated to the scalar field ψ in analogy with classical mechanics [9], i.e., by taking the difference of the kinetic and potential energy of the field, namely

$$\mathscr{L} = \frac{1}{2}(\partial_\mu \psi)(\partial^\mu \psi) - U(\psi), \tag{3.1}$$

where $U(\psi)$ is an arbitrary, but smooth, function of ψ. The first term is referred to as the kinetic energy of the field because it is essentially proportional (after integration by parts) to the square of the 4-momentum of the field, which determines its kinetic properties. To see this, let us first introduce the action functional, as the integral of the Lagrangian density, i.e.,

$$S = \int d^4x \; \mathscr{L}. \tag{3.2}$$

Let us know concentrate on the kinetic term and observe the following:

$$\int d^4x \, (\partial_\mu \psi)(\partial^\mu \psi) = -\int d^4x \psi(\partial_\mu \partial^\mu \psi) = -\int d^4x \; \psi\left(\nabla^2 - \frac{\partial^2}{\partial t^2}\right)\psi, \tag{3.3}$$

where we have applied partial integration (assuming that the field at infinity vanishes, i.e., $\psi(\mathbf{x}, t)\big|_{(\mathbf{x}, \, t)\to\pm\infty} \to 0$) to go from the first to the second expression. If we now apply the usual momentum-derivative duality, i.e., $\nabla^2 \to |\mathbf{p}|^2$, it is easy to see that the term $(\partial_\mu \psi)(\partial^\mu \psi)$ indeed contains the information on the kinetic energy of the field.

This can also be seen by deriving the equation of motion for the field, i.e., its wave equation. To do so, we use the Hamilton–Maupertius principle, i.e., we require that the variation of the action is minimal $\delta S = 0$. Since the Lagrangian density depends on both ψ, through the potential term, and $\partial_\mu \psi$, through the kinetic term, the variation of the action δS can be written, in general, as

$$\delta S = \int d^4x \left[\frac{\partial \mathscr{L}}{\partial \psi}\delta\psi + \frac{\partial \mathscr{L}}{\partial(\partial_\mu \psi)}\delta(\partial_\mu \psi) \right] = 0. \tag{3.4}$$

To proceed further, we need to express the second term in terms of the variation $\delta\psi$, rather than the variation of its derivative. To do so, we employ partial integration to get

$$\int d^4x \, \frac{\partial \mathscr{L}}{\partial(\partial_\mu \psi)}\delta(\partial_\mu \psi) = -\int d^4x \, \partial_\mu\left[\frac{\partial \mathscr{L}}{\partial(\partial_\mu \psi)} \right]\delta\psi. \tag{3.5}$$

Requiring now that $\delta S = 0$ is equivalent to requiring that the integrand in equation (3.4) (after performing the integration by part as shown above) vanishes. This leads to a set of differential equations the Lagrangian density needs to satisfy, known as the Euler–Lagrange equations, that have the following form

$$\partial_\mu \left[\frac{\partial \mathscr{L}}{\partial(\partial_\mu \psi)} \right] - \frac{\partial \mathscr{L}}{\partial \psi} = 0. \tag{3.6}$$

If we consider a free, scalar field with mass m, we can choose $U(\psi) = (m^2/2)\psi^2$ as the explicit form of the potential in equation (3.1). Substituting then the expression of the Lagrangian (3.1) with this choice of the potential in the Euler–Lagrange equations above leads to the following equation of motion for ψ

$$(\partial_\mu \partial^\mu - m^2)\psi = \left(\nabla^2 - \frac{\partial^2}{\partial t^2} - m^2 \right)\psi = 0, \tag{3.7}$$

which is the Klein–Gordon equation for a massive real scalar field with mass m [1]. Notice, that for $m = 0$ the Klein–Gordon equation reduces to the wave equation (2.3) for an electric field defined as $\mathbf{E}(\mathbf{x}, t) = \psi(\mathbf{x}, t)\hat{\mathbf{f}}$, where $\hat{\mathbf{f}}$ is a fixed polarisation vector. Therefore, the Lagrangian (3.1) with $m = 0$ can be used to describe the dynamics of a scalar electric field, i.e, an electromagnetic field with fixed polarisation.

3.1.2 Equivalent Lagrangians

We have seen above how it is possible, from the Lagrangian of a field, to derive the equations of motion, through the principle of least action. To find the correct expression for the Lagrangian of a given system, the only criterion we have implicitly used is that $\mathscr{L}$ must be expressed in terms of the field ψ and its first derivatives $\partial_\mu \psi$ only, and the Lagrangian must be (for the general case of a relativistic system) Lorentz invariant[1]. Nevertheless, the choice of such a Lagrangian is not unique, since any addition of a total derivative leaves the equations of motion unchanged [9]. To illustrate this, let us consider the following modified Lagrangian for a scalar, real field $\mathscr{L}' = \mathscr{L} + \partial_\mu \Lambda^\mu$, where $\mathscr{L}$ is given by equation (3.1) and Λ^μ is a smooth vector field, subject to the same vanishing conditions at infinity, as the field ψ. We can immediately see, using integration by parts, that the action remains unchanged by the addition of this term, since

$$\begin{aligned} S' &= \int d^4x \, \mathscr{L}' = \int d^4x \, \mathscr{L} + \int d^4x \, \partial_\mu \Lambda^\mu \\ &= S - \Lambda^\mu \big|_{-\infty}^\infty = S, \end{aligned} \tag{3.8}$$

---

[1] This is not always a required condition, especially if one is considering an effective theory, such as, for example, the paraxial approximation, where Lorentz invariance is not necessarily a symmetry of the system. More in general, in constructing the Lagrangian, one should make sure that it possesses the same symmetries of the system, or effective situation, one wants to describe.

since the vector field Λ^μ vanished at infinity. Therefore, adding a total derivative to the Lagrangian does not change its equations of motion, and, by consequence, its physics. This property of the Lagrangian in useful in a variety of situations in field theory, from the determination of the conserved currents through Noether's theorem (see below) to gauge fields, where this extra condition can be used to impose a particular choice of gauge to, for example, the electromagnetic field.

3.1.3 Lagrangian density for a scalar paraxial field

Starting from the Lagrangian density for a scalar electromagnetic field, it is also possible to write a Lagrangian density, whose Euler–Lagrange equations reduce to the paraxial equation (2.39). To do this, we assume that the field ψ is monochromatic and varies slowly along the z-direction, which is defined as the propagation direction of the field. This allows us to write

$$\psi(\mathbf{x}, t) = \phi(\mathbf{R}, z)\exp[i(kz - \omega t)]. \tag{3.9}$$

It is worth noticing, however, that in doing so we are forced to break Lorentz invariance, as we have specified a propagation direction and an oscillation frequency for the field ψ. The resulting Lagrangian density will then only be valid in the nonrelativistic limit[2].

The kinetic term appearing in equation (3.1) then becomes (dropping the exponential term after derivation)

$$
\begin{aligned}
(\partial_\mu\psi)(\partial^\mu\psi) &= (\partial_j\phi)(\partial^j\phi) + (\partial_z\phi + ik\phi)(\partial^z\phi + ik\phi) - (-i\omega\phi)(-i\omega\phi) \\
&= (\nabla_\perp\phi)\cdot(\nabla_\perp\phi) + 2ik\phi\partial_z\phi + \partial_z\phi\partial_z\phi - (k^2 - \omega^2)\phi^2 \\
&= \partial_\mu(\phi\partial_\mu\phi) - \phi(\partial_j\partial^j\phi - 2ik\partial_z\phi),
\end{aligned}
\tag{3.10}
$$

where $j = \{x, y\}$ and we have used the fact, that for spatial derivatives $\partial_{x_i} = \partial^{x_i}$. In going from the second to the third line we have used the vacuum dispersion relation of the electromagnetic field, i.e., $\omega = k$ to cancel the last term on the second line, and the following identity to transform the terms containing the spatial derivatives

$$\partial_\mu\phi\partial_\mu\phi = \partial_\mu(\phi\partial_\mu\phi) - \phi\partial_\mu\partial^\mu\phi, \tag{3.11}$$

and we have dropped terms proportional to $\partial_z^2\phi$, according to the paraxial equation. This last assumption, in particular, lets us write

---

[2] The use of the term 'nonrelativistic' in this context might sound a bit confusing, and it is worth an explanation. Usually, we refer to a nonrelativistic system, when its velocity is much smaller than the speed of light. In the context of photonics, a nonrelativistic electromagnetic field is instead understood as a paraxial field. Then, the term 'nonrelativistic' is simply a synonym of paraxiality, in the sense that the rate of change of the field (i.e., its velocity) along the propagation direction is much smaller than that along the transverse direction, i.e., it is a slowly varying field along that direction. The term 'nonrelativistic' is used because of the isomorphism existing between the paraxial equation and the Schrödinger equation. As the Schrödinger equation is the nonrelativistic limit of the relativistic Klein–Gordon equation, so is the paraxial equation the nonrelativistic (i.e., paraxial) limit of the relativistic (i.e., nonparaxial) Helmholtz equation.

$$\partial_z\phi\partial_z\phi = \partial_z(\phi\partial_z\phi) - \partial_z^2\phi \simeq \partial_z(\phi\partial_z\phi). \tag{3.12}$$

The first term in the last line of equation (3.10) amounts to a total derivative and can be neglected [10], since it doesn't contribute to the equations of motion. We are then left with the following form for the Lagrangian of a paraxial, scalar field

$$\mathscr{L}_{\text{paraxial}} = \phi(\partial_j\partial^i - 2ik\partial_z)\phi = \frac{1}{2}(\partial_j\phi)(\partial^j\phi) + ik\phi\partial_z\phi, \tag{3.13}$$

where for the second equality we have used partial integration to transform the term $\phi(\partial_j\partial^j\phi)$ into its usual kinetic form $(\partial_j\phi)(\partial^j\phi)$ for consistency of notation. Substituting this into the Euler–Lagrange equations (3.6) gives the paraxial equation

$$i\partial_z\phi = -\frac{1}{2k}\nabla_\perp^2\phi. \tag{3.14}$$

Notice, that the form of the paraxial Lagrangian given above is not unique, and the same result can be derived in many different ways, for example by considering a complex, rather than real, scalar field (see, e.g., reference [7] for details). However, all the formulations are equivalent and lead to the same paraxial equation above.

3.1.4 Hamiltonian dynamics

Another alternative, but equivalent, formulation of field theory to the Lagrangian is the so-called Hamiltonian formulation, which has a more direct connection with the total energy of the system, and it is historically the preferred formulation for quantum mechanics and light–matter interaction. The core of the Hamiltonian representation of a system is the definition of the so-called conjugated variables, often indicated as (x, p) for mechanical systems, or (ψ, Π) for fields. The fundamental property of these variables is to be, in a sense, Fourier pairs. In classical mechanics, canonically conjugated variables are defined by the following action of the Poisson brackets[3] of them [9]

$$\{q_i, p_j\} = \delta_{ij}, \tag{3.15}$$

which in quantum mechanics becomes the celebrated commutation relation between position and momentum of a quantum particle, i.e., $[\hat{x}_i, \hat{p}_j] = i\hbar\delta_{ij}$. The concept of canonically conjugated variables, Poisson brackets (and commutators) can be easily extended to fields. In this context, the canonically conjugated field (called, by analogy with mechanics, momentum) to the field ψ is the field $\Pi \equiv \Pi(\mathbf{x}, t)$ given by

$$\Pi = \frac{\partial\mathscr{L}}{\partial(\partial_t\psi)}, \tag{3.16}$$

---

[3] The Poisson brackets between two quantities f and g are defined as $f, g = \sum_{i=1}^{N}[\partial_{q_i}f\partial_{p_i}g - \partial_{p_i}f\partial_{q_i}g]$, where N is the number of components of the system.

which for a scalar field is given by $\Pi = \partial_t \psi$. Using the momentum, we can introduce the Hamiltonian density of the field by means of a Legendre transformation of its Lagrangian density as follows[4]

$$\mathscr{H} = \Pi \partial_t \psi - \mathscr{L}. \tag{3.17}$$

In doing so, we essentially go from a description of the system in terms of the field ψ and its derivatives $\partial_\mu \psi$ (i.e., the Lagrangian approach), to that of the field ψ and its conjugated momentum Π (i.e., the Hamiltonian approach). Substituting the Lagrangian density (3.1) for a scalar field into the expression above gives

$$\mathscr{H} = \frac{1}{2}\Pi^2 + \frac{1}{2}(\partial_i \psi)(\partial^i \psi) + U(\psi), \tag{3.18}$$

which is the Hamiltonian density for a real, scalar field. $\mathscr{H}$ represents the total energy contained in the field, as it is expressed as the sum of its kinetic energy $[\Pi^2 + (\partial_i \psi)^2]$ and potential energy $[U(\psi)]$. Analogously to the Lagrangian case, we can also define an action for the field in terms of its Hamiltonian density as

$$S = \int d^4x \, \mathscr{H}, \tag{3.19}$$

and use the principle of least action to derive the equations of motion. Notice, that while in the Lagrangian case the equations of motion are found by requiring that the action is minimal, in the Hamiltonian case they are instead found by requiring that the action is maximal. This is due to the presence of the minus sign in front of the Lagrangian in the definition of the Hamiltonian given by equation (3.17). However, since the minimum and maximum of a functional are both found by imposing the condition $\delta S = 0$, the two approaches are, de facto, equivalent and give rise to the same physics. An illustration of this difference is shown in figure 3.1. The variation of the action in this case is given by

$$\delta S = \int d^4x \left[\frac{\partial \mathscr{H}}{\partial \Pi}\delta\Pi + \frac{\partial \mathscr{H}}{\partial \psi}\delta\psi + \frac{\partial \mathscr{H}}{\partial(\partial_i \psi)}\delta(\partial_i \psi) \right], \tag{3.20}$$

which, upon partial integration, leads to the following Hamilton equations of motion for the field ψ:

$$\partial_t \psi = \frac{\partial \mathscr{H}}{\partial \Pi}, \tag{3.21a}$$

$$\partial_t \Pi = -\frac{\partial \mathscr{H}}{\partial \psi} + \partial_i \left[\frac{\partial \mathscr{H}}{\partial(\partial_i \psi)} \right]. \tag{3.21b}$$

---

[4] A Legendre transformation is analogous to a contact transformation in classical mechanics, i.e., a change of variable that links a pair of canonically conjugated variables to another pair of canonically conjugated variables.

Lagrangian Mechanics

Hamiltonian Mechanics

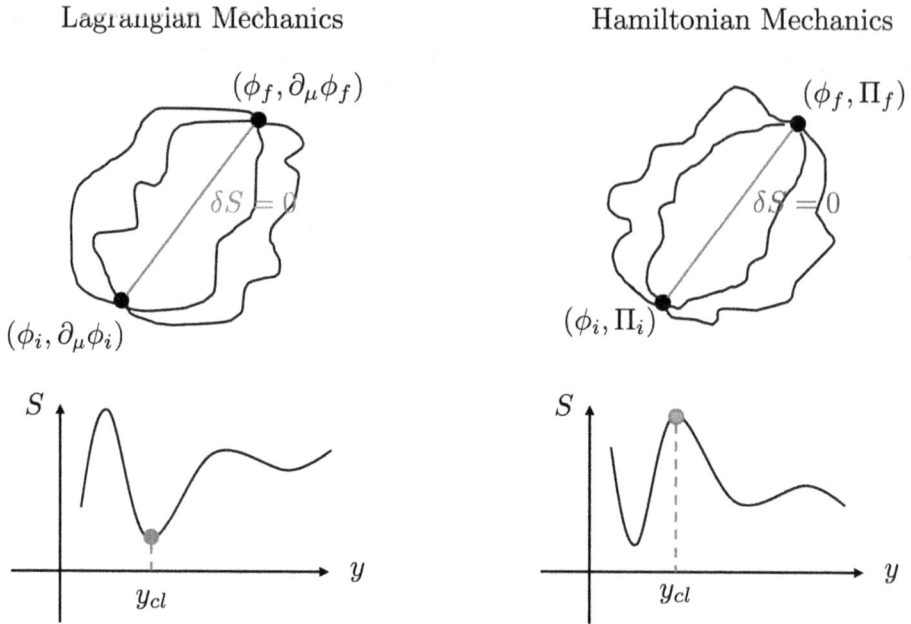

Figure 3.1. Pictorial representation of the conceptual difference between Lagrangian and Hamiltonian mechanics. In Lagrangian mechanics (left), the equations of motion are found by minimising the action. The classical path that realises this condition (y_{cl}, in red) is the classical trajectory that brings the system from the initial configuration (ϕ_i, $\partial_\mu \phi_i$) to its final configuration (ϕ_f, $\partial_\mu \phi_f$). The minimisation of the action can be seen in the action-path diagram on the left, for which $S(y_{cl}) < S(y)$, for any $y \neq y_{cl}$, meaning that the action corresponding to the classical path (in red) has the lowest value amongst all possible actions generated by other paths. For the case of Hamiltonian mechanics (right) instead, the equations of motion are found by maximising the action, and the classical trajectory y_{cl} (in blue) is then given as that with the maximum action. This classical path is the one that brings the system from the initial configuration (ϕ_i, Π_i) to the final configuration (ϕ_f, Π_f), corresponding to the path with the highest value of the action. This can be seen in the action-path diagram on the right, for which $S(y_{cl}) > S(y)$, for any $y \neq y_{cl}$. The two formulations are however equivalent, since the condition $\delta S = 0$, required to find the minimal or maximal action, doesn't distinguish between maxima and minima, but only finds extrema of the action functional.

It is left to the reader as an exercise to prove that by using the Hamiltonian density defined in equation (3.18) in the equations of motion above, and by choosing, again $U(\psi) = m^2 \psi^2 / 2$, one obtains the same Klein–Gordon equation for ψ as in equation (3.7)

3.1.5 Noether theorem and conserved quantities

We now go back to the Lagrangian description of a scalar field ψ and investigate what happens to the action, when a coordinate transformation is applied to the field ψ. In particular, we are interested in those transformations that do not change the physical content of the field, but just represent it in a different set of coordinates. In the language of special relativity, this means that the considered transformation must be compatible with Lorentz transformations. Amongst all the possible choices, here we focus on those transformations corresponding to particular symmetries of the fields, such as translations and rotations. Formally, a transformation of the type $\psi' = \hat{O}\psi$ is a symmetry for the field ψ if the action S associated with ψ is invariant under the effect of the transformation operator $\hat{O}$ [10]. In this section we will see, that requiring that a transformation leaves the action invariant corresponds to defining a current (and charge), linked to a conserved quantity of the field. This is the essence of Noether's theorem.

To illustrate how to derive Noether's current, let us assume that a coordinate transformation $x^\mu \to y^\mu = x^\mu + \delta x^\mu$ is performed on the field ψ, such that ψ undergoes a transformation of the type $\Psi = \hat{O}\psi = \psi + \delta\psi$, and that this transformation corresponds to a symmetry of the system. Since, in general, $\psi = \psi(\mathbf{x}, t)$, this coordinate transformation will also affect the Lagrangian, giving $\mathscr{L}' = \mathscr{L}(\Psi, \partial_\mu\Psi, y^\mu)$[5]. From the definition of symmetry of a system given above, the action must be left invariant, and therefore we must have that $\mathscr{L}(\Psi, \partial_\mu\Psi, y^\mu) = \mathscr{L}(\psi, \partial_\mu\psi, x^\mu)$ must hold. To find an explicit expression for this condition, we calculate the change $\delta\mathscr{L}$ induced to the Lagrangian by the transformation as

$$
\begin{aligned}
\delta\mathscr{L} &= \mathscr{L}(\Psi, \partial_\mu\Psi, y^\mu) - \mathscr{L}(\psi, \partial_\mu\psi, x^\mu) \\
&= \frac{\partial\mathscr{L}}{\partial\psi}\delta\psi + \frac{\partial\mathscr{L}}{\partial(\partial_\mu\psi)}\delta(\partial_\mu\psi) + \frac{\partial\mathscr{L}}{\partial x^\mu}\delta x^\mu \\
&= \partial_\mu\left[\frac{\partial\mathscr{L}}{\partial(\partial_\mu\psi)}\right]\delta\psi + \frac{\partial\mathscr{L}}{\partial(\partial_\mu\psi)}\delta(\partial_\mu\psi) + \frac{\partial\mathscr{L}}{\partial x^\mu}\delta x^\mu \qquad (3.22) \\
&= \partial_\mu\left[\frac{\partial\mathscr{L}}{\partial(\partial_\mu\psi)}\delta\psi\right] + \frac{\partial\mathscr{L}}{\partial x^\mu}\delta x^\mu,
\end{aligned}
$$

where to go from the second to the third line we made use of the Euler–Lagrange equations (3.6) to rewrite $\partial\mathscr{L}/\partial\psi$, and from the third to the fourth line we have used the product rule, i.e., $(\partial_\mu F)\delta\psi + F\delta(\partial_\mu\psi) = (\partial_\mu F)\delta\psi + F\partial_\mu(\delta\psi) = \partial_\mu(F\delta\psi)$. To ensure that the Lagrangian is left invariant by the coordiante transformation we need to set $\delta\mathscr{L} = 0$, which corresponds to

$$
\partial_\mu\left[\frac{\partial\mathscr{L}}{\partial(\partial_\mu\psi)}\delta\psi + \delta^\mu_\nu\mathscr{L}\delta x^\nu\right] \equiv \partial_\mu j^\mu = 0, \qquad (3.23)
$$

where we have defined the so-called Noether current as

$$
j^\mu = \frac{\partial\mathscr{L}}{\partial(\partial_\mu\psi)}\delta\psi + \delta^\mu_\nu\mathscr{L}\delta x^\nu. \qquad (3.24)
$$

---

[5] In writing this form of the transformed Lagrangian, we have made explicit the fact, that the Lagrangian could, in principle, also depend on the coordinates x^μ explicitly.

If we introduce the Noether charge associated with the Noether current, i.e.,

$$Q = \int d^3x \, j^0, \tag{3.25}$$

equation (3.23) can be written in a familiar form, namely that of a continuity equation involving charge and current, as

$$\partial_t Q - \nabla \cdot \mathbf{j} = 0. \tag{3.26}$$

From this result, we can see that a coordinate transformation of the field ψ generates a continuity equation, whose charge and current are reminiscent of the particular transformation operated on ψ, according to the definition of Noether current given in equation (3.24). It is worth pointing out that although this way of deriving Noether's theorem implicitly only handles continuous symmetries, Noether's theorem can also be applied in many other cases, including internal symmetries (normally linked to concepts of gauge invariance and isospin) [11], and discrete symmetries [12], and something similar to a Noether's theorem can also emerge from the topology of the field [13].

3.1.6 Conserved quantities and their continuity equations

From Noether's theorem, we can derive the conservation laws for momentum, energy, and angular momentum for the field, by simply imposing that the Lagrangian of the field is invariant under spatial translations, temporal translations, and rotations, respectively. To do so, we need to find the expressions for δx^μ in all the three mentioned cases, and express $\delta \psi$ in terms of them.

Let us begin with spacetime translations, for which $\delta x^\mu = a^\mu$, with a^μ being a constant vector, which implies $\delta \psi = \partial_\mu \psi \delta x^\mu = -(\partial_\mu \psi) \, a_\mu$, where the minus sign has been put for convention. Substituting this into equation (3.23) gives the following result

$$\partial_\mu \left[\frac{\partial \mathcal{L}}{\partial(\partial_\mu \psi)} \partial_\nu \psi - \delta_\nu^\mu \mathcal{L} \right] a^\nu \equiv \partial_\mu (T_\nu^\mu) a^\nu = 0, \tag{3.27}$$

where we have introduced the rank-two tensor T_ν^μ as the stress–energy tensor of the field [1, 10, 11], with which the Noether current can be written as $(j^\mu)_\nu = T_\nu^\mu$. We can then define the correspondent Noether charges as

$$Q_\nu = \int d^3x \, T_\nu^0 \equiv (E, \mathbf{p}) \tag{3.28}$$

$E = \int d^3x \, T_0^0$ is the energy density, and $p_i = \int d^3x \, T_i^0$ the momentum density. We then see that invariance under temporal or spatial translations corresponds to conservation of energy and momentum, respectively, which are governed by the continuity equation $\partial_\mu T_\nu^\mu = 0$, or, in terms of energy and momentum density

$$\frac{\partial E}{\partial t} - \nabla \cdot \mathbf{p} = 0. \tag{3.29}$$

For the case of rotations, instead, we have that $\delta x^\mu = M^\mu_\nu x^\nu$, where M^μ_ν is the rotation matrix. Consequently, $\delta\psi = (\partial_\mu\psi)M^\mu_\nu \delta x^\nu$, and equation (3.23) implies the following condition

$$\partial_\mu T^\mu_\nu M^{\nu\sigma} x_\sigma = 0. \tag{3.30}$$

To define the Noether current associated with this transformation, we need to rewrite the expression above in the following form first

$$
\begin{aligned}
\partial_\mu T^\mu_\nu M^{\nu\sigma} x_\sigma &= \frac{1}{2}\left(\partial^\mu T^\nu_\mu M_{\nu\sigma} x^\sigma + \partial^\mu T^\nu_\mu M_{\nu\sigma} x^\sigma\right) \\
&= \frac{1}{2}\left(\partial^\mu T^\nu_\mu M_{\nu\sigma} x^\sigma - \partial^\mu T^\nu_\mu M_{\sigma\nu} x^\sigma\right) \\
&= \frac{1}{2}\left(\partial^\mu T^\nu_\mu x^\sigma - \partial^\mu T^\sigma_\mu x^\nu\right)M_{\nu\sigma} \\
&= \frac{1}{2}\partial_\mu (T^{\mu\nu} x^\sigma - T^{\sigma\mu} x^\nu)M_{\nu\sigma},
\end{aligned}
\tag{3.31}
$$

where in passing from the second to the third row we have used the antisymmetry property of rotation matrices, i.e., $M_{\mu\nu} = -M_{\nu\mu}$, then in the third row we exchanged the dummy indices $\nu \leftrightarrow \sigma$ and finally we have raised the index μ for consistency. The Noether current associated with rotations can then be written as

$$(j^\mu)^{\nu\sigma} = T^{\mu\nu} x^\sigma - T^{\sigma\mu} x^\nu, \tag{3.32}$$

and the associated Noether charge reads

$$Q^{\nu\sigma} = \int d^3x \, (T^{0\nu} x^\sigma - T^{\sigma 0} x^\nu). \tag{3.33}$$

If we limit ourselves to consider spatial rotations, then the charge Q^{ij} defines a vector $L_i = (1/2)\varepsilon_{ijk} Q^{jk}$, which can be identified with the angular momentum of the field. On the other hand, for rotations mixing both spatial and temporal directions, the charge Q^{0i} accounts for the conservation of boost [1, 10]. For the former case, we can rewrite the conservation law $\partial_\mu (j^\mu)^{\nu\sigma} = 0$ in terms of the continuity equation involving the angular momentum density and flux as follows

$$\frac{\partial \mathcal{J}_k}{\partial t} + \frac{\partial \mathcal{M}_{jk}}{\partial x_j} = 0, \tag{3.34}$$

where

$$\mathcal{J}_i = \varepsilon_{ijk} \, T^{0j} \, x^k = \varepsilon_{ijk} \, p^j \, x^k, \tag{3.35}$$

is the ith component of the angular momentum (AM) density (and we have identified $T^{0j} = p^j$, as the jth component of the momentum) and

$$\mathcal{M}_{ij} = \varepsilon_{jkl} \, x^k \, T^l_{\ i}, \tag{3.36}$$

is the angular momentum flux across a surface oriented along the i-direction, and ε_{ijk} is the totally anti-symmetric Levi-Civita symbol [14].

The relations above suggest how it is possible to write a whole hierarchy of continuity equations for the conserved quantities of a field, each related to a specific symmetry. These relations have been investigated extensively for the electromagnetic field and generalised to an arbitrary hierarchy of continuity equations involving higher-order conserved quantities of the field [15, 16].

In the case of fields carrying a nonzero spin, or, more generally, when the spatial distribution of the field is nontrivial, spatial rotations can generate an extra term in the continuity equation, which is related to the spin angular momentum (SAM) density of the field. This can be calculated for the case of a simple scalar field by assuming that a spatial rotation transforms the field as follows $\delta\psi = \varepsilon_{\mu\nu} S^{\mu\nu}\psi + (\partial_\mu\psi)M_\nu^\mu \delta x^\nu$, i.e., it not only rotates the coordinates, but also rotates the field itself, in a coordinate-independent manner. In this case, while the second term of $\delta\psi$ gives the OAM derived above, the first term instead corresponds to the SAM, which is generated by the following conservation law

$$\partial_\mu \left[\frac{\partial \mathscr{L}}{\partial(\partial_\mu\psi)} \varepsilon_{\nu\sigma} S^{\nu\sigma}\psi \right] = 0, \tag{3.37}$$

and the correspondent Noether charge defines the SAM as

$$S_i = \frac{1}{2}\varepsilon_{ijk} \int d^3x \left[\frac{\partial \mathscr{L}}{\partial(\partial_0\psi)} S^{jk}\psi \right]. \tag{3.38}$$

3.2 Noether theorem for internal symmetries of the field

We conclude this chapter by mentioning another important application of Noether's theorem, i.e., the study of internal symmetries of the field. This is a rather important topic when dealing with gauge theories, fields with nontrivial internal symmetries (such as isospin or colour symmetry [4, 11]), or matter and gauge fields interacting, as in the case of quantum electrodynamics. In this latter example, in fact, the electric charge emerges as a conserved quantity correspondent to an internal symmetry of the light-plus-matter system, i.e., the gauge symmetry [17]. For the sake of completeness, and to provide a non-conventional example of application of Noether's theorem to a field theory, here we consider the simplest case of internal symmetry, i.e., a translation of the field itself, and show how this symmetry leads to the conservation of the canonically conjugated momentum density $\Pi = \partial\mathscr{L}/\partial(\partial_0\psi)$.

In this case, we do not assume any coordinate transformation, i.e., $\delta x^\mu = 0$, but allow the existence of a transformation of the sort $\psi \to \psi + \delta\psi$. To a certain extent, this is the translational analogue of the reference frame independent field rotation that leads to SAM. The Noether current can be derived directly from equation (3.23) by setting $\delta x^\mu = 0$, thus obtaining[6]

---

[6] We now indicate the Noether current as J^μ to differentiate it from that generated by external symmetries.

$$\partial_\mu \left[\frac{\partial \mathcal{L}}{\partial(\partial_\mu \psi)} \delta\psi \right] \equiv \partial_\mu J^\mu = 0, \qquad (3.39)$$

and the associated Noether charge reads

$$Q = \int d^3x \, J^0 \qquad (3.40)$$

Notice, that this expression is very general, and can be applied to any internal transformation of the field. As an example, let us consider the case of a translation of the field itself, described by the (infinitesimal) transformation[7] $\psi' = \psi - i\varepsilon$, with $\varepsilon \ll 1$. In this case, we get

$$\partial_\mu \left[\frac{\partial \mathcal{L}}{\partial(\partial_\mu \psi)} \right] = \partial_t \Pi + \nabla \cdot \mathbf{P} = 0, \qquad (3.41)$$

where we have introduced the conjugated momentum of the field as $\Pi = \partial \mathcal{L}/\partial(\partial_0 \psi)$ and defined $P_i = \partial \mathcal{L}/\partial(\partial_i \psi)$. If we integrate the above equation in an infinite volume V, we can use the divergence theorem to get rid of the ∇-operator on $\mathbf{P}$ and get

$$\int d^3x \, \nabla \cdot \mathbf{P} = \int_{\partial V} d^2x \, \mathbf{P} = 0, \qquad (3.42)$$

since $\mathbf{P}$ is defined in terms of fields that vanish on ∂V at infinity. Therefore, by identifying the integral of the conjugated momentum with the Noether charge Q, we get the following conservation law for this particular internal symmetry

$$\partial_t Q = \partial_t \Pi = 0. \qquad (3.43)$$

Invariance of the field upon internal translation symmetry, i.e., translation of the field itself, thus corresponds to conservation of the conjugated momentum $\Pi = \partial \mathcal{L}/\partial(\partial_0 \psi)$.

References

[1] Radmon P 1997 *Field Theory: A Modern Primer* 2nd edn (Boulder, CO: Westview Press)
[2] Barut O 1981 *Electrodynamics and Classical Theory of Fields and Particles* (Mineola, NY: Dover)
[3] Scheck F 2012 *Classical Field Theory* (Boston, MA: Springer)
[4] Frampton P H 2000 *Gauge Field Theories* (New York: Wiley)
[5] Das A 2006 *Field Theory: A Path Integral Approach* 2nd edn (Singapore: World Scientific)
[6] Baez J and Munian J P 1994 *Gauge Fields, Knots and Gravity* (Singapore: World Scientific)
[7] Landau L D and Lifshitz E M 1980 *The Classical Theory of Fields* 4th edn (Oxford: Butterworth-Heinemann)

---

[7] Notice, that we have not made any assumption on whether the field ψ takes real or complex values. Nevertheless, we implement the transformation as a complex translation. Although this can be formally justified, here we take it as a convention.

[8] Noether E 1918 Invariante variationsprobleme *Nachr. Ges. Wiss. Göttingen, Math.-Phys. Kl.* **1918** 235

[9] Arnold V I 1989 *Mathematical Methods of Classical Mechanics* (Berli: Springer)

[10] Maggiore M 2005 *A Modern Introduction to Quantum Field Theory* (Oxford: Oxford University Press)

[11] Itzykson C and Zuber J B 1980 *Quantum Field Theory* (New York: Dover)

[12] Seidl G 2023 Discrete noether currents arXiv:1405.1409v1

[13] Skopenkov M 2023 Discrete field theory: symmetries and conservation laws *Math. Phys. Anal. Geom.* **26** 19

[14] Byron F W and Fuller R W 1992 *Mathematics of Classical and Quantum Physics* (Mineola, NY: Dover)

[15] Cameron R P, Barnett S M and Yao A M 2012 Optical helicity, optical spin and related quantities in electromagnetic theory *New J. Phys.* **14** 053050

[16] Aghapour S, Andersson L and Rosquist K 2020 The zilch electromagnetic conservation law revisited *J. Math. Phys.* **61** 122902

[17] Coleman S 2019 *Quantum Field Theory* (Singapore: World Scientific)

IOP Publishing

A Field Theory Approach to Photonics

Marco Ornigotti

Chapter 4

Electromagnetic field theory

This chapter constitutes the core of this book. Here, in fact, we apply the concepts and methods briefly outlined in chapter 3 for a scalar field, to the case of the electromagnetic field and derive explicitly the general expressions of momentum, orbital angular momentum (OAM), spin angular momentum (SAM) and helicity density for the electromagnetic field using the Lagrangian formalism.

The first part of the chapter, encompassing sections 4.1–4.6, constitutes the entry-level application and the simplest example of usage of field theory methods in photonics, discussing basic properties such as momentum, angular momentum and polarisation of the field. It will also serve to fix notation and jargon between optics and field theory.

The rest of the chapter is devoted to developing a Lagrangian-based microscopic theory of light–matter interaction, based on the model proposed by Huttner and Barnett in 1992 [1]. Contrary to the work presented in reference [1], in this chapter we will limit ourselves to the classical case only, leaving the quantisation of the model for chapter 7. Section 4.7, where the Huttner–Barnett model is discussed in detail, also serves as a starting point for describing the quantised electromagnetic field in arbitrary media, which will be carried out in chapter 7 using path integrals, as a means to introduce a quantum field theory for nonlinear interaction of photons in arbitrary media.

Finally, we briefly touch into the topic of the electromagnetic field in curved spacetime, to give some insight into the recently introduced concept of transformation optics [2–4]. More on the topic of transformation optics and invisibility cloaks can be found in the book by Leonhardt [5]. Lastly, appendix B contains the detailed derivation of the Power–Zienau–Wooley transformation sketched in section 2.6.6.

The discussion on the momentum, OAM and SAM of the electromagnetic field is fairly standard, and further details can be found in any classical field theory book that treats electromagnetism, such as, for example Landau–Lifshitz [6], the book by

doi:10.1088/978-0-7503-5789-0ch4   4-1   © IOP Publishing Ltd 2025. All rights,

Barut [7] or that by Scheck [8]. On the other hand, the derivation of the helicity density from the Noether's theorem, presented in section 4.6 is based on the works by Cameron, Barnett and Yao [9] and Bliokh [10].

Within this chapter, if not specified otherwise, we omit the $(\mathbf{r},\ t)$-dependence from the various fields, for simplicity.

4.1 A preliminary discussion

Frequently, when one finds a reference to the Lagrangian or Hamiltonian density of the free electromagnetic field within the context of photonics, one finds the following expressions for the two functionals

$$\mathscr{L} = \frac{\varepsilon_0}{2}|\mathbf{E}|^2 - \frac{1}{2\mu_0}|\mathbf{B}|^2, \qquad (4.1a)$$

$$\mathscr{H} = \frac{\varepsilon_0}{2}|\mathbf{E}|^2 + \frac{1}{2\mu_0}|\mathbf{B}|^2. \qquad (4.1b)$$

Sometimes, in those references, the argument of gauge invariance, in combination with the electric and magnetic fields being the physical quantities (in opposition to the electromagnetic potentials) is brought up as a justification to write the Lagrangian (Hamiltonian) in the form above. While the definition of the Hamiltonian density of the electromagnetic field as sum of the electric and magnetic energy is consistent with Maxwell's equations, the Lagrangian (4.1a) is problematic, as Maxwell's equations cannot be derived directly from it. To see this, let us derive the Euler–Lagrange equations (3.6) for the Lagrangian density above, assuming that $\mathbf{E}$ ad $\mathbf{B}$ are independent fields. Since $\mathscr{L}$ does not depend on the derivative of either $\mathbf{E}$ of $\mathbf{B}$, they simply reduce to

$$\frac{\partial \mathscr{L}}{\partial \mathbf{E}} = 0 \ \rightarrow \ \mathbf{E} = 0, \qquad (4.2a)$$

$$\frac{\partial \mathscr{L}}{\partial \mathbf{B}} = 0 \ \rightarrow \ \mathbf{B} = 0, \qquad (4.2b)$$

which is in contrast with Maxwell's equations. The reason for this erroneous result can be explained in many different ways. One possible explanation could be that the assumptions of treating the electric and magnetic fields as independent quantities is wrong. In fact, Maxwell's equations require $\mathbf{E}$ and $\mathbf{B}$ to be interlinked (a variation in time of the electric field generates a magnetic field, and vice versa).

To find a more formal justification for why the result above is wrong, let us consider an easier example, i.e., that of a massless scalar field, and let us rewrite its Lagrangian density by introducing the auxiliary field $\phi^\mu = \partial_\mu \psi$. In terms of this new field, the Lagrangian (3.1) can be written as follows

$$\mathscr{L} = \frac{1}{2}(\partial_\mu \psi)(\partial^\mu \psi) = \frac{1}{2}\phi_\mu \phi^\mu. \qquad (4.3)$$

Practically, what we have done is that we have eliminated the derivatives of the field from $\mathscr{L}$. Using now this new expression to calculate the Euler–Lagrange equations, we get $\partial \mathscr{L} / \partial \phi_\mu = \phi_\mu = \partial_\mu \psi = 0$, which is not the (massless) Klein–Gordon equation obeyed by ψ.

In this example, the fact that we have introduced a transformation between the original configuration space $\{\psi, \partial_\mu \psi\}$ and the new configuration space $\{\phi^\mu, \partial_\nu \phi^\mu\}$ is the reason of this erroneous result. This change of configuration space, in fact, does not produce an equivalent Lagrangian to the starting one, and therefore the Euler–Lagrange equation for this new Lagrangian are not constrained to reproduce the same results as those of the original Lagrangian. Moreover, this is not an allowed change (in the sense that it doesn't generate an equivalent Lagrangian to the original one) essentially because it implicitly treats ψ and $\partial_\mu \psi$ as independent quantities, when in reality they are not independent from each other. This is the same problem that emerges when using the Lagrangian (4.1a) to describe the electromagnetic field: since it is not cast in the right configuration space, it cannot produce an exact result.

In general, therefore, if one wants to make a change in configuration variables, i.e., a change of fields, one must be careful in making a transformation that brings the Lagrangian into an equivalent one, so that the Euler–Lagrange equations (3.6) can be used correctly.

For the electromagnetic field, as it is well-known, the correct configuration space is the one spanned by the electromagnetic potentials $\{\phi, \mathbf{A}\}$, and not the one spanned by the electric and magnetic fields $\mathbf{E}$ and $\mathbf{B}$. Therefore, even though the Lagrangian (4.1a) is conveniently written in terms of electric and magnetic fields solely, it is not the correct one to use for deriving Maxwell's equations. Instead, from equation (4.1a) one can generate the correct Lagrangian by simply substituting the expression of the electric and magnetic fields in terms of the scalar and vector potential, i.e., $\mathbf{E}(\mathbf{r}, t) = -\partial_t \mathbf{A}(\mathbf{r}, t) - \nabla \phi(\mathbf{r}, t)$ and $\mathbf{B}(\mathbf{r}, t) = \nabla \times \mathbf{A}(\mathbf{r}, t)$, to obtain

$$\mathscr{L} = \frac{\varepsilon_0}{2} |\partial_t \mathbf{A}(\mathbf{r}, t) - \nabla \phi(\mathbf{r}, t)|^2 - \frac{1}{2\mu_0} | \nabla \times \mathbf{A}(\mathbf{r}, t)|^2. \tag{4.4}$$

Although in principle correct, this form is rather cumbersome to utilise. In the next section we introduce a much easier to handle form, i.e., the standard, relativistic electromagnetic Lagrangian written as a function of the electromagnetic tensor.

4.2 The Lagrangian for the electromagnetic field

The Lagrangian density of the electromagnetic field in the presence of external sources is easily written in terms of the 4-vector potential $A_\mu \equiv (\phi(\mathbf{x}, t), \mathbf{A}(\mathbf{x}, t))$ and the 4-current $J^\mu \equiv (\rho, \mathbf{J})$ as

$$\mathscr{L} = -\frac{1}{4} F_{\mu\nu} F^{\mu\nu} - A_\mu J^\mu, \tag{4.5}$$

where $F_{\mu\nu} = \partial_\mu A_\nu - \partial_\nu A_\mu$ is the Faraday electromagnetic tensor, whose explicit expression, in terms of electric and magnetic field components, is given by

$$F_{\mu\nu} = \begin{pmatrix} 0 & E_x & E_y & E_z \\ -E_x & 0 & -B_z & B_y \\ -E_y & B_z & 0 & -B_x \\ -E_z & -B_y & B_x & 0 \end{pmatrix}. \tag{4.6}$$

The Faraday tensor is anti-symmetric, i.e., $F_{\mu\nu} = -F_{\nu\mu}$, and therefore traceless, i.e., $\mathrm{Tr}\{F_{\mu\nu}\} = 0$. Notice, moreover, that the interaction between the electromagnetic field and the sources, represented by the term $A_\mu J^\mu$ is in minimal coupling form. Using the Lagrangian above, the equations of motion can be calculated directly from the Euler–Lagrange equations (3.6) by identifying $\partial_\mu \psi \rightarrow \partial_\mu A_\nu$ and $\psi \rightarrow A_\nu$, and can be written as

$$\partial_\mu F^{\nu\mu} = J^\nu. \tag{4.7}$$

It is worth noticing that one could have arrived to the same equation of motion also by considering $\{F_{\mu\nu}, A_\mu\}$ as configuration space, instead of $\{\partial_\mu A_\nu, A_\mu\}$, since

$$\frac{\partial \mathcal{L}}{\partial(\partial_\mu A_\nu)} = -\frac{1}{4}\frac{\partial}{\partial(\partial_\mu A_\nu)}[(\partial_\alpha A_\beta - \partial_\beta A_\alpha)(\partial^\alpha A^\beta - \partial^\beta A^\alpha)]$$

$$= -\frac{1}{4}\frac{\partial}{\partial(\partial_\mu A_\nu)}[(\partial_\alpha A_\beta - \partial_\beta A_\alpha)\eta^{\alpha\sigma}(\partial_\sigma A_\tau - \partial_\tau A_\sigma)\eta^{\beta\tau}] \tag{4.8}$$

$$= -\frac{1}{2}F^{\mu\nu}(\delta_{\mu\alpha}\delta_{\nu\beta} - \delta_{\nu\alpha}\delta_{\mu\beta}) = -F^{\mu\nu} = \frac{\partial \mathcal{L}}{\partial F_{\mu\nu}}.$$

Substituting equation (4.11) into equation (4.7) gives the two Maxwell's equations for the divergence of **E** and the curl of **B**, i.e., equations (2.1b) and (2.1d). The other two equations, i.e., the one for the divergence of **E** and the curl of **B**, are instead derived from the so-called Bianchi identity, namely

$$\varepsilon^{\mu\nu\sigma\tau}\partial_\nu F_{\sigma\tau} = 0. \tag{4.9}$$

We can rewrite the above equation in a more inspiring form by introducing the so-called dual electromagnetic tensor, also known as the Maxwell's tensor, as

$$G^{\mu\nu} = \frac{1}{2}\varepsilon^{\mu\nu\sigma\tau}F_{\sigma\tau}, \tag{4.10}$$

which, in terms of electromagnetic field components, is given by

$$G^{\mu\nu} = \begin{pmatrix} 0 & B_x & B_y & B_z \\ -B_x & 0 & -E_z & E_y \\ -B_y & E_z & 0 & -E_x \\ -B_z & -E_y & E_x & 0 \end{pmatrix}, \tag{4.11}$$

i.e., the Maxwell tensor is simply the Faraday tensor with the roles of **E** and **B** exchanged. As we will see later in this chapter, the Maxwell tensor plays an important role in determining the helicity of the electromagnetic field. The Bianchi identity in equation (4.9) can be then rewritten in the following form, using the Maxwell tensor

$$\partial_\nu G^{\mu\nu} = 0, \tag{4.12}$$

which is analogue to equation (4.7), but casted in terms of the Maxwell, rather than the Faraday, tensor. Notice, moreover, that the Maxwell tensor is nonzero only in an even number of spatial dimensions [7, 11]. To understand why this is true, let us consider the scalar product of the Faraday and Maxwell tensors, which constitutes one of the electromagnetic invariants (the other being $F_{\mu\nu}F^{\mu\nu}$) [6], namely

$$G_{\mu\nu}F^{\mu\nu} = \frac{1}{2}\varepsilon_{\mu\nu\sigma\tau}F^{\mu\nu}F^{\sigma\tau} = -4\mathbf{E}\cdot\mathbf{B}. \tag{4.13}$$

In regular three-dimensional space, the electric and magnetic fields are orthogonal to each other, so $G_{\mu\nu}F^{\mu\nu} = 0$. However, since $F_{\mu\nu} \neq 0$ because it appears in the field Lagrangian (see equation (4.5)), for the above expression to be zero, one must set $G_{\mu\nu} = 0$.

Using both the Faraday and Maxwell tensor, Maxwell's equations can then be written as the following set of tensorial equations

$$\partial_\mu F^{\nu\mu} = J^\nu, \tag{4.14a}$$

$$\partial_\mu G^{\nu\mu} = 0. \tag{4.14b}$$

Written in this form, it is not hard to see that electromagnetism is invariant under translations, rotations, special Lorentz transformations, parity inversion, and combinations of these transformations. Moreover, if we consider no sources (i.e., $J^\nu = 0$), then these equations are also invariant with respect to an internal symmetry, that amounts to an arbitrary rotation between the electric and magnetic fields. This internal symmetry, as we will see later in this chapter, is at the root of the helicity conservation of the electromagnetic field.

4.2.1 Gauge invariance and the choice of gauge

Because $F^{\mu\nu}$ is expressed in terms of the electromagnetic potentials, i.e., the 4-vector potential A^μ, the Lagrangian in equation (4.5) is not manifestly gauge invariant, and in order to describe physical electric and magnetic fields, it must be supported by a gauge fixing condition. The usual choice in relativistic (classical) field theory is the Lorentz gauge, for which $\partial_\mu A^\mu = 0$. However, since the gauge fixing operation is, in principle, arbitrary, there are many other possible choices, that have an impact in different fields of physics. Other useful choices in field theory, especially for quantising the electromagnetic field, is the so-called R_ξ-gauge. In this case, instead of fixing the gauge with an external constraint, one inserts the R_ξ gauge condition as a gauge-breaking term

$$\delta\mathcal{L} = \frac{(\partial_\mu A^\mu)^2}{2\xi}, \tag{4.15}$$

into the Lagrangian, and postpones the choice of gauge, i.e., the fixing of the parameter ξ, to the end of the calculations. Typical choices for ξ are the Landau gauge ($\xi \rightarrow 0$, which retrieves the Lorentz gauge), the Feynman gauge ($\xi = 1$), and the Yennie gauge ($\xi = 3$) [12, 13].

For the case of non-relativistic fields, the Coulomb (or, as often referred to, radiation) gauge reads $\nabla \cdot \mathbf{A} = 0$ and requires the electric and magnetic fields to be purely transverse. The Weyl (or temporal) gauge, setting $A^0 = 0$, is also another option, which requires a nonzero longitudinal component of the fields and a constraint on the electromagnetic state. The Weyl gague is commonly used, for example, when treating the propagation of electromagnetic fields inside a material in terms of vector potential, instead of Hertz potentials [14]. Moreover, the Weyl gauge is incomplete, in the sense that fixing $A^0 = 0$ leaves enough gauge freedom to impose other, more stringent, gauge conditions, such as the Coulomb gauge, for example. Finally, the multipolar gauge requires $\mathbf{r} \cdot \mathbf{A} = 0$. Notice, that the multipolar gauge is a special case of the more general Fock–Schwinger gauge $x^\mu A_\mu = 0$, where $A^0 = 0$ has been implicitly assumed.

4.3 The Hamiltonian for the electromagnetic field

In analogy with what we have done in chapter 3, we now apply a Legendre transformation to the electromagnetic Lagrangian density (4.5) to obtain the electromagnetic Hamiltonian density, i.e., the total electromagnetic energy density. First, we calculate the canonically conjugated momenta of the electromagnetic field, using the definition give by equation (3.16). However, for a reason that will become clear in a moment, we calculate the temporal and spatial canonically conjugated momenta separately, namely

$$\pi^0 = \frac{\partial \mathcal{L}}{\partial(\partial_0 A_0)} = 0, \tag{4.16a}$$

$$\pi^i = \frac{\partial \mathcal{L}}{\partial(\partial_0 A_i)} = \frac{\partial \mathcal{L}}{\partial F_{0i}} = -F^{0i} = E^i. \tag{4.16b}$$

The first thing we notice from this calculation is that there is no associated canonically conjugated momentum to the field A_0, i.e., to the scalar potential. Moreover, the electric field $\mathbf{E}$ acts as the (spatial) canonical momentum for the electromagnetic field. The Hamiltonian density for the electromagnetic field can be then written as

$$\begin{aligned} \mathcal{H} &= -\pi^i \partial_0 A^i - \mathcal{L} \\ &= -\mathbf{E} \cdot \partial_0 \mathbf{A} + \frac{1}{2}(|\mathbf{E}|^2 - |\mathbf{B}|^2) + J^\mu A_\mu \\ &= \frac{1}{2}(|\mathbf{E}|^2 + |\mathbf{B}|^2) + J^\mu A_\mu + \mathbf{E} \cdot \nabla A_0, \end{aligned} \tag{4.17}$$

where in passing from the second to the third line we have used the definition for the electric field in terms of the scalar and vector potentials, i.e.,

$\mathbf{E} - \partial_t \mathbf{A} - \nabla \phi = -\partial_0 \mathbf{A} - \nabla A_0$. Notice, that the last term above cancels with the temporal component of the 4-current, therefore leaving the (spatial) current as the only physically meaningful source of the electromagnetic field[1]. We can see this by integrating the Hamiltonian density above and applying the divergence theorem to the last term, obtaining

$$\int d^3x \, (j^0 A_0 + \mathbf{E} \cdot \nabla A_0) = \int d^3x \, [j^0 A_0 - (\nabla \cdot \mathbf{E}) A_0]$$
$$= \int d^3x \, [\rho(x) A_0(x) - \rho(x) A_0(x)] = 0. \tag{4.18}$$

We then get that the total Hamiltonian in the presence of field sources is given by

$$H = \int d^3x \, \mathscr{H} = \int d^3x \left[\frac{1}{2}(|\mathbf{E}|^2 + |\mathbf{B}|^2) + \mathbf{J} \cdot \mathbf{A} \right], \tag{4.19}$$

where $\mathbf{J}$ is the electromagnetic current. The above equality reduces to the well-known expression

$$H_{free} = \int d^3x \left[\frac{1}{2}(|\mathbf{E}|^2 + |\mathbf{B}|^2) \right], \tag{4.20}$$

in the absence of sources, i.e., for the free electromagnetic field propagating in vacuum.

4.4 Conserved quantities of the electromagnetic field

4.4.1 Electromagnetic stress–energy tensor

In this section, we apply the results we got in section 3.1.6 to the electromagnetic field. We start by deriving the expression of the stress–energy tensor and discussing the physical meaning of its components. From equation (3.27), we can directly derive the definition of the stress–energy tensor for the electromagnetic field by means of the substitution $\partial_\mu \psi \to \partial_\mu A_\nu$ and $\psi \to A_\mu$, which leads to

$$T^{\mu\nu} = \frac{\partial \mathscr{L}}{\partial(\partial_\mu A_\alpha)} \partial^\nu A_\alpha - \delta^{\mu\nu} \mathscr{L}$$
$$= \frac{\partial \mathscr{L}}{\partial F_{\mu\alpha}} F_\alpha{}^\nu + \frac{1}{4}\eta^{\mu\nu} F_{\alpha\beta} F^{\alpha\beta} \tag{4.21}$$
$$= F^{\mu\alpha} F_\alpha{}^\nu + \frac{1}{4}\eta^{\mu\nu} F_{\alpha\beta} F^{\alpha\beta},$$

where to pass from the first to the second line we have used equation (4.8) to write the derivative of the Lagrangian with respect to $\partial_\mu A_\alpha$ in terms of the derivative of the Lagrangian with respect to the Faraday tensor $F_{\mu\alpha}$, and we have applied the same

---

[1] Again, this underlines the role of the scalar potential ϕ as being connected to the electrostatic charge distribution, and thus for not contributing to the propagation dynamics of the electromagnetic field.

line of reasoning to the second term, to transform $\partial^\nu A_\alpha$ into $F_\alpha{}^\nu$, using the anti-symmetric property of the Faraday tensor. Then, we made use once again of equation (4.8) to pass from the second to the third line. One direct consequence of the form of the stress–energy tensor above is that its divergence is directly proportional to the external sources. In fact,

$$
\begin{aligned}
\partial_\mu T^{\mu\nu} &= \partial^\mu(F_{\mu\alpha}F^{\alpha\nu}) + \frac{1}{4}\partial^\nu(F_{\alpha\beta}F^{\alpha\beta}) \\
&= (\partial^\mu F_{\mu\alpha})F^{\alpha\nu} + F_{\mu\alpha}\partial^\mu F^{\alpha\nu} + \frac{1}{2}F_{\alpha\beta}\partial^\nu F^{\alpha\beta} \\
&= J_\alpha F^{\alpha\nu} + \frac{1}{2}F_{\mu\alpha}[\partial^\mu F^{\alpha\nu} + \partial^\nu F^{\mu\alpha} + \partial^\mu F^{\alpha\nu}].
\end{aligned}
\tag{4.22}
$$

In the first line, we have raised the index α by means of $\partial_\mu F^{\mu\alpha}F_\alpha{}^\nu = \partial^\mu F_{\mu\alpha}F^{\alpha\nu}$. To go from the second to the third line, for the first term we used Maxwell's equations to write $\partial^\mu F_{\mu\alpha} = J_\alpha$. For the second term in the third line, we first have written $F_{\mu\alpha}\partial^\mu F^{\alpha\nu} = (1/2)F_{\mu\alpha}\partial^\mu F^{\alpha\nu} + (1/2)F_{\mu\alpha}\partial^\mu F^{\alpha\nu}$, then we have renamed the dummy indices $\{\mu, \alpha\}$ as $\{\alpha, \beta\}$ to be able to collect a common factor of $(1/2)F_{\alpha\beta}$ between all three terms Finally, to go from the third to the fourth line we have made use of the Bianchi identity (4.9) to evaluate the term in brackets to zero. This leaves the continuity equation linking the electromagnetic stress–energy tensor to the external sources as

$$
\partial_\mu T^{\mu\nu} + F^{\nu\alpha}J_\alpha = 0.
\tag{4.23}
$$

Notice, moreover, that the stress–energy tensor is traceless, i.e., $\mathrm{Tr}\{T^\mu_\mu\} = 0$ but, contrary to the Faraday tensor, the stress–energy tensor is symmetric, i.e., $T^{\mu\nu} = T^{\nu\mu}$.

4.4.2 Physical meaning of the components of $T^{\mu\nu}$

Let us now discuss the physical meaning of the various components of the electromagnetic stress–energy tensor. From its definition in equation (4.21), we can readily see that, for $\mu = \nu = 0$ we get

$$
T^{00} = F_{0\alpha}F^0_\alpha + \frac{1}{4}\eta^{00}F_{\alpha\beta}F^{\alpha\beta} = \frac{1}{2}(|\mathbf{E}|^2 + |\mathbf{B}|^2) = \mathscr{H},
\tag{4.24}
$$

i.e., the 00-component of the stress–energy tensor is the total energy density of the field, i.e., its Hamiltonian density. Analogously, the components T^{0i} are given by

$$
T^{0i} = F_{0\alpha}F^i_\alpha + \frac{1}{4}\eta^{0i}F_{\alpha\beta}F^{\alpha\beta} = (\mathbf{E} \times \mathbf{B})^i \equiv P^i,
\tag{4.25}
$$

where $\mathbf{P} = \mathbf{E} \times \mathbf{B}$ is the Poynting vector. We can then use the continuity equation (4.23) to link the energy density and the Poynting vector as follows (choose $\nu = 0$ in equation (4.23))

$$
\frac{\partial \mathscr{H}}{\partial t} + \nabla \cdot \mathbf{P} + \mathbf{E} \cdot \mathbf{J} = 0,
\tag{4.26}
$$

which is the well-known Poynting theorem [11]. For the simple case $\mathbf{J} = 0$, we can reinterpret the equation above as the continuity equation for the Noether charge $Q^\mu = \int d^3x \ T^{0\mu}$, as done in section 3.1.6, and therefore interpret $\mathscr{H}$ as the energy density of the electromagnetic field, to which we associate the Poynting vector as the energy density flux (see equation (3.29)), i.e., the momentum density of the electromagnetic field. The term $\mathbf{E} \cdot \mathbf{J}$ is instead associated with work done (per unit volume and time) by the electromagnetic field on the sources, i.e., the Joule effect [11].

Proceeding further, we can see that the spatial components of the stress–energy tensor are defined as

$$-T^{ij} = E^i E^j + B^i B^j - \frac{\delta^{ij}}{2}(|\mathbf{E}|^2 + |\mathbf{B}|^2) \equiv \sigma^{ij}, \qquad (4.27)$$

where σ^{ij} is the so-called Maxwell stress tensor, describing the interaction between electromagnetic forces and mechanical momentum. Again, we can use equation (4.23) to write down a continuity equation linking the components of the Maxwell tensor to their respective 'energy density', which in this case is the momentum density. From equation (4.23) with $\mu \rightarrow i$ and $\nu \rightarrow j$ we then obtain[2]

$$\frac{\partial \mathbf{P}}{\partial t} - \nabla \cdot \boldsymbol{\sigma} + \rho \, \mathbf{E} + \mathbf{J} \times \mathbf{B} = 0, \qquad (4.28)$$

i.e., σ_{ij} can be interpreted as the flux of the momentum component P_i across a surface characterised by $x^j =$ const. Notice, moreover, the appearance of the Lorentz force density $\rho \mathbf{E} + \mathbf{J} \times \mathbf{B}$, which can be interpreted as the work done by the electromagnetic field to move a charged particle [11].

4.5 Orbital and spin angular momentum of the electromagnetic field

In the previous section we derived the expression for the linear momentum of the electromagnetic field, i.e., the Poynting vector $\mathbf{P} = \mathbf{E} \times \mathbf{B}$ and we have seen how it is related to the flow of energy and momentum flux.

Here, instead, we focus on the angular momentum carried by the electromagnetic field. In particular, the aim of this section is to derive the general expression of the AM flux of the electromagnetic field and discuss how this in general can be seen as the sum of two contributions, one proportional to the OAM and the other proportional to the SAM of the electromagnetic field.

From the definitions of AM and AM flux given by equations (3.35) and (3.36), we can immediately define the AM density associated with the electromagnetic field as

$$J_i = \varepsilon_{ijk} T^{0j} x^k \ \rightarrow \ \mathbf{J} = \mathbf{x} \times \mathbf{P} = \mathbf{x} \times (\mathbf{E} \times \mathbf{B}), \qquad (4.29)$$

---

[2] Since we are using the convention $c = 1$, in this book there is no formal difference between the Poynting vector and the electromagnetic momentum, i.e., $\mathbf{P} = \mathbf{S}$. However, in SI units, when c is restored we have that $\mathbf{P} = \mathbf{S}/c^2$.

and the AM flux as

$$\mathcal{M}_{ij} = \begin{pmatrix} \sigma_{xy}z - \sigma_{xz}y & \sigma_{xz}x - \sigma_{xx}z & -\sigma_{xy}x + \sigma_{xx}y \\ -\sigma_{yz}y + \sigma_{yy}z & \sigma_{yz}x - \sigma_{xy}z & -\sigma_{yy}x + \sigma_{xy}y \\ -\sigma_{zz}y + \sigma_{yz}z & \sigma_{zz}x - \sigma_{xz}z & -\sigma_{yz}x + \sigma_{xz}y \end{pmatrix}, \tag{4.30}$$

where the expression for σ_{ij} is given by equation (4.27). This expression of the AM flux density contains information on both the OAM and SAM of the electromagnetic fields. However, this information is not explicitly evident from the expression above. This is due to the fact, that the stress–energy tensor defined in equation (4.21) that we used to derive the expression of the AM flux density is (a) manifestly gauge invariant (since it is defined only in terms of the Faraday tensor $F_{\mu\nu}$) and (b) symmetric. This form of the stress–energy tensor (see equation (4.21)) is known in literature as the Belinfante–Rosenfeld tensor [15, 16].

To make the OAM and SAM appear explicitly in the definition of the AM flux density, we need to go a step back and rewrite the stress–energy tensor in a non-gauge invariant manner, i.e., as a function of the vector potential A_μ. This can be done by repeating the calculations in equation (4.21) without writing the term $\partial^\nu A_\alpha$ in terms of the Faraday tensor, thus obtaining the following definition of stress–energy tensor

$$\begin{aligned} T_{\text{can}}^{\mu\nu} &= \frac{\partial \mathscr{L}}{\partial(\partial_\mu A_\alpha)} \partial^\nu A_\alpha - \delta^{\mu\nu} \mathscr{L} \\ &= -F^{\mu\alpha} \partial^\nu A_\alpha + \frac{1}{4} \eta^{\mu\nu} F_{\alpha\beta} F^{\alpha\beta}. \end{aligned} \tag{4.31}$$

This is often called the *canonical* stress–energy tensor. One can readily prove, that $T_{\text{can}}^{\nu\mu} \neq T_{\text{can}}^{\mu\nu}$. Hence, the canonical stress–energy tensor is not symmetric. Notice, however, that the tensor

$$T^{\mu\nu} = T_{\text{can}}^{\mu\nu} + \partial_\sigma K^{\sigma\mu\nu}, \tag{4.32}$$

is symmetric, if $K^{\sigma\mu\nu}$ is anti-symmetric in the indices σ and μ. The tensor $T^{\mu\nu}$ defined above is, indeed, the Belinfante–Rosenfeld tensor defined in equation (4.21), as can be easily checked by choosing $K^{\sigma\mu\nu} = F^{\mu\sigma} A^\nu$. Moreover, since $\partial_\mu \partial_\sigma K^{\sigma\mu\nu} = 0$, both the Belinfante–Rosenfeld tensor and the canonical stress–energy tensor satisfy Noether's theorem and thus describe the same physics. The difference between the two tensors is in the way they define their conserved currents. While the conserved current associated with the Belinfante–Rosenfeld tensor is the *total* AM, the one associated with the canonical stress–energy tensor can be easily broken down into its SAM and OAM parts. This can be seen by substituting equation (4.32) into equation (4.29) to obtain

$$\begin{aligned} J_i &= \varepsilon_{ijk} T^{0j} k^k = \varepsilon_{ijk} \left[T_{\text{can}}^{0j} + \partial_\sigma K^{\sigma 0j} \right] x^k \\ &= \varepsilon_{ijk} T_{\text{can}}^{0j} x^k + \varepsilon_{ijk} \partial_\sigma K^{\sigma 0j} x^k \\ &\equiv L_i + S_i, \end{aligned} \tag{4.33}$$

where L_i is the OAM (the first term above, in fact, is in the form $\mathbf{r} \times \mathbf{p}$, typical of orbital contribution to the angular momentum) and we have defined

$$S_i = \varepsilon_{ijk} \partial_\sigma K^{\sigma 0j} x^k, \tag{4.34}$$

as the SAM associated with the electromagnetic field. With a little algebra and vector calculus identities, we can write the expressions above in the following, more familiar form

$$\mathbf{L} = \mathbf{E} \cdot (\mathbf{r} \times \nabla)\mathbf{A}, \tag{4.35a}$$

$$\mathbf{S} = \mathbf{E} \times \mathbf{A}, \tag{4.35b}$$

where $\mathbf{r} \times \nabla$ is the OAM operator [11, 17]. To obtain the explicit expression of the SAM, we have used the relation $K^{\sigma\mu\nu} = F^{\nu\sigma} A^\mu - F^{\nu\mu} A^\sigma$, which is equivalent to the definition given above, just easier to use for calculation. The tensor $K^{\sigma\mu\nu}$ is known in literature with the name of *spin tensor*, since it is essentially carrying information about the spin of the electromagnetic field [10]. With this result, we can then write, combining equations (4.29) and the results above, $\mathbf{J} = \mathbf{L} + \mathbf{S}$, which is the usual decomposition rule for the total AM of the electromagnetic field in its OAM and SAM components.

Notice, that while $\mathbf{L}$ is reference-frame-dependent, and therefore it accounts for the orbital part of the AM (i.e., that coming from rotations about a fixed axis, say, the propagation direction), $\mathbf{S}$ is instead reference-frame-independent, as it only depends on the electromagnetic field itself (and it therefore describes intrinsic rotations of the field, i.e., the spin). Notice, moreover, that the expressions above for OAM and SAM are intrinsically gauge dependent, since the vector potential appears explicitly in their definition.

To obtain a gauge invariant expression for the OAM and SAM it is then necessary to start from equation (4.29) and work out the separation there, as a function of the electric and magnetic field.

Instead of doing that, we use a different route, i.e., we start from the gauge invariant expression of the AM flux given in equation (3.36) and prove that, under reasonable assumptions, it can be divided into two contributions, one related to OAM and the other to SAM.

To make things easier, let us look at the specific example of light beams, i.e., electromagnetic fields that propagate mainly along a given direction, which, without loss of generality, we can assume being parallel to the z axis. We can then calculate the AM flux M_z across a plane perpendicular to the propagation direction of the field as

$$M_z = \int d^2r \, \mathcal{M}_{33} = \int d^2r \, [y(E_x E_z + B_x B_z) - x(E_y E_z + B_y B_z)]. \tag{4.36}$$

For the sake of simplicity, let us also assume that the electromagnetic field is monochromatic, so that we can represent electric and magnetic field components as $E_i \to E_i \exp(i\omega t) + \text{c.c.}$ and $B_i \to B_i \exp(i\omega t) + \text{c.c.}$, where now E_i and B_i are complex amplitudes. Let us moreover assume, for simplicity, that the field possesses cylindrical symmetry, i.e., that the intensity distribution of the electromagnetic field in the plane (x, y) perpendicular to the propagation direction is invariant under rotations about the z-axis.

As a first step towards calculating the integral in equation (4.36), we use Maxwell's curl equations (2.1a) and (2.1d) to write the z-component of the electric and magnetic fields in terms of their transverse components. Then, we use cylindrical coordinates to write $d^2r = RdRd\varphi$ (with $R = \sqrt{x^2 + y^2}$ and $\varphi = \arctan(y/x)$) and transform the derivatives $\partial/\partial x$ and $\partial/\partial y$, coming from Maxwell's equations, into their cylindrical counterparts, i.e., $\partial/\partial x = \cos\varphi(\partial/\partial R) - (\sin\varphi/R)(\partial/\partial\varphi)$ and $\partial/\partial y = \sin\varphi(\partial/\partial R) + (\cos\varphi/R)(\partial/\partial\varphi)$. Finally, we integrate M_z with respect to time over one period $T = 2\pi/\omega$ (i.e., we take its cycle average). After a straightforward calculation we arrive at the following result for the cycle-averaged AM flux [18]

$$M_z = \frac{1}{2\omega} \operatorname{Im} \left\{ \int d^2r \left[\left(E_x B_x^* + E_y B_y^* \right) \right. \right.$$
$$\left. \left. + \frac{1}{2}\left(-B_x^*\frac{\partial}{\partial\varphi}E_y + E_y\frac{\partial}{\partial\varphi}B_x^* - E_x^*\frac{\partial}{\partial\varphi}B_y + B_y^*\frac{\partial}{\partial\varphi}E_x \right) \right] \right\}, \quad (4.37)$$

where now $d^2r = RdRd\varphi$. The expression above can be naturally separated into two parts, which we define as the SAM and OAM part of the AM flux of a light beam as

$$M_z^{(\text{SAM})} = \frac{1}{2\omega} \operatorname{Im} \left\{ \int d^2r \left(E_x B_x^* + E_y B_y^* \right) \right\}, \quad (4.38a)$$

$$M_z^{(\text{OAM})} = \frac{1}{4\omega} \operatorname{Im} \left\{ \int d^2r \left[\left(-B_x^*\frac{\partial}{\partial\varphi}E_y + E_y\frac{\partial}{\partial\varphi}B_x^* - E_x^*\frac{\partial}{\partial\varphi}B_y + B_y^*\frac{\partial}{\partial\varphi}E_x \right) \right] \right\}. \quad (4.38b)$$

As can be seen, the SAM flux only depends on the electric and magnetic fields and it is therefore frame-independent, while the OAM flux depends on the particular choice of the reference frame, since the coordinates appear explicitly in its definition via the derivatives. Moreover, since only the electric and magnetic fields appear in the definitions above, both the OAM and SAM fluxes are manifestly gauge invariant. One could then use the results above to define the OAM and SAM of the electromagnetic field, starting directly from the Belinfante–Rosenfeld tensor, instead of passing through the canonical stress–energy tensor and the spin tensor.

To fix ideas, let us consider two explicit, easy examples of electromagnetic field carrying SAM and OAM. For the former, we choose a circularly polarised plane wave, while for the latter, we choose a Laguerre–Gaussian beam, such as the one defined in equation (2.49).

4.5.1 SAM and circular polarisation

The electric field of a circularly polarised plane wave can be written as $\mathbf{E} = (E_0/\sqrt{2})(\hat{\mathbf{x}} + i\,\sigma\,\hat{\mathbf{y}})\exp(ikz)$, with $\sigma = \pm 1$ indicate left-handed $(+1)$ and right-handed (-1) circular polarisation, respectively, and E_0 is the (constant) amplitude of the field. Using Maxwell's equations, the magnetic field can be instead written as $\mathbf{B} = (B_0/\sqrt{2})(i \, \sigma \, \hat{\mathbf{y}} + \hat{\mathbf{x}})\exp(ikz)$. Substituting these expressions into equations (4.38) gives

$$M_z^{(SAM)} = \frac{\sigma}{2\omega} \int d^2r \, E_0 B_0, \qquad (4.39a)$$

$$M_z^{(OAM)} = 0, \qquad (4.39b)$$

where the second result follows from the fact, that both the amplitude and phase distribution of a plane wave are constant in space. Notice, that the quantity $\mathscr{F} = (1/2) \int d^2 \, E_0 B_0$ appearing in the expression above is the energy flux of a plane wave across the plane (x, y) perpendicular to the propagation direction. If we normalise the SAM flux by the energy flux we get the following expression [18, 19]

$$\frac{M_Z^{(SAM)}}{\mathscr{F}} = \frac{\sigma}{\omega}. \qquad (4.40)$$

This result tells us that $M_z^{(SAM)}$ can be indeed interpreted as the flux of spin angular momentum (i.e., circular polarisation) of an electromagnetic field through a surface perpendicular to its propagation direction.

4.5.2 OAM and twisted phase fronts

As a second example, let us consider a linearly polarised Laguerre–Gaussian beam, whose electric field can be written as $\mathbf{E} = E(\mathbf{R}, z)\exp(i\ell\theta)\hat{\mathbf{x}}$, where we have arbitrarily chosen to polarise the electric field along the x-direction. For the purposes of this example, only the twisted phase front of the beam, i.e., the factor $\exp(i\ell\theta)$ is relevant, so the spatial dependence of the beam, as defined in equation (2.49) all contained in the term $E(\mathbf{R}, z)$. The magnetic field, using Maxwell's equations, can be defined as $\mathbf{B} = B(\mathbf{R}, z)\exp(i\ell\theta)\hat{\mathbf{y}}$. Substituting these expressions into equations (4.38) now gives

$$M_z^{(SAM)} = 0, \qquad (4.41a)$$

$$M_z^{(OAM)} = \frac{\ell}{2\omega}\text{Re}\left\{\int d^2r \, [E^*(\mathbf{R}, z)B(\mathbf{R}, z) + B^*(\mathbf{R}, z)E(\mathbf{R}, z)]\right\}, \qquad (4.41b)$$

where now the first term is zero because the field is linearly polarised. Once more, the quantity $\mathscr{F} = (1/2)\text{Re}\{\int d^2r \, [E^*(\mathbf{R}, z)B(\mathbf{R}, z) + B^*(\mathbf{R}, z)E(\mathbf{R}, z)]\}$ is the energy flux of the field across a surface orthogonal to the propagation direction, and if we normalise the OAM flux to it we get [18, 19]

$$\frac{M_z^{(OAM)}}{\mathscr{F}} = \frac{\ell}{\omega}, \qquad (4.42)$$

which allows us once more to interpret $M_z^{(OAM)}$ as the flux of OAM (generated by the twisted phase front $\exp(i\ell\theta)$ of the field) of an electromagnetic field through a surface perpendicular to its propagation direction.

These two simple examples show what is the physical origin of SAM and OAM (circular polarisation and twisted phase, respectively) for the electromagnetic field and illustrate how the separation between these two degrees of freedom can be useful to better understand the properties of the field. The problem of separating the OAM and SAM of light has been the subject of intense research since the very early days of electromagnetism [20] and has been investigated in many different contexts and from many different perspectives. The interested reader will find a good starting point to delve deeper into this topic in references [9, 10, 15, 18, 19, 21–29].

4.6 Helicity, chirality and spin angular momentum

4.6.1 The second electromagnetic potential

Let us consider an electromagnetic field in a region of space where there are no sources. In this case, Maxwell's equations (2.1) reduce to the following form

$$\nabla \times \mathbf{E} = -\partial_0 \mathbf{B}, \tag{4.43a}$$

$$\nabla \cdot \mathbf{E} = 0, \tag{4.43b}$$

$$\nabla \cdot \mathbf{B} = 0, \tag{4.43c}$$

$$\nabla \times \mathbf{B} = \partial_0 \mathbf{E}. \tag{4.43d}$$

It is then said, that since $\nabla \cdot \mathbf{B} = 0$, the magnetic field can be defined in terms of a (magnetic) vector potential through $\mathbf{B} = \nabla \times \mathbf{A}$, since $\nabla \cdot \nabla \times \mathbf{f} = 0$. From this assumption, it then follows that the electric field can also be defined in terms of the (magnetic) vector potential as $\mathbf{E} = -\partial_0 \mathbf{A}$. In doing so, however, one somehow breaks the inherent symmetry between Maxwell's equations for the electric and magnetic fields. For example, if $\nabla \cdot \nabla \times \mathbf{f} = 0$ for any vector field $\mathbf{f}$, why do we choose the magnetic field, and not the electric field, to define a vector potential? In vacuum $\nabla \cdot \mathbf{B} = 0 = \nabla \cdot \mathbf{E}$, so any choice would be as good as the one we made, so the natural question we could ask is why we prefer the magnetic field to be defined in terms of $\mathbf{A}$ and not the electric field instead? The answer is that, while the definition of the (magnetic) vector potential holds also in the presence of charges (because $\nabla \cdot \mathbf{B} = 0$ holds everywhere, since there are no magnetic monopoles), the definition of an electric vector potential becomes very complicated in the presence of charges (see, e.g., the work of Cameron [30], where he discusses the electric vector potential in the presence of charges, both in the classical and quantum domain). In vacuum, however, both choices are on the same footing, and there is no real reason to choose one in favour of the other one, rather than convention. This *electric–magnetic democracy* [31], i.e., the fact that vacuum Maxwell's equations favour neither the electric nor the magnetic nature of an electromagnetic wave, is often visualised by means of the so-called duality transformation [32–34]. Under an arbitrary rotation of the electric and magnetic fields by a (pseudo-scalar, time-odd) angle θ, i.e.,

$$\begin{pmatrix} \mathbf{E}' \\ \mathbf{B}' \end{pmatrix} = \begin{pmatrix} \cos\theta & \sin\theta \\ -\sin\theta & \cos\theta \end{pmatrix} \begin{pmatrix} \mathbf{E} \\ \mathbf{B} \end{pmatrix}, \tag{4.44}$$

in fact, equations (4.43) remain invariant, and so does the electromagnetic Lagrangian (4.5)[3]. Thanks to the duality transformation above, we can then decide to treat the electric and magnetic fields on equal footing, and therefore introduce an electric vector potential $\mathbf{C}$ such that $\mathbf{E} = \nabla \times \mathbf{C}$, so that we can symmetrise the definition of electric and magnetic fields in terms of their vector potentials and, most importantly, make it also that the electromagnetic potentials are invariant under dual transformations. The electric and magnetic fields can be then defined in terms of the magnetic ($\mathbf{A}$) and electric ($\mathbf{C}$) vector potential as follows

$$\mathbf{E} = -\partial_0\mathbf{A} + \nabla \times \mathbf{C}, \tag{4.45a}$$

$$\mathbf{B} = -\partial_0\mathbf{C} + \nabla \times \mathbf{A}. \tag{4.45b}$$

It is then not hard to prove that if $\mathbf{E}$ and $\mathbf{B}$ fulfil the duality transformation described by equation (4.44), so also do the electric and magnetic vector potentials.

4.6.2 Dual electromagnetism, helicity density and optical spin

The electromagnetic field Lagrangian defined in equation (4.5), however, is not invariant under the duality transformation defined by equation (4.44). Inverting equation (4.44) and substituting the expressions of the electric and magnetic fields as a function of the rotated ones back into the Lagrangian (4.5) leads, in fact, to

$$\mathscr{L}' = \frac{1}{2}[\cos 2\theta \ (|\mathbf{E}|^2 - |\mathbf{B}|^2) - 2\sin 2\theta \ \mathrm{Re}\{\mathbf{E} \cdot \mathbf{B}^*\}]$$
$$= \mathscr{L}\cos 2\theta - \frac{1}{4}G^{\mu\nu}F_{\mu\nu}\sin 2\theta \neq \mathscr{L}. \tag{4.46}$$

Notice the appearance of the Maxwell tensor $G^{\mu\nu}$, dual to the Faraday tensor $F^{\mu\nu}$ [35–37]. One possible explanation for this lack of dual symmetry for the Lagrangian (4.5) is that only the magnetic vector potential $\mathbf{A}$ appears in its definition, and there is no trace of the electric vector potential $\mathbf{C}$. This suggests the possibility of defining a dual-symmetric Lagrangian, where both $\mathbf{A}$ and $\mathbf{C}$ appear explicitly by firstly introducing two electromagnetic tensors, namely the usual (magnetic) Faraday tensor $F_{\mu\nu} = \partial_\mu A_\nu - \partial_\nu A_\mu$ and the (electric) Faraday tensor $D_{\mu\nu} = \partial_\mu C_\nu - \partial_\nu C_\mu$, so that the Lagrangian of the electromagnetic field in vacuum can be written as[4]

$$\mathscr{L} = -\frac{1}{8}[F_{\mu\nu}F^{\mu\nu} + D_{\mu\nu}D^{\mu\nu}], \tag{4.47}$$

with half of the energy stored in the magnetic Faraday tensor, and the other half in the electric one. The equations of motion associated with this Lagrangian separate into two families, one only depending on $F_{\mu\nu}$ and the other depending on $D_{\mu\nu}$, so that

---

[3] This symmetry has then, by virtue of Noether's theorem, a conserved quantity associated with it, which, as we will see, is the optical helicity, whose flux is then the optical spin.

[4] Note, that in SI the two terms have different weights. The magnetic part of the Lagrangian, proprtional to $F_{\mu\nu}F^{\mu\nu}$ is weighted by $1/\mu_0$, while the electric part, proportional to $D_{\mu\nu}D^{\mu\nu}$ is instead weighted by $1/\varepsilon_0$. Here, they have the same weight, since we have assumed $c = 1$.

$\partial_\mu F^{\mu\nu} = 0$ reproduces Maxwell's equations for the divergence of **E** and the curl of **B**, i.e., equations (2.1b) and (2.1d), while $\partial_\mu D^{\mu\nu}$ reproduces the equations for the divergence of **B** and the curl of **E**, i.e., equations (2.1a) and (2.1c). Moreover, the electric Faraday tensor $D_{\mu\nu}$ is related to the Maxwell tensor $G_{\mu\nu}$ by the relation

$$D_{\mu\nu} = G_{\mu\nu}. \tag{4.48}$$

Dual electromagnetism and its consequences in optics have been studied quite intensively during the last decade, as the proper formal ground, where to define quantities like helicity and chirality, and discuss their properties in various different physical contexts. Amongst the vast literature on the subject, a good starting point is the works of Stephen Barnett [9, 38, 39] and the field theory approach presented by Bliokh [10], as well as the older works by Rañada [40] and Teitelboim [34].

The duality transformation (4.44) is an internal symmetry for the electromagnetic field, since it only amounts to a rotation of the electric and magnetic degrees of freedom. To find the Noether current associated with such internal symmetry, therefore, we generalise equation (3.39) to the case of two vector fields, namely A_μ and C_μ, and we consider an infinitesimal version of the rotation described by equation (4.44), i.e., $A' \simeq A_\mu + \theta\, C_\mu$ and $C' \simeq -\theta\, A_\mu + C_\mu$. Therefore, we can set $\delta\, A_\mu = \theta\, C_\mu$ and $\delta\, C_\mu = -\theta\, A_\mu$. To make calculations simpler, we can also use equation (4.8) to express derivatives with respect to $\partial_\mu A_\nu$ (and, analogously with respect to $\partial_\mu C_\nu$ in terms of derivatives of the Lagrangian density (4.47) with respect to the magnetic and electric Faraday tensors, respectively. Equation (3.39) then reads

$$\partial_\mu \left(\frac{\partial \mathcal{L}}{\partial F_{\mu\nu}} \delta\, A_\nu + \frac{\partial \mathcal{L}}{\partial D_{\mu\nu}} \delta\, C_\nu \right) = \frac{\theta}{2}\, \partial_\mu (F^{\mu\nu}\, C_\nu - D^{\mu\nu}\, A_\nu) = 0, \tag{4.49}$$

which defines the Noether current associated with the duality transformation as being

$$J^\mu = \frac{1}{2}(F^{\mu\nu}\, C_\nu - D^{\mu\nu}\, A_\nu). \tag{4.50}$$

As we have seen in chapter 3, the temporal component of the Noether current, J^0, represents the conserved charge density, while the spatial component **J**, is related to the flux of said conserved charge density. For the case of the duality transformation we get

$$J^0 \equiv h = \frac{1}{2}(\mathbf{A} \cdot \mathbf{B} - \mathbf{C} \cdot \mathbf{E}), \tag{4.51a}$$

$$\mathbf{J} \equiv \mathbf{S} = \frac{1}{2}(\mathbf{B} \times \mathbf{C} + \mathbf{E} \times \mathbf{A}). \tag{4.51b}$$

The associated continuity equation then reads

$$\frac{\partial h}{\partial t} + \nabla \cdot \mathbf{S} = 0, \tag{4.52}$$

and the two quantities are known as the helicity density (h) and the flux of the helicity density, i.e., the optical spin (**S**). As the name suggests, the optical spin is related to the spin of the photon, which is essentially the degree of circular polarisation of an electromagnetic field [11]. However, formally, the helicity is well defined only for plane wave fields, i.e., in momentum space representation, and it is not a good quantity to use for expressing the polarisation properties of general electromagnetic fields in real space [41, 42]. A connected quantity, well defined in real space and for arbitrary polarisation, is instead the optical chirality, introduced by Lipkin in 1964 [43], revisited by Tang and Cohen in 2010 in connection with light–matter interaction [44], and brought to the domain of optical beams by Andrews and co-workers [45–47], whose definition is as follows

$$\chi = \frac{1}{2}[\mathbf{E} \cdot (\nabla \times \mathbf{E}) + \mathbf{B} \cdot (\nabla \times \mathbf{B})], \qquad (4.53)$$

to which, by virtue of Noether's theorem, we can associate the chirality flow

$$\varphi = \frac{1}{2}[\mathbf{E} \times (\nabla \times \mathbf{B}) - \mathbf{B} \times (\nabla \times \mathbf{E})]. \qquad (4.54)$$

As can be seen, while the helicity is defined as a function of the electric and magnetic vector potentials (and, therefore, not gauge invariant), the chirality is instead defined as a function of the electric and magnetic fields solely, and it is therefore manifestly gauge invariant. The interested reader can refer to the aforementioned works, or to references [48–51] and references therein for a comprehensive discussion on helicity, chirality, and their interplay in electrodynamics.

4.7 Field theory description of light–matter interaction

In this section we construct a field theory of light–matter interaction, based on the work of Huttner and Barnett in 1992 [1]. Here, we only consider a linear interaction between light and matter, and we limit ourselves to considering the classical interaction between them. In chapter 7, after we have introduced the path integral formalism, we will come back to this model and look at its quantum properties and we will also extend it to the case of nonlinear optical properties of materials.

The basic assumption for developing this theory is to assume the standard form of the electromagnetic Lagrangian given by equation (4.5), and to model the medium as a harmonic oscillator, in accordance with the Lorentz model of dispersion [11]. Using the same idea, one could also model the lossess of this material as a continuous collection of harmonic oscillators, interacting with the material and taking away energy from it. This is the framework of the Huttner–Barnett model, and the one we want to explore in this section.

The most general field theory describing light–matter interaction, which encompasses all the relevant material properties, including losses (and can be readily quantised, if necessary) is given by the following Lagrangian density [1][5]

---

[5] Here, we use **E**, **B**, and **A** in an interchangeable and mixed way, to remain faithful to the notation introduced in the work by Huttner and Barnett. However, as usual, $c = 1$ is implicitly assumed.

$$\mathscr{L} = \frac{1}{2}(|\mathbf{E}|^2 - |\mathbf{B}|^2) + \frac{\rho}{2}\left(|\partial_0 \mathbf{X}|^2 - \omega_0^2 |\mathbf{X}|^2\right)$$
$$+ \frac{\rho}{2}\int_0^\infty d\omega \ (|\partial_0 \mathbf{Y}(\omega)|^2 - \omega^2 |\mathbf{Y}(\omega)|^2)$$
$$- \alpha(\mathbf{A} \cdot \partial_0 \mathbf{X} + \phi \ \nabla \cdot \mathbf{X}) - \int_0^\infty d\omega \ f(\omega)\mathbf{X} \cdot \partial_0 \mathbf{Y}(\omega) \tag{4.55}$$
$$\equiv \mathscr{L}_{\text{em}} + \mathscr{L}_{\text{mat}} + \mathscr{L}_{\text{res}} + \mathscr{L}_{\text{int}} + \mathscr{L}_{\text{mr}}.$$

The first term is the free electromagnetic Lagrangian ($\mathscr{L}_{\text{em}}$), with $\mathbf{E} = -\partial_0 \mathbf{A} - \nabla \phi$ and $\mathbf{B} = \nabla \times \mathbf{A}$. The second term is the free Lagrangian for the matter polarisation $\mathbf{P}(\mathbf{r}, t)$ ($\mathscr{L}_{\text{mat}}$), which is modelled as a harmonic oscillator with characteristic frequency ω_0. The third term is the free Lagrangian for the reservoir field $\mathbf{Y}(\mathbf{r}, t; \omega) \equiv \mathbf{Y}(\omega)$ ($\mathscr{L}_{\text{res}}$), which models the losses and it is represented by a continuous collection of harmonic oscillators, each with its own characteristic frequency ω. The fourth and fifth terms are, respectively, the interaction of the electromagnetic field with the matter ($\mathscr{L}_{\text{int}}$) and the matter–reservoir interaction ($\mathscr{L}_{\text{mr}}$). The quantities ρ, α and $f(\omega)$ are the mass density (per unit frequency), the light–matter coupling constant, and the reservoir (spectral) coupling constant, respectively.

To comply with reference [1], we assume that $f(\omega)$ is non-singular, square-integrable, i.e., $\int d\omega \ |f(\omega)|^2 < \infty$, and, most importantly, we assume that (i) $f(\omega) \neq 0$ if $\omega \neq 0$, and (ii) the analytical continuation of $|f(\omega)|^2$ on the negative frequency axis is an odd function. These assumptions will be useful later, when deriving the dielectric constant from this model.

The Lagrangian above is not gauge invariant, due to the explicit appearance of $\mathbf{A}$ and ϕ. Without loss of generality, we can then fix the gauge to be the Coulomb gauge, i.e., $\nabla \cdot \mathbf{A} = 0$ and also assume that there are no free charges in the material we are considering, which allows us to set $\phi = 0$. A more formal way to eliminate ϕ from the Lagrangian would be to notice that it is not a dynamical variable, since $\mathscr{L}$ does not contain a term proportional to $\dot{\phi}$, and use the equations of motion for ϕ to eliminate it from the expression of $\mathscr{L}$.[6] This is indeed the approach taken in reference [1] to deal with ϕ. Here, for the sake of simplicity, we assume to simply put $\phi = 0$ and because we have selected the Coulomb gauge, we consider all fields to be transverse, since only their transverse components participate in describing the propagation properties of light inside the medium. Formally, we can do this by invoking Helmholtz theorem, and write every vector field appearing in equation (4.55) as the sum of its transverse and longitudinal components as $\mathbf{V} = \mathbf{V}_\perp + \mathbf{V}_\parallel$. This procedure allows us to separate the Lagrangian in its transverse and longitudinal parts, and we can focus on the transverse part solely.

Since we are considering the linear interaction between light and matter, the quantity we are interested in is the permittivity $\varepsilon(\omega)$. From electrodynamics, we know that the permittivity links the polarisation to the electric field through the

---

[6] Another way to say this is that ϕ acts as a Lagrange multiplier.

constitutive relation $\mathbf{D} = \varepsilon_0\mathbf{E} + \mathbf{P} = \varepsilon_0(1 + \chi)\mathbf{E} = \varepsilon_0\varepsilon\mathbf{E}$. This tells us that we have essentially two ways to calculate the permittivity from our model. One possibility would be to derive the expression of the displacement vector $\mathbf{D}$ from the Lagrangian above through the relation $\mathbf{D} = \partial\mathscr{L}/\partial\mathbf{E}^*$ [6] and then use $\mathbf{D} = \varepsilon\mathbf{E}$ to extract directly the permittivity. The other possibility is to instead derive first the relation between polarisation and electric field, i.e., $\mathbf{P} = \varepsilon_0\chi\mathbf{E}$ and then use the susceptibility to derive the permittivity as $\varepsilon = 1 + \chi$. In this section, we adopt the second approach, for which the Hamiltonian formulation of the problem will be more suited. We will show how to derive the permittivity using the first approach in chapter 7, when we will get back to this problem from a path integral perspective.

Our goal is then to derive an expression for $\varepsilon(\omega)$. To obtain such a quantity from the Lagrangian (4.55) it is more convenient to switch to the momentum space representation of the fields by means of the 3D Fourier transform

$$\mathbf{V}(\mathbf{x}, t) = \left(\frac{1}{2\pi}\right)^{3/2} \int d^3k \; \tilde{\mathbf{V}}(\mathbf{k}, t)\, e^{i\mathbf{k}\cdot\mathbf{r}}. \tag{4.56}$$

Substituting the above expression for all the vector fields into equation (4.55), and imposing that $\tilde{\mathbf{E}}(-\mathbf{k}, t) = \tilde{\mathbf{E}}^*(\mathbf{k}, t)$[7], we get the following Lagrangian density in momentum space

$$\begin{aligned}\mathscr{L}_k &= (|\partial_0\tilde{\mathbf{A}}|^2 - k^2|\tilde{\mathbf{A}}|^2) + \rho\left(|\partial_0\tilde{\mathbf{X}}|^2 - \omega_0^2|\tilde{\mathbf{X}}|^2\right)\\ &+ \rho\int_0^\infty d\omega\,(|\partial_0\tilde{\mathbf{Y}}(\omega)|^2 - \omega^2|\tilde{\mathbf{Y}}(\omega)|^2)\\ &- \alpha(\tilde{\mathbf{A}}^*\cdot\partial_0\tilde{\mathbf{X}}+\tilde{\mathbf{A}}\cdot\partial_0\tilde{\mathbf{X}}^*) - \int_0^\infty d\omega\, f(\omega)[\tilde{\mathbf{X}}^*\cdot\partial_0\tilde{\mathbf{Y}}(\omega) + \tilde{\mathbf{X}}\cdot\partial_0\tilde{\mathbf{Y}}^*(\omega)],\end{aligned} \tag{4.57}$$

where the subscript k indicates that this is the Lagrangian density in momentum space. Notice, that while the Lagrangian in direct space is obtained by integrating the Lagrangian density (4.55) in the whole space, i.e., $L = \int d^3x\,\mathscr{L}$, to obtain the same result using the Lagrangian density in momentum space we need to integrate only over positive frequencies (because of the imposed constraint $\tilde{\mathbf{E}}(-\mathbf{k}, t) = \tilde{\mathbf{E}}^*(\mathbf{k}, t)$), i.e., $L = \int_0^\infty d^3k\,\mathscr{L}_k$[8].

To derive the equations of motion in an easy manner, we need to switch to the Hamiltonian representation of this system. To do so, we first calculate the canonically conjugated momenta for all the fields ($\tilde{\mathbf{A}},\tilde{\mathbf{X}}$, and $\tilde{\mathbf{Y}}(\omega)$) and their conjugates[9]

---

[7] This constraint is necessary to keep the Lagrangian a real functional. With this constraint, in fact, we are ensuring that no frequency component is overcounted, and that all quantities appearing in the Lagrangian are real, but represented in terms of complex numbers.

[8] To do so, remember that $\int_{-\infty}^\infty d^3k = 2\int_0^\infty d^3k$, and this explains the lack of factors $1/2$ in $\mathscr{L}_k$.

[9] The Lagrangian density (4.57) is the vector analog of the Lagrangian density for a complex scalar field, where the two independent degrees of freedom are the field and its conjugate. To each of these degrees of freedom, we can then associate a canonically conjugated momentum, defined according to equation (3.16), with just the difference that the derivative with respect to the conjugated field will give the momentum for the field and vice versa. See, for example, reference [52] for more details.

$$\frac{\partial \mathscr{L}_k}{\partial(\partial_0 \tilde{\mathbf{A}}^*)} = \partial_0 \tilde{\mathbf{A}} = -\tilde{\mathbf{E}}, \tag{4.58a}$$

$$\frac{\partial \mathscr{L}_k}{\partial(\partial_0 \tilde{\mathbf{X}}^*)} = -\rho \partial_0 \tilde{\mathbf{X}} - \alpha \tilde{\mathbf{A}} \equiv \tilde{\mathbf{P}}, \tag{4.58b}$$

$$\frac{\partial \mathscr{L}_k}{\partial(\partial_0 \tilde{\mathbf{Y}}^*(\omega))} = \rho \partial_0 \tilde{\mathbf{Y}}(\omega) - f(\omega)\tilde{\mathbf{X}} = \tilde{\mathbf{Q}}(\mathbf{r}, t; \omega) \equiv \tilde{\mathbf{Q}}(\omega), \tag{4.58c}$$

and the same result can be obtained by taking the derivative of the Lagrangian with respect to the time derivative of the fields, which will give the complex conjugate momenta. To derive the Hamiltonian, we then need to apply a Legendre transformation to the Lagrangian (4.57). It is instructive to make this calculation explicitly, at least for some fields, to have an idea on how these transformations work. Let us start from the reservoir first. The Legendre transformation for the reservoir field reads

$$
\begin{aligned}
\mathscr{H}_{\text{res}}(\omega) &= \frac{\partial \mathscr{L}_k}{\partial(\partial_0 \tilde{\mathbf{Y}}^*(\omega))}\partial_0 \tilde{\mathbf{Y}}^*(\omega) + \frac{\partial \mathscr{L}_k}{\partial(\partial_0 \tilde{\mathbf{Y}}(\omega))}\partial_0 \tilde{\mathbf{Y}}(\omega) - \mathscr{L}_{\text{res}} \\
&= \tilde{\mathbf{Q}}(\omega)\partial_0 \tilde{\mathbf{Y}}^*(\omega) + \tilde{\mathbf{Q}}^*(\omega)\partial_0 \tilde{\mathbf{Y}}(\omega) - \rho \partial_0 \tilde{\mathbf{Y}}^*(\omega)\partial_0 \tilde{\mathbf{Y}}(\omega) + \rho\omega^2|\tilde{\mathbf{Y}}(\omega)|^2 \quad (4.59) \\
&= \frac{|\tilde{\mathbf{Q}}(\omega)|^2}{\rho} + \rho\omega^2|\tilde{\mathbf{Y}}(\omega)|^2 - \frac{f^2(\omega)}{\rho}|\tilde{\mathbf{X}}|^2,
\end{aligned}
$$

where to go from the second to the third line we have used the definition of $\tilde{\mathbf{Q}}(\omega)$ given by equation (4.58c) to first remove the term $\rho\partial_0\tilde{\mathbf{Y}}^*(\omega)\partial_0\tilde{\mathbf{Y}}(\omega)$, and then the same definition a second time to write $\partial_0\tilde{\mathbf{Y}}(\omega)$ in terms of $\tilde{\mathbf{Q}}(\omega)$. This procedure will be repeated on all the other terms below. Notice, moreover, how calculating the Hamiltonian for the reservoir field gives a contribution proportional to $|\tilde{\mathbf{X}}|^2$, which acts as a correction term for the characteristic frequency of the matter oscillator. Next, we can calculate the Hamiltonian for the matter field, to get

$$
\begin{aligned}
\mathscr{H}_{\text{mat}}^{(0)} &= \frac{\partial \mathscr{L}_k}{\partial(\partial_0 \tilde{\mathbf{X}})^*}\partial_0 \tilde{\mathbf{X}}^* + \frac{\partial \mathscr{L}_k}{\partial(\partial_0 \tilde{\mathbf{X}})}\partial_0 \tilde{\mathbf{X}} - \mathscr{L}_{\text{mat}} \\
&= \tilde{\mathbf{P}}\partial_0 \tilde{\mathbf{X}}^* + \tilde{\mathbf{P}}^*\partial_0 \tilde{\mathbf{X}} - \rho\partial_0 \tilde{\mathbf{X}}^*\partial_0 \tilde{\mathbf{X}} + \rho\omega_0^2|\tilde{\mathbf{X}}|^2 \quad (4.60) \\
&= \frac{|\tilde{\mathbf{P}}|^2}{\rho} + \rho\omega_0^2|\tilde{\mathbf{X}}|^2 - \frac{\alpha^2}{\rho}|\tilde{\mathbf{A}}|^2,
\end{aligned}
$$

where the superscript $^{(0)}$ indicates that we haven't considered any correction term to it, as we will do that later, at the end of the calculation. Notice, again, how the matter Hamiltonian naturally generates a correction term for the electromagnetic field. The Hamiltonian of the electromagnetic field can then be calculated as follows (again, we ignore any correction term, for the moment)

$$\mathcal{H}_{\text{em}}^{(0)} = \frac{\partial \mathcal{L}_k}{\partial(\partial_0 \tilde{\mathbf{A}}^*)} \partial_0 \tilde{\mathbf{A}}^* + \frac{\partial \mathcal{L}_k}{\partial(\partial_0 \tilde{\mathbf{A}})} \partial_0 \tilde{\mathbf{A}} - \mathcal{L}_{\text{em}}$$

$$= 2|\partial_0 \tilde{\mathbf{A}}|^2 - |\partial_0 \tilde{\mathbf{A}}|^2 + k^2|\tilde{\mathbf{A}}|^2 = |\tilde{\mathbf{E}}|^2 + k^2|\tilde{\mathbf{A}}|^2, \tag{4.61}$$

Finally, we can calculate the interaction Hamiltonian. To do so, we rewrite the time derivatives in terms of the correspondent canonically conjugated momenta using equations (4.58) to obtain

$$\mathcal{H}_{\text{int}} = \frac{\alpha^2}{\rho}|\tilde{\mathbf{A}}|^2 + \frac{\alpha}{\rho}(\tilde{\mathbf{A}}^*\cdot\tilde{\mathbf{P}} + \tilde{\mathbf{A}}\cdot\tilde{\mathbf{P}}^*)$$

$$+ 2\frac{f^2(\omega)}{\rho}|\tilde{\mathbf{X}}|^2 + \int_0^\infty d\omega \, \frac{f(\omega)}{\rho}[\tilde{\mathbf{X}}^*\cdot\tilde{\mathbf{Q}}(\omega) + \tilde{\mathbf{X}}\cdot\tilde{\mathbf{Q}}^*(\omega)] \tag{4.62}$$

A closer inspection of these results allows us to define the following quantities

$$\Omega_0^2 = \omega_0^2 + \int_0^\infty d\omega \, \frac{f^2(\omega)}{\rho^2}, \tag{4.63a}$$

$$\tilde{k}^2 = k^2 + \frac{\alpha^2}{\rho}, \tag{4.63b}$$

which can be interpreted as the renormalised characteristic frequency of the material (by means of the reservoir field) and the renormalised k-vector of the electromagnetic field, by means of the matter field. Putting everything together and including these definitions we can finally arrive at the following expression of the light–matter Hamiltonian

$$\mathcal{H} = |\tilde{\mathbf{E}}|^2 + \tilde{k}^2|\tilde{\mathbf{A}}|^2 + \frac{|\tilde{\mathbf{P}}|^2}{\rho} + \rho\Omega_0^2|\tilde{\mathbf{X}}|^2 + \int_0^\infty d\omega \left[\frac{|\tilde{\mathbf{Q}}(\omega)|^2}{\rho} + \rho\omega^2|\tilde{\mathbf{Y}}(\omega)|^2 \right]$$

$$+ \frac{\alpha}{\rho}(\tilde{\mathbf{A}}^*\cdot\tilde{\mathbf{P}} + \tilde{\mathbf{A}}\cdot\tilde{\mathbf{P}}^*) + \int_0^\infty d\omega \frac{f(\omega)}{\rho}[\tilde{\mathbf{X}}^*\cdot\tilde{\mathbf{Q}}(\omega) + \tilde{\mathbf{X}}\cdot\tilde{\mathbf{Q}}^*(\omega)]. \tag{4.64}$$

To derive the equations of motion, we can either use the Lagrangian (4.57) or the Hamiltonian above. We choose the second approach because it allows us to have an easier access to the dielectric permittivity, which is the ultimate goal of this section. Using equations (3.21) for each pair of canonically conjugated pairs appearing in equation (4.64), i.e., $\{\tilde{\mathbf{E}}, \tilde{\mathbf{A}}\}$, $\{\tilde{\mathbf{P}}, \tilde{\mathbf{X}}\}$, and $\{\tilde{\mathbf{Q}}(\omega), \tilde{\mathbf{Y}}(\omega)\}$ we get the following set of coupled differential equations

$$\frac{\partial \tilde{\mathbf{E}}}{\partial t} = \tilde{k}^2\tilde{\mathbf{A}} + \frac{\alpha}{\rho}\tilde{\mathbf{P}}, \tag{4.65a}$$

$$\frac{\partial \tilde{\mathbf{A}}}{\partial t} = -\tilde{\mathbf{E}}, \tag{4.65b}$$

$$\frac{\partial \tilde{\mathbf{P}}}{\partial t} = \rho \Omega_0^2 \tilde{\mathbf{X}} + \int_0^\infty d\omega \, \frac{f(\omega)}{\rho} \tilde{\mathbf{Q}}(\omega), \tag{4.65c}$$

$$\frac{\partial \tilde{\mathbf{X}}}{\partial t} = -\frac{1}{\rho}\tilde{\mathbf{P}} - \frac{\alpha}{\rho}\tilde{\mathbf{A}}, \tag{4.65d}$$

$$\frac{\partial \tilde{\mathbf{Q}}(\omega)}{\partial t} = \rho \omega^2 \tilde{\mathbf{Y}}(\omega), \tag{4.65e}$$

$$\frac{\partial \tilde{\mathbf{Y}}(\omega)}{\partial t} = -\frac{1}{\rho}\tilde{\mathbf{Q}}(\omega) - \frac{f(\omega)}{\rho}\tilde{\mathbf{X}}. \tag{4.65f}$$

Notice, that we can eliminate $\tilde{\mathbf{Q}}(\omega)$ from equation (4.65c) using equations (4.65e) and (4.65f). If we take the time derivative of equation (4.65e) and use equation (4.65f) to substitute $\partial \tilde{\mathbf{Y}}(\omega)/\partial t$, we get an equation of a driven oscillator for $\tilde{\mathbf{Q}}(\omega)$,i.e.,

$$\left(\frac{\partial^2}{\partial t^2} + \omega^2\right)\tilde{\mathbf{Q}}(\omega) = -\omega^2 f(\omega) \, \tilde{\mathbf{X}}, \tag{4.66}$$

whose solution can be written using the Green's function for the harmonic oscillator [53] as

$$\begin{aligned}
\tilde{\mathbf{Q}}(\omega) &= -\omega^2 f(\omega) \int_0^t d\tau G(t - \tau)\tilde{\mathbf{X}}(\tau) \\
&= -\omega^2 f(\omega) \int_0^t d\tau \int \frac{d\Omega}{2\pi} \frac{\exp[-i\Omega(t - \tau)]}{\Omega^2 - \omega^2}\tilde{\mathbf{X}}(\tau).
\end{aligned} \tag{4.67}$$

Substituting this result into equation (4.65c) and defining (in accordance with Huttner and Barnett [1] and to get a similar result than the one presented in Dutra *et al* [54]) $V(\omega) = (f(\omega)/\rho)\sqrt{\Omega_0/\omega}$, gives the following result for the second term in equation (4.65c)

$$\int_0^\infty d\omega \, \frac{f(\omega)}{\rho}\tilde{\mathbf{Q}}(\omega) = -\frac{i\rho\Omega_0}{2}\int_0^t d\tau \, \tilde{\mathbf{X}}(\tau) \int d\omega \, |V(\omega)|^2 \exp[-i\omega(t - \tau)]. \tag{4.68}$$

Details of the derivation of this equation are given in the appendix at the end of this chapter. To get the expression for the permittivity from the system of equations (4.65), we can first take the Fourier transform with respect to time of all of equations (4.65), using equation (4.68) to eliminate $\tilde{\mathbf{Q}}(\omega)$ from equation (4.65c), and then try to obtain an equation linking the matter polarisation $\tilde{\mathbf{P}}$ with the electric field $\tilde{\mathbf{E}}$, since we are looking for a relation of the kind $\mathbf{P} = \varepsilon_0 \chi(\omega)\mathbf{E}$ [11]. We can then use this to extract the permittivity, which has the following explicit form

$$\varepsilon(\omega) = 1 - \frac{\omega_c^2}{\omega^2 - \Omega_0^2 + (\Omega_0/2)F(\omega)}, \tag{4.69}$$

where, again, $\omega_c^2 = \tilde{k}^2$[10] and

$$F(\omega) = \int d\Omega \frac{|V(\Omega)|^2}{\Omega - \omega}, \tag{4.70}$$

is the contribution of the spectral coupling, essentially deriving from the Fourier transform of the convolution integral appearing in equation (4.67)[11]. It is worth noticing that although the quantity $\varepsilon(\omega)$ reported above can be physically interpreted as the permittivity of the medium, some dielectric permittivities cannot be described by this microscopic model. In fact, following Dutra *et al* [54], if we would like to derive from equation (4.69) the permittivity of the Lorentz oscillator model, we would have to require that $F(\omega) = 4i\kappa\omega/\Omega_0$ (where κ is the rate of absorption), which would require the function $V(\omega)$ to simultaneously fulfil the following constraints

$$\begin{cases} |V(\omega)|^2 = \dfrac{4\kappa\omega}{\pi\Omega_0}, \\[2mm] \mathscr{P} \displaystyle\int d\Omega \dfrac{|V(\omega)|^2}{\Omega - \omega} = 0, \end{cases} \tag{4.71}$$

where $\mathscr{P}$ stands for the Cauchy principal value. The conditions above, however, are incompatible with each other, and therefore the permittivity given by equation (4.69) cannot reproduce the Lorentz model. However, it can reproduce other models, like the Drude model for metals and epsilon-near-zero materials [55, 56]. A possible solution of this issue is discussed at the end of the Huttner and barnett paper [1], and their argument, based on the quantisation of the electromagnetic field, essentially amounts to abandoning the microscopic model and starting directly with an effective light–matter Hamiltonian, carefully chosen to avoid the aforementioned problem. A classical discussion of this assumption is also briefly reported in the work by Dutra *et al* [54]. We refer the interested reader to these references, and also to reference [57] for more details on this discussion.

4.8 Electrodynamics in curved spacetime and transformation optics

Let us now turn our attention to a more general framework, i.e., that of electrodynamics in curved spacetime. In this section, we will show how an electromagnetic field in vacuum on a curved background can be interpreted as propagating in an effective medium, whose permittivity and permeability are defined by the metric of the underlying curved spacetime. We will do so by looking at the field equations for the electromagnetic field inside a medium and propagating in vacuum on a curved background and establish the correspondence by comparison. A more rigorous derivation of the field equations for curved spacetime can be found, for example, in

---

[10] Remember we are working with natural units, i.e, $c = 1$.
[11] Because of the properties of the function $V(\omega)$, $F(\omega)$ is not allowed to have zeros in the upper-half complex plane. Moreover, it can also be shown, that the zeros of $F(\omega)$ all lie on the negative imaginary axis [55].

reference [6, 58]. To start with, let us write down the constitutive relations for the displacement vector $\mathbf{D}(\mathbf{r}, t)$ and the magnetic field $\mathbf{H}(\mathbf{r}, t)$ as

$$\mathbf{D}(\mathbf{r}, t) = \varepsilon_0 \mathbf{E}(\mathbf{r}, t) + \mathbf{P}(\mathbf{r}, t) \equiv \varepsilon_o \varepsilon\, \mathbf{E}(\mathbf{r}, t) \tag{4.72a}$$

$$\mathbf{H}(\mathbf{r}, t) = \frac{1}{\mu_0}\mathbf{B}(\mathbf{r}, t) - \mathbf{M}(\mathbf{r}, t) \equiv \frac{1}{\mu}\mathbf{B}(\mathbf{r}, t), \tag{4.72b}$$

where $\mathbf{P}(\mathbf{r}, t) = \varepsilon_0 \chi\, \mathbf{E}(\mathbf{r}, t)$ is the polarisation, $\mathbf{M}(\mathbf{r}, t) = \chi_m\, \mathbf{H}(\mathbf{r}, t)$ is the magnetisation, and we have made explicit reference to the vacuum permittivity (ε_0) and permeability (μ_0) for clarity, and we assumed we are dealing with a linear dispersive material, to write the polarisation and magnetisation in terms of electric and magnetic fields through their respective susceptibilities. In the equations above, we have introduced ε and μ as the permittivity and permeability of the material.

Following the original work of Minkowski [59], we can rewrite the vectors $\mathbf{D}$ and $\mathbf{H}$ in a rank-two tensor, called electromagnetic displacement tensor, as

$$\mathscr{D}^{\mu\nu} = \begin{pmatrix} 0 & -D_x & -D_y & -D_z \\ D_x & 0 & -H_z & H_y \\ D_y & H_z & 0 & -H_x \\ D_z & -H_y & H_x & 0 \end{pmatrix}, \tag{4.73}$$

and analogously, we can introduce the magnetisation-polarisation tensor [60]

$$\mathscr{M}^{\mu\nu} = \begin{pmatrix} 0 & P_x & P_y & P_z \\ -P_x & 0 & -M_z & M_y \\ -P_y & M_z & 0 & -M_x \\ -P_z & -M_y & M_x & 0 \end{pmatrix}. \tag{4.74}$$

With these definitions at hand, it is not hard to prove that the constitutive relations above can be rewritten in tensor form as

$$\mathscr{D}^{\mu\nu} = F^{\mu\nu} - \mathscr{M}^{\mu\nu}, \tag{4.75}$$

and the Lagrangian density for the electromagnetic field in the presence of matter can be written as

$$\mathscr{L} = -\frac{1}{4}F_{\mu\nu}F^{\mu\nu} - A_\mu J^\mu_{\text{free}} + \frac{1}{2}F_{\mu\nu}\mathscr{M}^{\mu\nu}, \tag{4.76}$$

where J^μ_{free} is the free 4-current, containing information on the free charges and the unbound currents. Using equations (4.72), the equations of motions associated with the above Lagrangian can be written as

$$\partial_\nu \mathscr{D}^{\mu\nu} = J^\mu_{\text{free}}, \tag{4.77}$$

which are fully equivalent to Maxwell's equations in the presence of matter [11, 60], i.e., equations (2.1).

Let us now focus our attention on the constitutive relation (4.75). In vacuum, $\mathcal{M}^{\mu\nu} = 0$ and the electromagnetic displacement tensor coincides with the Faraday tensor. We can then rewrite equation (4.75) in a more convenient way by lowering the indices of the Faraday tensor, so that the Minkowski metric tensor appears explicitly, i.e.,

$$\mathcal{D}^{\mu\nu} = \eta^{\mu\alpha}F_{\alpha\beta}\eta^{\beta\nu}. \tag{4.78}$$

This way of writing the electromagnetic displacement is very useful when switching to curved spacetime. When doing so, one needs to change every instance of the Minkowski metric $\eta^{\mu\nu}$ with the (curved) metric $g^{\mu\nu}$, and all the instances of partial derivatives $\partial_\mu V^\nu$ with covariant derivatives $\nabla_\mu V^\nu = \partial_\mu V^\nu + \Gamma^\nu_{\mu\sigma}V^\sigma$, where $\Gamma^\nu_{\mu\sigma}$ are the Christoffel symbols that account for the effect of gravity (i.e., spacetime curvature) [58]. In curved spacetime, then, the equation above is then rewritten as

$$\mathcal{D}^{\mu\nu} = g^{\mu\alpha}F_{\alpha\beta}g^{\beta\nu}\sqrt{-g}, \tag{4.79}$$

where $g = \det(g^{\mu\nu})$ and the term $\sqrt{-g}$ is introduced so that the Lagrangian density in curved spacetime can still be interpreted as an energy density [58], and $F_{\alpha\beta} = \nabla_\alpha A_\beta - \nabla_\beta A_\alpha = \partial_\beta A_\alpha - \partial_\alpha A_\beta$, i.e., the definition of the Faraday tensor is not changed by the presence of gravity[12].

To understand how the presence of a curved spacetime changes the constitutive relations in vacuum and allows us to draw a parallelism with an effective dielectric permittivity, let us now have a look at, for example, the electric field component of the constitutive relation above, i.e., let us set $F_{\alpha\beta} = F_{0\beta} = E_i$ and $\mathcal{D}^{\mu\nu} = \mathcal{D}^{0\nu} = D^j$. With this substitution, we then want to reduce the constitutive relation above into something like $\mathbf{D} = \varepsilon\,\mathbf{E}$. Substituting this into leads to equation (4.79) leads to

$$D^j = \mathcal{D}^{0\nu} = g^{0\alpha}F_{\alpha\beta}g^{\beta\nu}\sqrt{-g} = g^{00}E_i\,g^{i\nu}\sqrt{-g} = E_i\,g^{ij}\sqrt{-g}, \tag{4.80}$$

where at the end of the calculation we have set, without loss of generality, $g^{00} = 1$. If we compare this result with the first of equations (4.72) we get the following result

$$D^j = \sqrt{-g}\,g^{ji}E_i = \varepsilon^{ji}E_i. \tag{4.81}$$

This is the main result of this section: an electromagnetic field propagating in curved spacetime can be thought of being completely analogous to an electromagnetic field propagating into a medium characterised by a permittivity $\varepsilon^{\mu\nu} = \sqrt{-g}\,g^{\mu\nu}$. With an analogous calculation, but on the magnetic field instead, one also arrives at the result $B^i = \sqrt{-g}\,g^{ij}H_j = \mu^{ij}H_j$. As one can immediately see by comparing the electric and magnetic results, a metric $g^{\mu\nu}$ defines a material whose electric and magnetic properties are fully determined by the metric and, moreover, are linked by the

---

[12] The proof of this can be found in any gravity textbook and it practically boils down to the fact that the two Christoffel symbols coming from the covariant derivatives cancel out, because the Christoffel symbol is symmetric in its lower indices, i.e., $\Gamma^\alpha_{\beta\gamma} = \Gamma^\alpha_{\gamma\beta}$.

relation $\varepsilon^{\alpha\beta} = \mu^{\alpha\beta}$. This result is at the foundation of transformation optics. By looking at the dynamics of the electromagnetic field in different curved spaces, such as a sphere, for example, one can, by virtue of the correspondence established above, design an optical material with the same characteristics as those imposed by the curved space under investigation (often referred to as metamaterials, as they are materials that cannot be found in Nature), thus achieving new ways to control and shape the propagation of light in matter. The most striking example of this correspondence is certainly the so-called invisibility cloak, demonstrated for the first time by Leonhardt [2] and Pendry [3] in 2006. An extensive set of examples of different metrics $g^{\mu\nu}$, the materials they generate, and the electromagnetic properties of such materials is contained in the book by Leonhardt and Philbin [5].

Appendix A: Derivation of equation (4.68)

Substituting directly equation (4.67) into equation (4.65c) gives

$$
\begin{aligned}
\int_0^\infty d\omega \, \frac{f(\omega)}{\rho} \tilde{\mathbf{Q}}(\omega) &= -\frac{1}{\rho} \int_0^\infty \omega^2 f(\omega)^2 \int_0^t d\tau \tilde{\mathbf{X}}(\tau) \int \frac{d\Omega}{2\pi} \frac{\exp[-i\Omega(t-\tau)]}{\Omega^2 - \omega^2} \\
&= \frac{1}{\rho} \int_0^t d\tau \tilde{\mathbf{X}}(\tau) \int \frac{d\Omega}{2\pi} \int_0^\infty d\omega \frac{\omega^2 |f(\omega)|^2}{\omega^2 - \Omega^2},
\end{aligned}
\tag{4.82}
$$

where in the last line we have analytically continued the function $f(\omega)$ to the complex plane. If we add and subtract $\Omega^2 |f(\omega)|^2 / (\omega^2 - \Omega^2)$ in the last integral, we can split it into two parts as

$$
\int_0^\infty d\omega \frac{\omega^2 |f(\omega)|^2}{\omega^2 - \Omega^2} = \int_0^\infty d\omega |f(\omega)|^2 + \Omega^2 \int_0^\infty , d\omega \frac{|f(\omega)|^2}{\omega^2 - \Omega^2}.
\tag{4.83}
$$

We can use the residue theorem to calculate both integrals. Before doing that, however, let us use the relation $V(\omega) = (f(\omega)/\rho)\sqrt{\Omega_0/\omega}$ in the first integral, to transform it into a Cauchy integral, and let us calculate the two integrals separately. First we have

$$
\int_0^\infty d\omega |f(\omega)|^2 = \rho\Omega_0 \int_0^\infty d\omega \frac{|V(\omega)|^2}{\omega} = 2\pi i \rho \Omega_0 |V(0)|^2 = 0,
\tag{4.84}
$$

since we have initially assumed that $f(\omega)$ is only zero at $\omega = 0$, and the same must also be assumed for $V(\omega)$ in order to avoid divergences. The second integral instead gives

$$
\Omega^2 \int_0^\infty d\omega \frac{|f(\omega)|^2}{\omega^2 - \Omega^2} = -\frac{i\pi}{\Omega^2} |V(\Omega)|^2,
\tag{4.85}
$$

where we have chosen a path in the complex plane that only encloses the pole at $\omega = -\Omega$. Substituting these two results into equation (4.82) gives equation (4.68).

Appendix B: Derivation of the Power–Zienau–Wooley transformation

In section 2.6.6 we have introduced a transformation, called Power–Zienau–Wooley transformation, that allows us to change the representation of the light–matter interaction, from the minimal coupling, involving the vector potential, to the electric dipole approximation. Here we present a detailed derivation of this transformation, using the tools of field theory. A more in-depth discussion on this topic can be found in the book by Cohen-Tannoudji [61].

Let us start by writing the Lagrangian for the light–matter interaction system described in chapter 3, i.e., a classical electromagnetic field interacting with a collection of atoms, as follows

$$\mathscr{L} = \sum_i \frac{m_i \dot{\mathbf{x}}_i}{2} - U_{\text{Coulomb}} + \int d^3x [|\mathbf{E}|^2 - |\mathbf{B}|^2] + \int d^3 \, \mathbf{J} \cdot \mathbf{A}$$
$$\equiv \mathscr{L}_{\text{mat}} + \mathscr{L}_{\text{field}} + \mathscr{L}_{\text{int}}, \tag{4.86}$$

where the dot indicates time derivation and the first two terms are the kinetic energy of the collection of atoms (the sum running over all the atoms in the system) and the collective single-particle and two-particle Coulomb potential, i.e.,

$$U_{\text{Coulomb}} = \sum_{i<j} \frac{q_i q_j}{|\mathbf{x}_i - \mathbf{x}_j|} + V_{\text{Coulomb}}, \tag{4.87}$$

with V_{Coulomb} being the standard, single-particle Coulomb potential. Note, moreover, that we have introduced the matter, field and interaction Lagrnangian for future convenience. The light–matter interaction is introduced in the minimal coupling way by the last term, where the current generated by the moving atoms is coupled to the electromagnetic field via the vector potentials. Remember, moreover, that we are working in the Coulomb gauge, so the appropriate gauge condition for this system is given by $\nabla \cdot \mathbf{A} = 0$. Let us now define a function F as

$$F = -\int d^3x \, \mathbf{P} \cdot \mathbf{A}, \tag{4.88}$$

where $\mathbf{P}$ is the matter polarisation, and add the time derivative of this expression to the Lagrangian above obtaining

$$\mathscr{L}' = \mathscr{L} + \frac{dF}{dt} = \mathscr{L} - \int d^3x \, [\dot{\mathbf{P}} \cdot \mathbf{A} + \mathbf{P} \cdot \dot{\mathbf{A}}]$$
$$= \mathscr{L}_{\text{mat}} + \mathscr{L}_{\text{field}} + \int d^3x \, (\mathbf{J} - \dot{\mathbf{P}}) \cdot \mathbf{A} - \int d^3x \, \mathbf{P} \cdot \dot{\mathbf{A}}. \tag{4.89}$$

As we can see, the addition of the total derivative dF/dt (which leaves the equations of motion invariant) has resulted in splitting the interaction Lagrangian into two terms, one containing only the matter polarisation (the last term above), and the other one containing both current and polarisation. Intuitively, we could expect the term containing only the polarisation to give rise to the electric dipole interaction, while the term involving the current should somehow take care of the magnetic dipole interaction. In fact, if we recall the definition of the electric field in terms of the vector potential, i.e., $\mathbf{E} = -\dot{\mathbf{A}}$, we can immediately see that the last term is just the electric dipole term in disguise, i.e.,

$$-\int d^3x \ \mathbf{P} \cdot \dot{\mathbf{A}} = \int d^3x \ \mathbf{P} \cdot \mathbf{E}, \tag{4.90}$$

To transform the other interaction term, which we deem intuitively responsible for the magnetic interaction, we use the second of equations (2.64), i.e., $\mathbf{J} = \dot{\mathbf{P}} + \nabla \times \mathbf{M}$, where $\mathbf{M}$ is the magnetisation of the material. Substituting then gives

$$\int d^3x \ (\mathbf{J} - \dot{\mathbf{P}}) \cdot \mathbf{A} = \int d^3x \ (\nabla \times \mathbf{M}) \cdot \mathbf{A}$$
$$= \int d^3x \ \mathbf{M} \cdot \mathbf{B}, \tag{4.91}$$

where to get the result in the last line we have integrated by parts and used the fact that $\nabla \times \mathbf{A} = \mathbf{B}$. Putting everything together, the transformed Lagrangian is then given by

$$\mathscr{L}' = \mathscr{L}_{\text{mat}} + \mathscr{L}_{\text{field}} + \int d^3x \ \mathbf{P} \cdot \mathbf{E} + \int d^3x \ \mathbf{M} \cdot \mathbf{B}. \tag{4.92}$$

If we now perform a Legendre transformation, after having defined the canonically conjugated momenta to the position of the atoms $(\mathbf{p}_i)$ and the electromagnetic field $(\dot{\mathbf{E}})$, we arrive at the expression of the Power–Zienau–Wooley Hamiltonian

$$\mathscr{H}_{\text{PZW}} = \mathscr{H}_{\text{mat}} + \mathscr{H}_{\text{field}} - \int d^3x \ \mathbf{P} \cdot \mathbf{E} - \int d^3x \ \mathbf{M} \cdot \mathbf{B}$$
$$\simeq \mathscr{H}_{\text{mat}} + \mathscr{H}_{\text{field}} - \int d^3x \ \mathbf{P} \cdot \mathbf{E}, \tag{4.93}$$

where $\mathscr{H}_{\text{mat}}$ and $\mathscr{H}_{\text{field}}$ are the Hamiltonians of the bare atoms and free electromagnetic field, respectively, and in the last line we have used the fact, that the magnetic dipole interaction is approximately α times smaller than the electric dipole (see equation (2.72)), and therefore it can be neglected in first approximation.

References

[1] Huttner B and Barnett S M 1992 Quantization of the electromagnetic field in dielectrics *Phys. Rev.* A **46** 4306
[2] Pendry J B, Schurig D and Smith D R 2006 Controlling electromagnetic fields *Science* **312** 1780
[3] Leonhardt U 2006 Optical conformal mapping *Science* **312** 1777
[4] Leonhardt U and Philbin T G 2006 General relativity in electrical engineering *New J. Phys.* **8** 247
[5] Leonhardt U 2012 *Geometry and Light: The Science of Invisibility* (New York: Dover)

[6] Landau L D and Lifshitz E M 1980 *The Classical Theory of Fields* 4th edn (Oxford: Butterworth-Heinemann)

[7] Barut O 1981 *Electrodynamics and Classical Theory of Fields and Particles* (Mineola, NY: Dover)

[8] Scheck F 2012 *Classical Field Theory* (Boston, MA: Springer)

[9] Cameron R P, Barnett S M and Yao A M 2012 Optical helicity, optical spin and related quantities in electromagnetic theory *New J. Phys.* **14** 053050

[10] Bliokh K Y, Bekshaev A Y and Nori F 2013 Dual electromagnetism: helicity, spin, momentum and angular momentum *New J. Phys.* **15** 033026

[11] Jackson J D 1998 *Classical Electrodynamics* (New York: Wiley)

[12] Das A 2006 *Field Theory: A Path Integral Approach* 2nd edn (Singapore: World Scientific)

[13] Coleman S 2019 *Quantum Field Theory* (Singapore: World Scientific)

[14] Hatfield B 1992 *Quantum Field Theory of Point Particles and Strings* (Boston, MA: Addison-Wesley)

[15] Belinfante F J 1940 On the current and the density of the electric charge, the energy, the linear momentum and the angular momentum of arbitrary fields *Physica* **7** 449

[16] Rosenfeld L 1940 Sur le tenseur daimpulsion-énergie *Mem. Acod. Roy. Bel.* **18** 1

[17] Messiah A 2014 *Quantum Mechanics* (New York: Dover)

[18] Barnett S M 2002 Optical angular-momentum flux *J. Opt. B: Quantum Semiclass. Opt.* **4** S7

[19] van Enk S J and Nienhuis G 1994 Spin and orbital angular momentum of photons *Europhys. Lett.* **25** 497

[20] Darwin C G 1932 Notes on the theory of radiation *Proc. R. Soc. Lond.* A **136** 36

[21] Barnett S M and Allen L 1994 Orbital angular momentum and nonparaxial light beams *Opt. Commun.* **110** 670

[22] van Enk S J and Nienhuis G 1994 Commutation rules and eigenvalues of spin and orbital angular momentum of radiation fields *J. Mod. Opt.* **41** 963

[23] Berry M V 2009 Optical currents *J. Opt. A: Pure Appl. Opt.* **11** 094001

[24] Li C-F 2009 Spin and orbital angular momentum of a class of nonparaxial light beams having a globally defined polarization *Phys. Rev.* A **80** 063814

[25] Barnett S M 2010 Rotation of electromagnetic fields and the nature of optical angular momentum *J. Mod. Opt.* **57** 1339

[26] Bialynicki-Birula I and Bialynicki-Birula Z 2011 Canonical separation of angular momentum of light into its orbital and spin parts *J. Opt.* **13** 064014

[27] Bliokh K Y, Dressel J and Nori F 2014 Conservation of the spin and orbital angular momenta in electromagnetism *New J. Phys.* **16** 093037

[28] Aghapour S, Andersson L and Rosquist K 2020 The zilch electromagnetic conservation law revisited *J. Math. Phys.* **61** 122902

[29] Ornigotti M and Aiello A 2014 Surface angular momentum of light beams *Opt. Express* **22** 6586

[30] Cameron R P 2014 On the second potential in electrodynamics *ZJ. Opt.* **16** 015708

[31] Berry M V 2009 Optical currents *J. Opt. A: Pure Appl. Opt.* **11** 094001

[32] Heaviside O 1892 On the forces, stresses and fluxes of energy in the electromagnetic field *Phil. Trans. R. Soc.* A **183** 423

[33] Larmor J 1897 Dynamical theory of the electric and luminiferous medium iii *Phil. Trans. R. Soc.* A **190** 205

[34] Deser S and Teitelboim C 1976 Duality transformations of Abelian and non-Abelian gauge fields *Phys. Rev.* D **13** 1592

[35] Radmon P 1997 *Field Theory: A Modern Primer* 2nd edn (Boulder, CO: Westview Press)

[36] Frampton P H 2000 *Gauge Field Theories* (New York: Wiley)

[37] Itzykson C and Zuber J B 1980 *Quantum Field Theory* (New York: Dover)

[38] Crimin F, Mackinnon M, Götte J P and Barnett S M 2019 Optical helicity and chirality: conservation and sources *Appl. Sci.* **9** 828

[39] Barnett S M, Cameron R P and Yao A M 2012 Duplex symmetry and its relation to the conservation of optical helicity *Phys. Rev.* A **86** 013845

[40] Trueba J L and Rañada A F 1996 The electromagnetic helicity *Eur. J. Phys.* **17** 141

[41] Landau L D and Lifshitz E M 1980 *Quantum Electrodynamics* 4th edn (Oxford: Butterworth-Heinemann)

[42] Bialynicki-Birula I 1996 The photon wave function *Prog. Opt.* **36** 245

[43] Lipkin D M 1964 Existence of a new conservation law in electromagnetic theory *J. Math. Phys.* **5** 696

[44] Tang Y and Cohen A E 2010 Optical chirality and its interaction with matter *Phys. Rev. Lett.* **104** 163901

[45] Coles M M and Andrews D L 2012 Chirality and angular momentum in optical radiation *Phys. Rev.* A **85** 063810

[46] Forbes K A and Andrews D L 2018 Optical orbital angular momentum: twisted light and chirality *Opt. Lett.* **43** 435

[47] Forbes K A and Andrews D L 2021 Orbital angular momentum of twisted light: chirality and optical activity *J. Phys. Photon.* **3** 022007

[48] Bliokh K Y and Nori F 2011 Characterizing optical chirality *Phys. Rev.* A **83** 021803(R)

[49] Caloz C and Sihvola A 2019 Electromagnetic chirality arXiv:1903.09087

[50] Mun J, Minkyung Kim M, Yang Y, Badloe T, Jincheng Ni J, Yang Chen Y, Qiu C-W and Rho Y 2020 Electromagnetic chirality: from fundamentals to nontraditional chiroptical phenomena *Light: Sci. Appl.* **9** 139

[51] Philbin T G 2013 Lipkin's conservation law, noether's theorem, and the relation to optical helicity *Phys. Rev.* A **87** 043843

[52] Maggiore M 2005 *A Modern Introduction to Quantum Field Theory* (Oxford: Oxford University Press)

[53] Byron F W and Fuller R W 1992 *Mathematics of Classical and Quantum Physics* (Mineola, NY: Dover)

[54] Dutra S M and Furuya K 1998 The permittivity in the Huttner-Barnett theory of QED in dielectrics *Europhys. Lett.* **43** 13

[55] Bechler A 1999 Quantum electrodynamics of the dispersive dielectric medium–a path integral approach *J. Mod. Opt.* **46** 901

[56] Tamashevich Y, Shubitidze T, Dal Negro L and Ornigotti M 2024 Field theory description of the non-perturbative optical nonlinearity of epsilon-near-zero media *APL Photon.* **9** 016105

[57] Dutra S M and Furuya K 1997 Macroscopic averages in QED in material media *Phys. Rev.* A **55** 3832

[58] Misner C W, Thorne K S and Wheeler J A 2017 *Gravitation* (Princeton, NJ: Princeton University Press)

[59] Minkowski H 1908 Die grundgleichungen far die elektromagnetischen vorgänge in bewegten körpern *Nachr. Kgl. Ges. Wiss. Goett.* **53** 111

[60] Vanderlinde J 2004 *Classical Electromagnetic Theory* (Berlin: Springer)

[61] Cohen-Tannoudji C 1989 *Photons and Atoms: Introduction to Quantum Electrodynamics* (New York: Wiley)

Part II

Quantum theory of light

IOP Publishing

A Field Theory Approach to Photonics

Marco Ornigotti

Chapter 5

Quantum field theory in a nutshell

This chapter marks the start of the second part of this book, dedicated to the quantisation of the electromagnetic field, and the construction of a quantum theory for linear and nonlinear optics in arbitrary media. Here, we first present a brief review of classical Hamiltonian mechanics in section 5.1, then we continue using the simple case of a real scalar field to discuss field quantisation, both in the Hamiltonian and Lagrangian (i.e., path integral) formalism.

We first look at the *canonical* quantisation of the free field in section 5.2, which is the usual and most common quantisation procedure starting from the field Hamiltonian and introducing the creation and annihilation operators and the particle (Fock) space. Then, in section 5.3 we introduce interactions and discuss how to treat them within the canonical formalism. Here, we will be mostly looking at self-interactions of the field, since they are the most common in photonics, through nonlinear optics. We then conclude our analysis of quantum fields in canonical formalism by discussing Feynman diagrams and giving Feynman rules for canonical quantum fields. This is done in section 5.4.

Section 5.5 is then dedicated to the quantisation of a free scalar field in the Lagrangian formalism, and is devoted to introduce path integrals. This is complemented by interacting fields in path integral formalism, discussed in section 5.6.

Most of the material presented in this chapter follows from standard quantum field theory and can be found presented in different references from different perspectives. Some easily accessible references for quantum field theory (QFT) are the books by Maggiore [1] and Srednicki [2]. The latter can also be used as a good reference for path integrals. Alternatively, the books by Brown [3], Das [4] and the classic book by Feynman and Hibbs [5] (although introducing path integrals for quantum mechanics, and not for quantum fields) are also excellent references for learning path integrals in the classical and quantum domain. The interested reader can delve deeper into the subject of QFT by looking at the books by Peskin and Schroeder [6], and the three volumes by Weinberg [7], while for path integrals, an excellent reference is the book by Kleinert [8], for a general overview of the many fields of physics, where path integrals find applications, or the book by Rivers [9].

5.1 Hamiltonian mechanics revisited

In section 3.1.4 we have introduced the Hamiltonian for a scalar field and derived the equations of motion from the action principle. For a free field, we can derive the Hamiltonian density by setting $U(\psi) = m^2\psi^2$ in equation (3.18) to get

$$\mathscr{H} = \frac{1}{2}[\Pi^2 + (\nabla\psi)^2 + m^2\psi^2], \tag{5.1}$$

where m is the mass of the field. The equations of motion can be then derived from the action (3.19) to get equations (3.21). Let us, however, try to rewrite those equations in a more compact form, more natural for quantisation, by making explicit use of the fact, that ψ and Π are canonically conjugated variables[1] and can therefore be used to define the Poisson brackets [10]

$$\{F, G\} = \int d^4x \left(\frac{\partial F}{\partial \psi}\frac{\partial G}{\partial \Pi} - \frac{\partial F}{\partial \Pi}\frac{\partial G}{\partial \psi}\right), \tag{5.2}$$

where F and G are functionals of the fields ψ and Π. Notice, that the Poisson brackets form a Lie algebra, and as such they represent an anti-symmetric operation, subject to the Jacobi identity

$$\{X, \{Y, Z\}\} + \{Y, \{Z, X\}\} + \{Z, \{X, Y\}\} = 0. \tag{5.3}$$

In particular, since ψ and Π are canonically conjugated, the following relations hold

$$\{\psi(\mathbf{x}), \psi(\mathbf{y})\} = 0 = \{\Pi(\mathbf{x}), \Pi(\mathbf{y})\}, \tag{5.4a}$$

$$\{\psi(\mathbf{x}), \Pi(\mathbf{y})\} = \delta(\mathbf{x} - \mathbf{y}), \tag{5.4b}$$

so that the equations of motion (3.21) can be rewritten in the following compact form

$$\dot{\psi} = \{\psi, \mathscr{H}\}, \tag{5.5a}$$

$$\dot{\Pi} = \{\Pi, \mathscr{H}\}. \tag{5.5b}$$

Moreover, the evolution of a functional $\rho \equiv \rho(\psi, \Pi, t)$ can be easily written with the help of Poisson brackets as

$$\frac{d\rho}{dt} = \{\rho, \mathscr{H}\} + \frac{\partial \rho}{\partial t}, \tag{5.6}$$

---

[1] In a more optics-friendly jargon, canonically conjugated variables are linked via Fourier transformation, i.e., they are Fourier pairs

and if A is a conserved quantity of the system, then

$$\{A, \mathcal{H}\} = 0. \tag{5.7}$$

Poisson brackets derive directly from the symplectic nature of Hamiltonian mechanics, i.e., the fact that a system possessing a Hamiltonian can also be endowed with a phase space spanned by its canonically conjugated variables[2]. The experienced reader might recognise in the structure above a certain similarity with that of a pair of non-commuting operators, where the Poisson brackets play essentially the role of the Lie bracket, commonly referred to as commutator. This similarity is not accidental. In classical physics, all observables commute (i.e., they can be measured simultaneously), and there is no need for a commutator. However, when coming to canonically conjugated variables a similar relation to that of commutators exists in the form of Poisson brackets. Although the underlying symplectic structure is preserved between the classical and quantum case, the algebra changes from that of the Poisson brackets in the classical case, to that of commutators in the quantum case. This change is necessary, since the fundamental objects of the theory are changing: classical fields are replaced with linear operators, which generally do not commute. This correspondence, which is even more corroborated by the fact that the classical limit of quantum mechanics in the Heisenberg representation is precisely Hamiltonian mechanics in the Poisson bracket formulation, led Dirac to propose that the quantum counterparts $\hat{A}$ and $\hat{B}$ of classical observables A and B should be found by imposing the following equality

$$[\hat{A}, \hat{B}] = i\hbar\widehat{\{A, B\}}, \tag{5.8}$$

where the ^ symbol indicates that the quantities underneath it have to be considered as operators, and $[\hat{A}, \hat{B}] = \hat{A}\hat{B} - \hat{B}\hat{A}$ is the usual (quantum) commutator. It is worth noticing, however, that a general systematic correspondence between quantum commutators and Poisson brackets cannot be established, as stated by Groenewold in his 1946 paper on the matter [11]. However, such a correspondence exists between a deformed version of the Poisson bracket (called Moyal bracket), and it constitutes the starting point for an alternative quantisation scheme called *deformation quantisation*, and based, essentially, on a generalisation of the concept of Wigner function and phase space of a quantum system. Here, however, we will accept that the correspondence outlined in equation (5.8) is a bona fide classical-to-quantum mapping, and we will use it to quantise the field. The interested reader can find more information of Groenewold's theorem and deformation quantisation in reference [12].

5.2 Canonical quantisation

Before quantising the Hamiltonian (5.1), let us first rewrite it in k-space, by introducing the Fourier transform of the field ψ and the momentum Π as follows

---

[2] In classical mechanics, for example, the phase space of a system is spanned by the position q and its canonically conjugated momentum p.

$$\psi = \int^{\cdot} d^3\tilde{k}[a(\mathbf{k})\exp(ikr) + a^*(\mathbf{k})\exp(-ikr)], \tag{5.9a}$$

$$\Pi = \partial_0 \psi = \int d^3\tilde{k}\ [-i\omega(\mathbf{k})][a(\mathbf{k})\exp(ikr) - a^*(\mathbf{k})\exp{-(ikr)}], \tag{5.9b}$$

where $d^3\tilde{k} = d^3k/[2\omega(\mathbf{k})(2\pi)^3]$ is the Lorentz invariant measure, $kr = k_\mu r^\mu = \mathbf{k} \cdot \mathbf{r} - \omega\, t$ and $\omega(k) = \sqrt{|\mathbf{k}|^2 + m^2}$. The spectral amplitudes $a(\mathbf{k})$ and $a^*(\mathbf{k})$ will become, upon quantisation, the creation and annihilation operators of the field, in analogy with the harmonic oscillator (see appendix A).

Substituting the expressions above into equation (5.1) and integrating over the whole space we get the following form for the Hamiltonian in k-space representation

$$\begin{aligned}
\hat{H} &= \int d^3x\ \hat{\mathscr{H}} = \frac{1}{2}\int d^3x d^3k d^3k'\{[-\omega(\mathbf{k})\omega(\mathbf{k}') - \mathbf{k} \cdot \mathbf{k}' + m^2] \\
&\quad \times [a(\mathbf{k})a(\mathbf{k}')\exp[i(k + k')r]+\text{c.c.}] \\
&\quad + [\omega(\mathbf{k})\omega(\mathbf{k}') + \mathbf{k} \cdot \mathbf{k}' + m^2][a(\mathbf{k})a^*(\mathbf{k}')\exp[i(k - k')r]+\text{c.c.}]\} \\
&= \frac{1}{2}\int \frac{d^3k}{2\omega(\mathbf{k})}\{[\omega^2(\mathbf{k}) + |\mathbf{k}|^2 + m^2][a^*(\mathbf{k})a(\mathbf{k}) + a(\mathbf{k})a^*(\mathbf{k})] \\
&\quad + [-\omega^2(\mathbf{k}) + |\mathbf{k}|^2 + m^2][a(\mathbf{k})a(-\mathbf{k})\exp[-2i\omega(\mathbf{k})t] + a^*(\mathbf{k})a^*(-\mathbf{k})\exp[2i\omega(\mathbf{k})t]] \\
&= \frac{1}{2}\int d^3k\ \omega(\mathbf{k})[a^*(\mathbf{k})a(\mathbf{k}) + a(\mathbf{k})a^*(\mathbf{k})]\},
\end{aligned} \tag{5.10}$$

where we have used the spectral property $a(-\mathbf{k}) = a^*(\mathbf{k})$ and $\omega(-\mathbf{k}) = \omega(\mathbf{k})$, since $\omega(\mathbf{k}) \in \mathbb{R}$. To get the second identity, we have integrated over d^3x and used the Dirac delta identity

$$\int d^3x\ \exp[\pm i(k \pm k')r] = (2\pi)^3 \exp[\mp i(\omega \pm \omega')t]\delta(\mathbf{k} \pm \mathbf{k}'). \tag{5.11}$$

Notice, that when integrating the Dirac delta with respect to d^3k', there is an extra factor $1/2\omega(\mathbf{k})$ remaining from the Lorentz measure. To get the third equality, we have used the fact, that $\omega^2(\mathbf{k}) = |\mathbf{k}|^2 + m^2$. We have left the sum $a^*(\mathbf{k})a(\mathbf{k}) + a(\mathbf{k})a^*(\mathbf{k})$ in this expanded way, instead of summing the two terms, for later convenience. Notice, moreover, that if we use the Fourier representation of ψ and Π, the Poisson brackets in equations (5.4) can be written in terms of the spectral amplitudes $a(\mathbf{k})$ and $a^*(\mathbf{k})$ as follows

$$\{a(\mathbf{k}), a(\mathbf{k}')\} = 0 = \{a^*(\mathbf{k}), a^*(\mathbf{k}')\}, \tag{5.12a}$$

$$\{a(\mathbf{k}), a^*(\mathbf{k}')\} = \delta(\mathbf{k} - \mathbf{k}'). \tag{5.12b}$$

In doing so, we need to assume that $a(\mathbf{k})$ and $a^*(\mathbf{k})$ are independent quantities. If we do so, they can basically be considered as a couple of canonically conjugated variables in Fourier domain.

We are now ready to quantise the free scalar field. We can use the correspondence (5.8), and promote the field ψ and momentum Π or, equivalently, promote the quantities $a(\mathbf{k})$ and $a^*(\mathbf{k})$, to operators. By doing so, we need to change the Poisson brackets into commutators to obtain the commutation rules for the field operators, i.e.,

$$[\hat{\psi}(\mathbf{x}), \hat{\psi}(\mathbf{y})] = 0 = [\hat{\Pi}(\mathbf{x}), \hat{\Pi}(\mathbf{y})], \tag{5.13a}$$

$$[\hat{\psi}(\mathbf{x}), \hat{\Pi}(\mathbf{y})] = \delta(\mathbf{x} - \mathbf{y}), \tag{5.13b}$$

and, analogously, for the operators $\hat{a}(\mathbf{k})$ and $\hat{a}^{\dagger}(\mathbf{k})$

$$[\hat{a}(\mathbf{k}), \hat{a}(\mathbf{k}')] = 0 = [\hat{a}^{\dagger}(\mathbf{k}), \hat{a}^{\dagger}(\mathbf{k}')], \tag{5.14a}$$

$$[\hat{a}(\mathbf{k}), \hat{a}^{\dagger}(\mathbf{k}')] = \delta(\mathbf{k} - \mathbf{k}'). \tag{5.14b}$$

The operators $\hat{a}^{\dagger}(\mathbf{k})$ and $\hat{a}(\mathbf{k})$ are known as creation and annihilation operators, respectively, since their action onto the state of the system creates or destroys an excitation of the field, i.e., a particle associated with the quantum field. With these prescriptions, the Hamiltonian operator can be easily written from equation (5.10) as

$$\hat{H} = \int d^3k \; \hbar\omega(\mathbf{k})\left[\hat{a}^{\dagger}(\mathbf{k})\hat{a}(\mathbf{k}) + \frac{1}{2}\right]. \tag{5.15}$$

Notice how the Hamiltonian above can be interpreted as a collection of uncoupled harmonic oscillators, one for each frequency component $\mathbf{k}$ of the field (see appendix A).

To derive the equations of motion for the field operators, we can use the correspondence (5.8) again, this time to transform the equation of motion for an arbitrary observable of the system, i.e., equation (5.6) into its quantum version, known as the Heisenberg equation of motion[3]

$$i\hbar\frac{d\hat{\rho}}{dt} = [\hat{\rho}, \hat{H}]. \tag{5.16}$$

Moreover, if $\hat{A}$ is an operator associated with an observable of the system, then $A = \langle\hat{A}\rangle$ is a conserved quantity of the system if, in analogy with equation (5.7)

$$[\hat{A}, \hat{H}] = 0, \tag{5.17}$$

i.e., if $\hat{A}$ commutes with the Hamiltonian of the system.

The quantisation procedure highlighted above is completely independent on the state of the system. This is related to the fact that canonical quantisation stems from the Hamiltonian formulation of mechanics, where the state of the system is determined by its configuration in phase space (i.e., position and momentum at each time), rather than its trajectory. This is then a different perspective than regular quantum mechanics, where the state of the system, i.e., its wave function, evolves in time according to the Schrödinger equation, while the operators acting on the wave

---

[3] This way of quantising the field leads directly and naturally to the Heisenberg representation of quantum mechanics, where the state is time-independent, while the operators acting on the state evolve in time according to the Heisenberg equation.

function are time-independent. Here, instead, the opposite is true: the state of the system does not evolve in time and remains therefore constant, while the operators acting on the system are evolving in time according to equation (5.16). If we apply this equation to the creation and annihilation operators, we in fact see that they evolve in time with frequency $\omega(\mathbf{k})$, i.e.,

$$i\hbar\frac{d\hat{a}(\mathbf{k})}{dt} = [\hat{a}(\mathbf{k}), \hat{H}] = \hbar\omega(\mathbf{k})\hat{a}(\mathbf{k}), \tag{5.18a}$$

$$-i\hbar\frac{d\hat{a}^{\dagger}(\mathbf{k})}{dt} = [\hat{a}^{\dagger}(\mathbf{k}), \hat{H}] = \hbar\omega(\mathbf{k})\hat{a}^{\dagger}(\mathbf{k}). \tag{5.18b}$$

This representation of quantum systems is known in quantum mechanics as the Heisenberg picture, in which, as explained above, the state of the system does not evolve in time, while the operators acting on it do.

5.2.1 Fock space for quantum fields

QFT uses the Heisenberg representation of quantum mechanics, where the state of the system does not evolve in time, but operators do, according to equation (5.16). This means that the focus of QFT is the study of the interaction, time evolution, and action of operators on the state of a system, rather than the evolution of the state of the system itself, as is the case for the traditional Schrödinger picture of quantum mechanics.

The state of the system is therefore always regarded as being the vacuum state $|0\rangle$, i.e., the state where no exitation of the field is present, and whose energy is given by the zero point, or vacuum, energy of the field. The vacuum state is defined as the state that is annihilated by the annihilation operator $\hat{a}(\mathbf{k})$, i.e., $\hat{a}(\mathbf{k})|0\rangle = 0$. Any other state of the system can be then fully generated by acting on the vacuum state with a number of creation operators equal to the number of particles (i.e., field excitation) in the field at a given instant in time. This can be done in two ways: first, if we fix $\mathbf{k}$ and assume that the field has only one single spectral mode, then the subsequent action of $\hat{a}^{\dagger}(\mathbf{k})$ on the vacuum state generates more particles in the same state $\mathbf{k}$, i.e.,

$$\frac{(\hat{a}^{\dagger}(\mathbf{k}))^{N}}{\sqrt{N!}}|0\rangle \equiv |N\rangle, \tag{5.19}$$

where $|N\rangle$ indicates a state (characterised by the momentum $\mathbf{k}$) containing N particles (field excitations), and the term $1/\sqrt{N!}$ is just a normalisation factor deriving from the action of the creation and annihilation operators on quantum states, namely

$$\hat{a}(\mathbf{k})|n\rangle = \sqrt{n}\,|n-1\rangle, \tag{5.20a}$$

$$\hat{a}^{\dagger}(\mathbf{k})|n\rangle = \sqrt{n+1}\,|n+1\rangle, \tag{5.20b}$$

in order to guarantee that $\langle n|n \rangle = 1$ for any n. The state in equation (5.19) is a *single-mode, N particle state*, and is commonly referred to as a number, or Fock, state. One important characteristic of Fock states is that they possess a *well-defined* number of particles n. Moreover, they define an orthonormal basis, that can be used to represent any quantum state in Fock space, i.e., the Hilbert space of particle states. Fock states can also be defined as the eigenstates of the so-called particle number operator $\hat{N}(\mathbf{k}) = \hat{a}^{\dagger}(\mathbf{k})\hat{a}(\mathbf{k})$ through the eigenvalue equation

$$\hat{N}(\mathbf{k})|n\rangle = \hat{a}^{\dagger}(\mathbf{k})\hat{a}(\mathbf{k})|n\rangle = n|n\rangle. \qquad (5.21)$$

The second possibility to generate an N particle state is to lift the assumption that the field only has a single spectral mode and regard a collection of spectral modes, each characterised by its own wave vector $\mathbf{k}$. In this case, each wave vector has its own set of creation and annihilation operators and harmonic oscillator states associated with it, and the multi-particle state can be written as follows

$$\hat{a}^{\dagger}(\mathbf{k})\hat{a}^{\dagger}(\mathbf{k}') \cdots \hat{a}^{\dagger}(\mathbf{k}^{(n)})|0_{\mathbf{k}}\rangle|0_{\mathbf{k}'}\rangle \cdots |0_{\mathbf{k}^{(n)}}\rangle \equiv |1_{\mathbf{k}}, 1_{\mathbf{k}'},\ldots, 1_{\mathbf{k}^{(n)}}\rangle \qquad (5.22)$$

where $|0_{\mathbf{k}}\rangle$ indicates the vacuum state pertaining to the wave vector $\mathbf{k}$. In general, however, a state containing N particles can have them distributed in any combination across all the possible states of the system, labelled by their wave vector $\mathbf{k}$. For example, assuming a system characterised by three wave vectors $\mathbf{k}_{1,2,3}$, then there will be three single-particle states, i.e., $\{|1, 0, 0\rangle, |0, 1, 0\rangle, |0, 0, 1\rangle\}$, six two-particle states, i.e., $\{|2, 0, 0\rangle, |0, 2, 0\rangle, |0, 0, 2\rangle, |1, 1, 0\rangle, |1, 0, 1\rangle, |0, 1, 1\rangle\}$ and so on. In general, if there are n particles and m available states, the number of different n particle states that can be generated is $\binom{n+m-1}{m-1}$. All these states are orthogonal to each other, since

$$\langle n|m \rangle \equiv \langle n_{\mathbf{k}}^{(1)}, \ldots n_{\mathbf{k}^{(p)}}^{(p)}|m_{\mathbf{k}}^{(1)}, \ldots m_{\mathbf{k}^{(q)}}^{(q)}\rangle = \langle n_{\mathbf{k}}^{(1)}|m_{\mathbf{k}}^{(1)}\rangle \cdots \langle n_{\mathbf{k}^{(p)}}^{(p)}|m_{\mathbf{k}^{(q)}}^{(q)}\rangle$$
$$= \delta_{n^{(1)},m^{(1)}} \cdots \delta_{n^{(p)},m^{(q)}}, \qquad (5.23)$$

and they then define a different orthonormal set of multi-particle states for each value of N. The number operator for multi-particle Fock states is also readily obtained by simply summing the number of operators of each individual state as $\hat{N} = \sum_{i=1}^{p} \hat{N}(\mathbf{k}_i)$, where p is the number of spectral states of the field.

5.3 Interacting field—Hamiltonian formalism

We now turn our attention to the case in which the field is not free, but it is interacting, in particular, since this is frequently the situation in photonics, we consider only self-interacting fields, which are a good model to describe optical nonlinearities [13, 14]. As usual, we limit our description to a scalar field, and we tackle the problem of an interacting field theory from a Hamiltonian point of view. The methods and techniques developed in this section can also, however, be readily applied to vector, tensor, and spinor fields, without much effort.

The full Hamiltonian for an interacting field theory can be written as $\hat{H} = \hat{H}_0 + \hat{H}_{\text{int}}$, where $\hat{H}_0$ is the free Hamiltonian derived in the previous section, and $\hat{H}_{\text{int}}$ is the interaction Hamiltonian, whose explicit form varies depending on the kind of interaction one takes into account. A very common form for the interaction Hamiltonian used in QFT, which is also very convenient for nonlinear optical purposes, is a polynomial function of the field ψ, i.e.,

$$\hat{H}_{\text{int}} = \frac{g}{n!}\psi^n, \tag{5.24}$$

where g is the coupling constant characteristic of the interaction, and typically $g \ll 1$ is assumed, so that the interaction Hamiltonian above can be treated using perturbation theory.

Before delving into solving the quantum problem, it is instructive to think about the correspondent classical one. The total Hamiltonian $H = H_0 + H_{\text{int}}$ generates a set of nonlinear equations of motion, whose complexity depends on the explicit form of H_{int}. In general, however, we can say that because of the complexity of the equations of motion, the field ψ does not admit a simple representation in terms of plane waves, and therefore we cannot use the expansion in equation (5.9). At the quantum level, this means that we cannot introduce the creation and annihilation operators for the field ψ, since we do not have access to its plane wave expansion (i.e., Fourier transform). To circumvent this problem, we would like to have access to a field ψ_I, instead, that evolves only according to the free, rather than total, Hamiltonian. If we were able to find such field, we could describe the whole system in terms of ψ_I rather than ψ and be able to introduce a Fourier representation for the field and, consequently, creation and annihilation operators. The framework of introducing an operator ψ_I that evolves only with the free part of the Hamiltonian is known in literature as the interaction (or Dirac) picture, and will be the subject of this section.

To change from the Heisenberg to the Dirac picture, we define a quantum field $\psi_I(\mathbf{r}, t)$, such that its evolution in time from the initial time $t = t_0$ to a generic time t is given by

$$\psi_I(\mathbf{r}, t) = \exp[i\hat{H}_0(t - t_0)]\psi_I(\mathbf{r}, t_0)\exp[-i\hat{H}_0(t - t_0)]. \tag{5.25}$$

By construction, ψ_I is a free field, and can be therefore represented in terms of creation and annihilation operators following equations (5.9). We can take advantage of this fact and express the full field ψ in terms of the field ψ_I as follows[4]

$$
\begin{aligned}
\psi(\mathbf{r}, t) &= \exp(i\hat{H}t)\psi(\mathbf{r}, 0)\exp(-i\hat{H}t) \\
&= \exp(i\hat{H}t)\exp(-i\hat{H}_0 t)\exp(i\hat{H}_0 t)\psi(\mathbf{r}, 0)\exp(-i\hat{H}_0 t)\exp(i\hat{H}_0 t)\exp(-i\hat{H}t) \quad (5.26)\\
&\equiv U^\dagger(\mathbf{r}, t)\psi_I(\mathbf{r}, t)U(\mathbf{r}, t),
\end{aligned}
$$

where to go from the second to the third line we have first assumed that at $t = 0$ $\psi(\mathbf{r}, 0) = \psi_I(\mathbf{r}, 0)$ and then used the definition of ψ_I given by equation (5.25), and then we have defined the evolution operator

---

[4] Without loss of generality, we have assumed $t_0 = 0$ for simplicity.

$$U(\mathbf{r}, t) = \exp(i\hat{H}_0 t)\exp(-i\hat{H}t). \tag{5.27}$$

Notice, that since $\hat{H}_0$ and $\hat{H}$ do not commute (because of the presence of the interaction Hamiltonian), one cannot write that $\exp(i\hat{H}_0 t)\exp(-i\hat{H}t) = \exp(-i\hat{H}_{\text{int}} t)$. However, if we introduce the interaction picture Hamiltonian as

$$\hat{H}_I(t) = \exp(i\hat{H}_0 t)\hat{H}_{\text{int}} \exp(-i\hat{H}_0 t), \tag{5.28}$$

then we can write the evolution operator in the interaction picture representation as simply

$$U(t, t_0) = \hat{T} \exp\left[-i \int_{t_0}^{t} d\tau\, \hat{H}_I(\tau)\right], \tag{5.29}$$

together with the initial condition $U(t_0, t_0) = 1$, where $\hat{T}$ is the temporal ordering operator [15]. This can be derived by looking at the equation of motion for $U(t, t_0)$ and realise that the evolution of $U(t, t_0)$ is indeed generated by $\hat{H}_I$.

The next step in solving the interacting field theory is to compute the n-point Green's function, defined as

$$G_n(x_1, x_2, \dots, x_n) = \langle 0|\psi(x_1)\psi(x_2) \cdots \psi(x_n)|0\rangle, \tag{5.30}$$

with $x_n = (\mathbf{r}, t_n)$, and $t_{n-1} > t_n$, so that time ordering is automatically accounted for. This quantity, essentially, contains all the information about the (self-)interaction of the field ψ, i.e., it corresponds to the probability amplitude for the interaction Hamiltonian in equation (5.24) to give a non-negligible contribution to the energy of the quantum system at hand. We can calculate the n-point Green's function by using the interaction picture representation developed above, to write the field $\psi(x_n)$ in terms of its interaction picture counterpart $\psi_I(x_n)$. Substituting equation (5.26) into the expression above, and taking care of respecting time ordering, we get the following expression for the interaction picture n-point Green's function

$$
\begin{aligned}
G_n(x_1, x_2, \dots, x_n) &= \langle 0|[U^\dagger(t_1, t_0)\psi_I(x_1)U(t_1, t_0)] \cdots [U^\dagger(t_n, t_0)\psi_I(x_n)U(t_n, t_0)]|0\rangle \\
&= \langle 0|U^\dagger(t_1, t_0)[\psi_I(x_1)U(t_1, t_2)\psi_I(x_2) \cdots U(t_{n-1}, t_n)\psi_I(x_n)U(t_n, -t)]U(-t, t_0)|0\rangle \\
&= \langle 0|U^\dagger(t, t_0)\hat{T}\left\{\psi_I(x_1) \cdots \psi_I(x_n)\exp\left[-i \int_{-t}^{t} d\tau\, \hat{H}_I(\tau)\right]\right\}U(-t, t_0)|0\rangle,
\end{aligned} \tag{5.31}
$$

where to go from the first to the second line we have first reordered the evolution operators according to the chain rule $U(t_1, t_0)U^\dagger(t_2, t_0) = U(t_1, t_0)U(t_0, t_2) = U(t_1, t_2)$ and then introduced the variable $t \gg t_1 > t_2 > \cdots > t_n \gg -t$ to use the same chain rule in reverse and write $U(t_n, t_0) = U(t_n, -t)U(-t, t_0)$. Then, to pass from the second to the third line it is enough to observe that the products appearing inside the square brackets are automatically time-ordered, and therefore the various evolution operators can be collected together and combined to give rise to $U(-t, t)$, which we have written in the third line in its explicit form, according to equation (5.29).

To complete our calculation, we observe that if we now choose $t_0 = -t \rightarrow \infty$, we have that $U(t_0, -t) = 1$ and $U^\dagger(t, t_0) = U^\dagger(-\infty, \infty)$, which we need to calculate, and it will serve as normalisation factor for the n-point Green's function. To do so, we can notice that the term $U(-\infty, \infty)|0\rangle$ (and its hermitian conjugate $\langle 0|U^\dagger(-\infty, \infty)$) are the same state as the vacuum state $|0\rangle$, up to a global phase, i.e.,

$$U(-\infty, \infty)|0\rangle = \exp(i\alpha)|0\rangle. \tag{5.32}$$

This means, essentially, that evolving the vacuum state from $-\infty$ to $+\infty$ doesn't change the vacuum state at all (apart from the global phase factor), if the vacuum state of the field we are considering is stable, and does not have any means of decay, i.e., if there is no spontaneous breaking of any symmetry of the vacuum state, or a background accelerating spacetime [16]. However, we can take account of the result above when normalising $G_n(x_1, \ldots, x_n)$, so that for $t_0 = -t \to \infty$ it gives the correct result of the normalised vacuum state. If we then multiply the left- and right-hand-side of the expression above by $\langle 0|$, and using the fact that $\langle 0|0\rangle = 1$ we arrive at the expression for the phase factor α, which has the following explicit form

$$\exp(i\alpha) = \langle 0|\hat{T}\left\{\exp\left[-i\int_{-\infty}^{\infty} d\tau\ \hat{H}_I(\tau)\right]\right\}|0\rangle, \tag{5.33}$$

which allows us to write the complete expression of the (normalised) n-point Green's function as [1]

$$G_n(x_1, x_2, \ldots, x_n) = \frac{\langle 0|\hat{T}\left\{\psi_I(x_1)\cdots\psi_I(x_n)\exp\left[-i\int_{-t}^{t} d\tau\ \hat{H}_I(\tau)\right]\right\}|0\rangle}{\langle 0|\hat{T}\left\{\exp\left[-i\int_{-\infty}^{\infty} d\tau\ \hat{H}_I(\tau)\right]\right\}|0\rangle}. \tag{5.34}$$

5.4 Feynman diagrams

5.4.1 The Feynman propagator

A very important case of the n-point Green's function is that of $n = 2$, which is often known in the QFT literature as the propagator [1, 2, 15, 17, 18]. In particular, defined starting from equation (5.34), the propagator is automatically causal (because it is time-ordered) and it is known also with the name of Feynman propagator, and it is defined as

$$D_F(x - y) = \langle 0|\hat{T}\{\psi_I(x)\psi_I(y)\}|0\rangle. \tag{5.35}$$

To calculate it, then, we need to find the explicit expression of the two-point Green's function. First, we notice that since $\psi_I(x)$ is by definition a free field, we can write it in terms of creation and annihilation operators following equations (5.9). In particular, we can write it as a sum of its creation (proportional to $\hat{a}^\dagger(\mathbf{k})$) and annihilation (proportional to $\hat{a}(\mathbf{k})$) parts as $\psi_I(x) = \psi^+(x) + \psi^-(x)$. Notice that we have removed the subscript $_I$, for simplicity of notation, and also because it is clear from the context that these fields are the interaction picture fields.

The time-ordered product appearing above can then be expanded and calculated as follows, assuming $x^0 > y^0$

$$\hat{T}\{\psi(x)\psi(y)\} = \; :\psi(x)\psi(y): \; + [\psi^+(x),\,\psi^-(y)], \qquad (5.36)$$

where the colon denotes normal ordering, i.e., all the creation operators are ordered on the left of all the annihilator operators (so, for example $:\hat{a}^\dagger(\mathbf{k})\hat{a}(\mathbf{k}): \; = \hat{a}^\dagger(\mathbf{k})\hat{a}(\mathbf{k})$ is automatically normal ordered, but $:\hat{a}(\mathbf{k})\hat{a}^\dagger(\mathbf{k}): \; = \hat{a}^\dagger(\mathbf{k})\hat{a}(\mathbf{k})$). An analogous result for $y^0 > x^0$ holds, with $x \leftrightarrow y$ inside the commutator. We can use this result, and the fact that $\langle 0|:\psi(x)\psi(y): |0\rangle = 0$, since normal ordering imposes putting all annihilation operators on the right (and they therefore act on the vacuum, annihilating it), to write the Feynman propagator as

$$D_F(x-y) = \Theta(x^0 - y^0)[\psi^+(x),\,\psi^-(y)] + \Theta(y^0 - x^0)[\psi^+(y),\,\psi^-(x)]. \qquad (5.37)$$

For practical use, it is better to have the momentum space representation of the Feynman propagator, i.e., its Fourier transform. To obtain this, we substitute the expression of $\psi^\pm(x)$ with its Fourier transform and calculate the commutator. After some algebra, and having transformed the three-dimensional k-integral into a four-dimensional one using the relation

$$\Theta(x^0 - y^0)\frac{1}{2\omega(\mathbf{k})}\exp[-i\omega(\mathbf{k})(x^0 - y^0)] = \int \frac{d\Omega\,\exp[-i\Omega(x^0 - y^0)]}{\Omega^2 - \omega^2(\mathbf{k}) + i\varepsilon}, \qquad (5.38)$$

and a similar expression for the term containing $\Theta(y^0 - x^0)$, we get the following result for the Feynman propagator in momentum space

$$D_F(\mathbf{k}) = \frac{i}{|\mathbf{k}|^2 - \omega^2(k) + i\varepsilon}, \qquad (5.39)$$

and, in real space

$$D_F(x-y) = \int \frac{d^4 k}{(2\pi)^4} \frac{i}{|\mathbf{k}|^2 - \omega^2(\mathbf{k}) + i\varepsilon} \exp[-ik(x-y)]. \qquad (5.40)$$

Notice that the (Feynman) propagator is called in this way because its physical meaning on the system is to propagate the (quantum) field, from the point $x = (\mathbf{r}, t)$ to the point $y = (\mathbf{r}', t')$. For this reason, as we will see in the following subsection, the Feynman diagram associated with the propagator is simply a straight line joining the point x with the point y.

5.4.2 Wick's theorem and the Feynman rules

The knowledge of the n-point Green's function is very important to calculate scattering amplitudes in QFT. However, the presence of the time-ordering operator in its definition (see equation (5.34)) makes it difficult to explicitly evaluate all the possible combinations of fields and times that make up $G_n(x_1,\dots, x_n)$, especially when the interacting field theory is characterised by a high polynomial order, as for example a ψ^4 field theory (typical of $\chi^{(3)}$ nonlinear processes, for example [13]). To avoid this situation, we can make use of an important theorem in QFT, called Wick's theorem, which gives us a practical tool to write n-point Green's function as combinations of propagators. Practically, Wick's theorem is a generalisation of equation (5.36) to $G_n(x_1,\ldots, x_n)$, and can be cast as follows

$$G_n(x_1,\ldots, x_n) = \;:\psi(x_1)\;\cdots\;\psi(x_n):\; + \text{ all possible combinations of propagators and normal ordered products}. \tag{5.41}$$

For example, we can calculate the four-point Green's function using Wick's theorem as follows

$$
\begin{aligned}
G_4(x_1, x_2, x_3, x_4) &= \langle 0|[:\psi(x_1)\psi(x_2)\psi(x_3)\psi(x_4): \; + D_F^{12}:\; \psi(x_3)\psi(x_4): \\
&\quad + D_F^{13}:\; \psi(x_2)\psi(x_4): \; + D_F^{14}:\; \psi(x_2)\psi(x_3): \; + D_F^{23}:\; \psi(x_1)\psi(x_4): \\
&\quad + D_F^{34}:\; \psi(x_1)\psi(x_2): \; + D_F^{12}D_F^{34} + D_F^{13}D_F^{24} + D_F^{14}D_F^{23}]|0\rangle \\
&\equiv D_F^{12}D_F^{34} + D_F^{13}D_F^{24} + D_F^{14}D_F^{23}
\end{aligned}
\tag{5.42}
$$

where we have used the shorthand $D_F^{ij} = D_F(x_i - x_j)$ and to pass from the third to the fourth line we have used the fact that $\langle 0|:\psi(x)\psi(y):|0\rangle = 0$, since the normal ordered product $:\psi(x)\psi(y):$ applied to the vacuum state always annihilates it. This result has a very nice and intuitive physical interpretation. The term $D_F^{12}D_F^{34}$, for example, describes the propagation of a particle from x_1 to x_2, while another particle propagates from x_3 to x_4, without interacting with the first one. We can use this visual picture to also represent more complicated situations, for example those generated by the interaction Hamiltonian $\hat{H}_I(t)$ in equation (5.34), by introducing the following rules for constructing the so-called Feynman diagrams:

(1) A line joining the initial (x) and final (y) point of a free evolution of the field ψ represents the propagator $D_F(x - y)$;

(2) A filled dot joining n lines is called a *vertex* and it represents the interaction between fields. The number of lines that can join at a vertex is determined by the ψ^n nature of the interaction. The number of vertices in a diagram indicates the expansion order of the interaction;

(3) To each vertex, moreover, we associate the quantity $ig \int d^4x$, which contains the coupling constant g defining the interaction. In addition, we require that energy and momentum conservation holds at each vertex;

(4) Account for all the equivalent contractions with a factor $1/n!$, and also remember to take into account the numerical factors coming from the interaction term (typically also of the form $1/n!$);

(5) Momenta appearing in a loop must only appear as integration variables.

To understand how this works, let us first look at the Taylor series expansion of the evolution operator appearing in equation (5.34), with respect to the coupling constant g appearing in the interaction Hamiltonian, which reads

$$
\begin{aligned}
\exp\left[-i \int d^4x \; \hat{H}_I\right] &= 1 - \frac{ig}{n!} \int d^4x \; \psi^n(x) \\
&\quad + \left(-\frac{ig}{n!}\right)^2 \int d^4x \; d^4y \; \psi^n(x)\psi^n(y) + \mathcal{O}(g^3),
\end{aligned}
\tag{5.43}
$$

where, for convenience of notation, we have replaced the time integral of the interaction Hamiltonian $\int_{-t}^{t} d\tau \, \hat{H}_I$ with the integral over d^4x of the interaction Hamiltonian density $\hat{\mathcal{H}}_I$. Let's see how this works with the explicit example of a ψ^4-theory, i.e., we set $n = 4$.

At order zero of the expansion above, equation (5.34) gives simply the n-point Green's function for a free field, which can only connect the initial and final point of the field evolution through field propagation. At this level, then, the only quantities that can appear are propagators, and the correspondent Feynman diagrams are depicted in figure 5.1(a).

At order one of the expansion above, we will have to calculate expectation values of the form $\langle 0|\hat{T}\{\psi(x_1)\psi(x_2)\psi(x_3)\psi(x_4)\psi^4(x)\}|0\rangle$. To calculate this, we can use Wick's theorem again and realise that we get a nonzero result only when we consider propagators between one of the x_i points and the point x, i.e., only when we can construct terms of the form $D_F(x_1 - x)D_F(x_2 - x)D_F(x_3 - x)D_F(x_4 - x)$, and we can do this in $4! = 24$ different ways, by considering all possible permutations of the indices appearing in this expression. The point x represents the point in which the four fields interact, and therefore it hosts a vertex. An example of this is given in figure 5.1(b).

At order two, something new happens. The expectation value we have now to calculate is of the form $\langle 0|\hat{T}\{\psi(x_1)\psi(x_2)\psi(x_3)\psi(x_4)\psi^4(x)\psi^4(y)\}|0\rangle$. We can handle this with Wick's theorem again, and with the same line of reasoning as before we now obtain a term containing $D_F(x - y)D_F(x - y)$, which corresponds to the loop shown in figure 5.1(c).

Following these simple rules, then, we can create Feynman diagrams for the interacting field at any order of perturbation theory, and then easily reconstruct the corresponding form of the amplitude integral to be calculated. We will give an

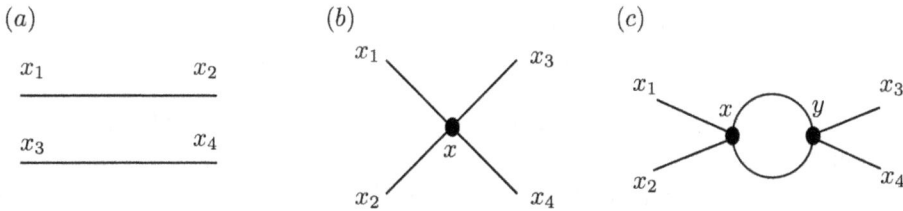

Figure 5.1. An example of Feynman diagrams for a ψ^4 theory. (a) Example of a Feynman diagram for the interaction at order $\mathcal{O}(1)$. The diagrams are depicting the propagators, connecting pairs of points without interaction (crossing of the lines). This panel, for example, shows the term $D_F(x_1 - x_2)D_F(x_3 - x_4)$ appearing in equation (5.42). The other two terms can be obtained via permutations of the indices, and give rise to similar diagrams. (b) Example of a Feynman diagram for the interaction at order $\mathcal{O}(g)$. Notice the presence of one vertex (filled dot), denoting that the order of the interaction is $\mathcal{O}(g)$. This panel shows one of the possible 4! combinations of propagators, i.e., $D_F(x_1 - x)D_F(x_2 - x)D_F(x_3 - x)D_F(x_4 - x)$. (c) Example of a Feynman diagram for the interaction at order $\mathcal{O}(g^2)$. Notice the presence of two vertices (filled dots), denoting that the order of interaction is $\mathcal{O}(g^2)$. The loop in the diagram is generated by the presence, in the expectation value, of terms proportional to $D_F(x - y)D_F(x - y)$, which correspond to the propagators from the dummy variable x to the dummy variable y.

example on how to use this formalism to calculate the probability amplitude of different nonlinear optical phenomena in chapter 6.

We conclude this section, by making the following observation: although in this section we have used the Hamiltonian formalism to introduce Wick's theorem, and derive the rules for Feynman diagrams, the same line of reasoning also applies for the case of path integrals described below. In that case, one needs to expand the interaction Lagrangian in a power series and look at the various orders of the partition function Z and its derivatives, instead of considering the n-point Green's function directly. Apart from this technical difference, the whole argument of Feynman diagrams and rules can be then transferred 1:1 to the case of path integrals as well.

5.5 Path integral quantisation

An alternative method to quantise a field, based on the Lagrangian, rather than Hamiltonian, formalism, is the path integral quantisation. In this quantisation scheme, the basic idea is to define a quantity, that plays a similar role to that of the partition function in statistical mechanics. Once the partition function has been calculated, all the relevant quantities, like transition and scattering amplitudes, can be calculated as derivatives of the partition function, or functionals derived from it [8]. Although there are many caveats when introducing path integrals, from using a proper definition of the integration measure, to the Trotter formula, in this section, to keep the description short and simple, but nevertheless insightful, we omit these mathematical details, addressing the interested reader to references [5, 8, 9, 19] for a more thorough explanation of path integral from the physical point of view, and to the book by Albeverio and co-workers [20], for a more mathematical approach to the topic.

5.5.1 Path integral for a free field

Contrary to the canonical formalism described above, which bases its premises on the concept of Hamiltonian and, therefore, total energy of the system, the basics for path integral formalism are rooted in the Lagrangian of a field, and, ultimately, its action. The basic idea behind this approach is to recuperate the concept of path in quantum mechanics, and in particular the choice of classical path as the one that minimised the action, and reinterpret it in terms of interference of all the possible paths, that a quantum system undertakes simultaneously. To do so, and following the ideas of Dirac [21] and Feynman [22, 23], one can interpret the quantity $S/\hbar$ as a measure of the (classical) path of the system, normalised by a fundamental 'wavelength' $\hbar$. Then, one should assign a plane wave of the form $\exp(iS/\hbar)$ to each of the (classical) paths the system can take. Then, the probability amplitude for a quantum system to evolve from the initial state $\psi(x)$ to the final state $\psi(y)$ will be given by the interference of all the waves propagating along all the possible paths the field can take[5] and calculate the

---

[5] Notice that here we are making a small abuse of language. We refer in fact to paths a field can take, meaning paths the quantum particle associated to the field can take in real space. To be more precise, we should refer to all the possible configurations in space one field can have, and associate with each of these configurations the plane wave weight $\exp(iS/\hbar)$, and then look at the most probable configuration of the field as emerging from the interference of all the possible configurations, each weighted by its own action-values phase factor.

(a)

Classical Mechanics

(b)

Quantum Mechanics

(c)

Quantum Field Theory

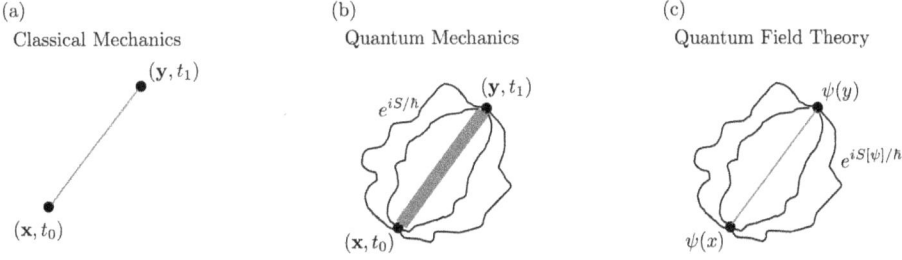

Figure 5.2. Pictorial representation of the different interpretations of path integrals (and the concept of trajectory) in classical mechanics (a), quantum mechanics (b), and quantum field theory (c). In classical mechanics (panel (a)), the least action principle $\delta S = 0$ gives as a result a single trajectory (in red, in panel (a)), joining the initial state of the system at time t_0, characterised by position $\mathbf{x}$, with the final state of the system at time t_1, characterised by position $\mathbf{y}$. All the other possible trajectories correspond to higher values of the action, and are therefore discarded and play no role in the dynamic of the system. For a (nonrelativistic) quantum mechanics system (panel (b)) the evolution from the initial state $(\mathbf{x}, t_0)$ to the final state $(\mathbf{y}, t_1)$ is probabilistic, and every possible path connecting these two points contributes it and is weighted by a factor $\exp(iS/\hbar)$, where S is the action corresponding to the quantum particle. Paths that are closer to the classical solution (in red) are contributing more than the ones further away. The blue area close to the classical solution, moreover, represents the area of uncertainty (given by Heisenberg's principle) inside which sits the classical path (this represents the quantum corrections imposed by quantum mechanics). The key concept here is that every path can be seen as a wave, and the overall transition probability from the initial to the final state is computed as the interference of all these waves. Finally, for quantum field theory (panel (c)), the situation is similar to that described in panel (b), with the difference that the paths are understood in field configuration space, and not in actual space, like in panels (a) and (b). This means, that each path in (c) corresponds to a different configuration of the field (a different value, a different shape, etc) and the 'classical path' (in red) represents the field configuration solving the equations of motion of the field (i.e., the Euler–Lagrange equations). The path integral in QFT is then built as the sum of all these different configurations, each weighted, as in panel(b), by a factor $\exp(iS[\psi]/\hbar)$, where now $S[\psi]$ is the action of the quantum field.

interference effects deriving by taking into account all the possible contributions given by all the possible paths the system can take to evolve from point x to point y. A pictorial representation of this is reported in figure 5.2.

Following these ideas, we can then define the central object of the path integral formalism, namely the *partition function* for the free field $Z_0(J)$ as follows[6]

$$Z_0(J) = \int \mathscr{D}\psi \, \exp(iS_0) = \int \mathscr{D}\psi \, \exp\left\{i \int d^4x \, [\mathscr{L}(\psi, \partial_\mu \psi) + J(x)\psi(x)]\right\}, \quad (5.44)$$

where S_0 is the action of the free field, $\mathscr{D}\psi$ is the so-called integration measure [8] and indicates that the integral is to be interpreted as running over all possible configurations of the field ψ, and $J(x)$ is the source term, which physically accounts for the eventual presence of sources for the field ψ, but here it can be thought as a convenient mathematical tool, inserted to make the calculations easier. In fact, the partition function is a function of the source term $J(x)$ solely, since the field ψ has been integrated away by means of the path integration $\int \mathscr{D}\psi$.

---

[6] Again, we set $\hbar = 1$, for simplicity.

Notice, moreover, that the above definition of partition function does not distinguish between ψ being a classical or a quantum field. For a classical field, the partition function $Z_0(J)$, in analogy with statistical mechanics, represents the generating function for all relevant quantities of the field, such as energy, momentum, and propagator. For a quantum field, on the other hand, after one has equipped the field ψ with a set of suitable commutation relations, $Z_0(J)$ represents the vacuum-to-vacuum correlation function, from which one can derive all the relevant transition probabilities and correlation functions of the field operator with respect to the ground state (i.e., the vacuum state), which is also in this case assumed to be a stable state. This means, that $Z_0(J)$ can only describe QFTs at equilibrium. To deal with out-of-equilibrium quantum systems, the path integral in equation (5.44) needs to be refined by introducing in a suitable way the information about the temperature of the system[7].

For a free, scalar field, the Lagrangian is quadratic in the field ψ (see equation (3.1) with $U(\psi) = m^2\psi$), and the path integral in equation (5.44) becomes Gaussian with respect to the field ψ and can then be computed analytically. This is better done in Fourier space, so we first introduce the Fourier transform of the field ψ as

$$\psi(x) = \int \frac{d^4k}{(2\pi)^4}\Psi(k)\exp(ikx), \tag{5.45}$$

and substitute it in the action, so that the exponential of equation (5.44) becomes

$$S_0 = \int \frac{d^4k}{(2\pi)^4}\left[-\frac{1}{2}\Psi(k)(|\mathbf{k}|^2 - \omega^2 + m^2)\Psi(-k) + \tilde{J}(k)\Psi(-k) + \tilde{J}(-k)\Psi(k)\right], \tag{5.46}$$

where $\tilde{J}(k)$ is the Fourier transform of the source term. If we then perform the change of variable

$$\xi(k) = \Psi(k) - \frac{\tilde{J}(k)}{|\mathbf{k}|^2 - \omega^2 + m^2}, \tag{5.47}$$

such that $\mathscr{D}\xi = \mathscr{D}\psi$, we can reduce the action above to a pure Gaussian integral, i.e.,

$$S_0 = \frac{1}{2}\int \frac{d^4k}{(2\pi)^4}[-\xi(k)(|\mathbf{k}|^2 - \omega^2 + m^2)\xi(-k)] + \frac{1}{2}\int \frac{d^4k}{(2\pi)^4}\frac{\tilde{J}(k)\tilde{J}(-k)}{|\mathbf{k}|^2 - \omega^2 + m^2}. \tag{5.48}$$

Substituting the expression above into equation (5.44), implementing the change of variables, performing the Gaussian integration with respect to the field ξ (see appendix B for details), and transforming back to real space, gives the following, final result

$$Z_0(J) = \exp(iW) = \exp\left[\frac{i}{2}\int d^4x \, d^4y \, J(x)D_F(x - y)J(y)\right], \tag{5.49}$$

---

[7] The interested reader can check out the book by Calzetta and Hu [24].

where $D_F(x - y)$ is the Feynman propagator, whose Fourier representation is given by equation (5.40).

The partition function for a free field contains therefore information about the propagator of the field, which is the only meaningful information one could extract from a free field, i.e., how it can propagate from a point x to a point y. However, the propagator appears at the exponent of the partition function, and it is not directly accessible by just reading the value of Z_0. However, if we take the derivative of the partition function with respect to the source terms $J(x)$ and $J(y)$, we get

$$\frac{\partial^2 Z_0(J)}{\partial J(x_1)\partial J(x_2)}\bigg|_{J=0} = \left[i\frac{\partial^2 W}{\partial J(x_1)\partial J(x_2)}Z_0 - \frac{\partial W}{\partial J(x_1)}\frac{\partial W}{\partial J(x_2)}Z_0\right]\bigg|_{J=0}$$

$$= [i\ D_F(x_1 - x_2)Z_0 - \text{terms containing}\ J(x_{1,2})Z_0]|_{J=0}$$

$$= D_F(x_1 - x_2), \tag{5.50}$$

which is the propagator of the free field. In deriving the result above we have made use of the following property for the functional derivative

$$\frac{\partial f(x)}{\partial f(y)} = \delta(x - y), \tag{5.51}$$

and we have used the fact that

$$\frac{\partial W}{\partial J(x_1)} = \frac{1}{2}\int d^4x\ d^4y\left[\frac{\partial J(x)}{\partial J(x_1)}D_F(x - y)J(y) + J(x)D_F(x - y)\frac{\partial J(y)}{\partial J(x_1)}\right]$$

$$= \frac{1}{2}\left[\int d^4y\ D_F(x_1 - y)J(y) + \int d^4x\ J(x)D_F(x - x_1)\right]. \tag{5.52}$$

As can be seen, $\partial W/\partial J$ is proportional to J, and therefore $(\partial W/\partial J(x_1))(\partial W/\partial J(x_2))$ will also contain only terms proportional to J, which go to zero once $J = 0$ is set at the end of the calculation. On the other hand,

$$\frac{\partial^2 W}{\partial J(x_1)\partial J(x_2)} = \frac{1}{2}\left[\int d^4y\ D_F(x_1 - y)\frac{\partial J(y)}{\partial J(x_2)}\right.$$

$$\left. + \int d^4x\ \frac{\partial J(x)}{\partial J(x_2)}D_F(x - x_1)\right] \tag{5.53}$$

$$= D_F(x_1 - x_2),$$

gives the free propagator $D_F(x_1 - x_2)$ as a result.

This calculation also reveals another important property of the partition function, i.e., its ability to generate the n-point Green's function by simply taking derivatives of the partition function Z_0 with respect to the source term $J(x)$. There is, in fact, a useful correspondence between the field ψ and the functional derivative $\partial/\partial J$, which allows for a simple expression for the n-point Green's function. To understand this, let us observe that the expectation value of the field ψ, which in path integral formalism is written as

$$\langle \psi \rangle = \langle 0 | \psi(x) | 0 \rangle = \int \mathcal{D}\psi \ \psi(x) \exp(i \ S_0), \qquad (5.54)$$

can be rewritten, using equation (5.44), as

$$
\begin{aligned}
\langle \psi \rangle &= \int \mathcal{D}\psi \ \frac{1}{i} \frac{\partial}{\partial J(x)} [\exp(i \ S_0)] \\
&= \frac{1}{i} \frac{\partial}{\partial J(x)} \left[\int \mathcal{D}\psi \ \exp(i \ S_0) \right] \qquad (5.55) \\
&= \frac{1}{i} \frac{\partial Z_0}{\partial J(x)},
\end{aligned}
$$

where to pass from the second to the third line we have used the fact that $\frac{1}{i} \frac{\partial}{\partial J(x)}$ does not depend on the field $\psi(x)$ and therefore it can be brought outside the path integral. In practice, what we obtain with this result is the possibility to enforce the substitution $\psi \to (1/i)\partial/\partial J$ whenever we need to calculate expectation values of field operators using the partition function. In particular, this result can be then immediately generalised to the case of n field, i.e., to the n-point Green's function, as

$$G_n(x_1,\ldots, x_n) = \frac{\partial^n Z_0}{\partial J(x_1) \cdots \partial J(x_n)} \Big|_{J=0}, \qquad (5.56)$$

where the constraint $J = 0$ at the end of the calculation is basically an implementation of Wick's theorem [2]. For $n = 2$, we obtain the result in equation (5.50), i.e., the field propagator.

5.6 Interacting fields—path integral formalism

To introduce interaction in the path integral picture, we need to introduce an interaction Lagrangian term $\mathcal{L}_{\text{int}}$ in the action, so that the partition function in equation (5.44) now becomes

$$Z = \int \mathcal{D}\psi \ \exp \left\{ i \int d^4x \ [\mathcal{L}(\psi, \partial_\mu \psi) + \mathcal{L}_{\text{int}}(\psi) + J(x)\psi(x)] \right\}. \qquad (5.57)$$

Contrary to the Hamiltonian case, where to properly account of the interaction term we had to introduce the interaction picture to circumvent the problem of not being able to represent the fields in terms of plane waves (and, therefore, creation and annihilation operators), in this case handling the interaction term is much simpler, since we can first factor the exponential into a free part, containing $\mathcal{L} + J(x)\psi(x)$, and an interacting part, containing $\mathcal{L}_{\text{int}}$. Then, we can use the correspondence $\psi \to (1/i)\partial/\partial J$ to rewrite the interacting Lagrangian as

$$\mathcal{L}_{\text{int}}(\psi) \ \to \mathcal{L}_{\text{int}}\left(\frac{1}{i} \frac{\partial}{\partial J(x)} \right), \qquad (5.58)$$

which we can use to rewrite the partition function for the interacting field appearing in equation (5.57) in the following form[8]

$$Z = \exp\left[i \int d^4x \; \mathscr{L}_{\text{int}}\left(\frac{1}{i}\frac{\partial}{\partial J(x)}\right)\right] Z_0. \tag{5.59}$$

Let us now assume that the interaction Lagrangian is of the polynomial type, i.e., $\mathscr{L}_{\text{int}} = (g/n!)\psi^n(x)$, where g is the coupling constant. If $g \ll 1$ we can expand the exponential appearing above in power series with respect to g (truncated at the desired order of accuracy), then expand $Z_0[J]$ as well in a power series to rewrite the partition function for the interacting theory in the following way

$$Z = \sum_{V=0}^{\infty} \frac{1}{V!}\left[\frac{ig}{n!} \int d^4x \left(\frac{1}{i}\frac{\partial}{\partial J(x)}\right)^n\right]^V$$
$$\times \sum_{P=0}^{\infty}\left[\frac{i}{2} \int d^4y \; d^4z \; J(y)D_F(y - z)J(z)\right]^P, \tag{5.60}$$

This expression allows us to treat path integrals for interacting fields in a perturbative manner to an arbitrary order V, which is taken as the perturbation order of the path integral. In addition to that, the equation above is also written in a very convenient form to be represented in terms of Feynman diagrams. We can do that by associating a vertex with each perturbation order, so that the index V does represent the number of vertices in the diagram (i.e., to which order the perturbative expansion of the interaction has been truncated). The index P, on the other hand, regulates the number of propagators appearing in the diagram. The rules for creating Feynman diagrams out of equation (5.60) are the same as those described in section 5.4.2.

One could then check, for example, that for a ψ^4 theory with zero vertices ($V = 0$, i.e., no interaction) and two propagators ($P = 2$), equation (5.60) reduces to equation (5.42), i.e., to the Feynman diagram for the free propagators depicted in figure 5.1(a), or that a ψ^4 theory with one vertex ($V = 1$) and four propagators ($P = 4$), one gets the diagram in figure 5.1(b).

Appendix A: The quantum harmonic oscillator

The canonical quantisation of fields has much in common with that of a quantum harmonic oscillator, in the sense that any free field can be written as a sum of uncoupled harmonic oscillators, i.e., one oscillator per point in space (or per value of $\mathbf{k}$ in k-space). It is therefore beneficial to review the canonical quantisation of a single harmonic oscillator, since it will give us insight on the quantum properties and dynamics of the field. Let us then start with a classical, one-dimensional harmonic oscillator, whose Hamiltonian is given by

---

[8] Details on the derivation of this formula, for the case of the electromagnetic field, are given in appendix 7.2.10. The explicit calculation for an interacting, scalar field can instead by found, for example, in references [2, 25].

$$H = \frac{p^2}{2m} + \frac{m\omega^2}{2}x^2 = \omega\left(\frac{p^2}{2m\omega} + \frac{m\omega}{2}x^2\right), \tag{5.61}$$

where $\{x, p\}$ are the canonically conjugated position and momentum of the oscillator. The volume (or area, in the case of a one-dimensional oscillator) occupied by the harmonic oscillator in phase space is given as $\Omega = \int dx\, dp\, H$. Since the harmonic oscillator is a closed system, energy is conserved, and by virtue of Liouville's theorem [10], the volume in phase space of the oscillator must remain constant. This means that

$$\Omega = \int dp\, dx\, H = \omega \int dp\, dx \left(\frac{p^2}{2m\omega} + \frac{m\omega}{2}x^2\right) \equiv \omega\, J, \tag{5.62}$$

where

$$J = \int dp\, dx \left(\frac{p^2}{2m\omega} + \frac{m\omega}{2}x^2\right) = \text{const}, \tag{5.63}$$

must have the dimensions of an energy per second, i.e., an action ($[J] = Js$), in order for E to have the dimensions of an energy. The relation above is a way to formulate Liouville's theorem and states that the area of the ellipse representing the harmonic oscillator (see figure 5.3(a)) is constant at all times, i.e., the ellipse can be deformed

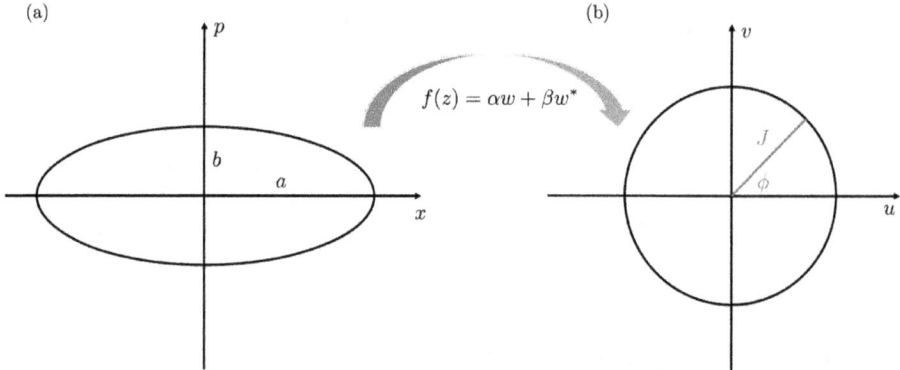

Figure 5.3. (a) Phase space representation of a one-dimensional harmonic oscillator defined by the Hamiltonian in equation (5.61), with respect to the canonically conjugated pair $\{x, p\}$. For a given set of values of m and ω, the trajectory in phase space traced by the oscillator is an ellipse with the major semiaxis a aligned along the x-direction, and the minor semiaxis b along the p-direction. (b) Phase space representation of the same harmonic oscillator described by equation (5.61) (and same values of the parameters m and ω), but with respect to the pair of canonically conjugated variables $\{u, v\}$. The two pairs of conjugated variables (or equialently, the two reference frames describing the oscillator) are related through a conformal mapping defined by the function $f(z) = \alpha w + \beta w^*$, which is pictorially represented here by a blue arrow linking the two representations. By virtue of the conformal mapping, the areas of the two curves are preserved during the transformation. Panel (b) also shows (in red) the action ($J = |w|$) and angle ($\phi = \text{arg}(w)$) variables, corresponding to polar coordinates in the phase space spanned by $\{u, v\}$. Because $|w|^2 = 1$ holds, $J = 1$ is enforced as an extra constraint, which renders the Hamiltonian independent of u and v.

during the evolution of the system, but its area remains constant. We can then calculate J by simply taking the area of the ellipse

$$\frac{p^2}{2m\omega} + \frac{m\omega}{2}x^2 = \kappa \qquad (5.64)$$

where κ is a constant with the dimension of an action, that will be determined later. The area of the ellipse is then given by $J = \pi ab$, where a and b are the major and minor axis of the ellipse in its canonical form, i.e., the coefficients of x^2 and p^2, which are readily calculated from equation (5.64) to be $a = \sqrt{2\kappa/m\omega}$ and $b = \sqrt{2m\omega\kappa}$. Substituting gives $\Omega = \omega J = 2\pi\omega\kappa$, for the volume, and, most importantly, $H = \kappa\omega$ for the Hamiltonian (i.e., the total energy of the oscillator).

This relation is already pretty telling, since it essentially says that the energy of the oscillator is proportional to its frequency. To have an even better feeling about this result, the reader should compare it with the quantum mechanical relation between energy and frequency, i.e., $E = \hbar\omega$, which would lead to the identification $\kappa = \hbar$, i.e., the reduced Planck's constant [26][9].

Introducing the action integral is quite useful to quantise the harmonic oscillator in terms of the creation and annihilation operators, and to show how they emerge from classical Hamiltonian mechanics. To do so, we basically seek to simplify the problem of calculating the integral in equation (5.63) by introducing a coordinate transformation that maps the ellipse into the unit circle. This can be done via conformal mapping, by introducing two sets of complex variables $z = x + ip$ and $w = u + iv$ (with the constraint $|w|^2 = u^2 + v^2 = 1$, so that the new variables will be constrained on a circle) and a holomorphic function $f: \mathbb{C} \to \mathbb{C}$ such that $f(z) = \alpha w + \beta/w = \alpha w + \beta w^*$ [28], which corresponds to the coordinate transformation

$$x + ip = \alpha(u + iv) + \beta(u - iv) = (\alpha + \beta)u + i(\alpha - \beta)v. \qquad (5.65)$$

From this relation we immediately get, by comparing real and imaginary parts on both sides of the equality, that $u = x/(\alpha + \beta)$ and $v = p/(\alpha - \beta)$. To find the connection between (α, β) and (a, b) we use $x^2/a^2 + p^2/b^2 = 1$ and $|w|^2 = u^2 + v^2 = 1$ (the former is the ellipse in phase space described by the harmonic oscillator, while the latter is the circle representing the conformally transformed ellipse in the new coordinates) to get the relations $u = x/(\alpha + \beta) = x/a$ and $v = p/(\alpha - \beta) = p/b$, so that the new coordinates on the circle are given by

$$w = u + iv = \frac{x}{a} + i\frac{p}{b} = \sqrt{\frac{m\omega}{2\kappa}}\left(x + \frac{i}{m\omega}p\right), \qquad (5.66a)$$

---

[9] This is essentially the way Born and Sommerfeld took to propose their quantisation rule at the dawn of quantum theory [27]. Their argument was based on the fact that orbits of quantum systems in phase space must be discrete, instead of continuous, which ultimately leads to the action J of the system being quantised in units of the fundamental area of phase space, i.e., $\hbar$.

$$\frac{1}{w} = w^* = u - iv = \frac{x}{a} - i\frac{p}{b} = \sqrt{\frac{m\omega}{2\kappa}}\left(x - \frac{i}{m\omega}p\right). \tag{5.66b}$$

Notice, that while x and p are real variables, w and w^* are instead complex variables[10]. We can then rewrite the Hamiltonian (5.61) in terms of the new variables w and w^* to obtain

$$H = \frac{1}{2}\kappa\omega(ww^*+w^*w) = \kappa\omega\,|w|^2 = \kappa\omega, \tag{5.67}$$

where in the first equality we have emphasised the way the two coordinates w and w^* multiply together (to already foreshadow quantisation), and to get the last equality we have used the fact that $|w|^2 = 1$. The phase space volume of the oscillator in the new coordinates can be calculated from the action integral (5.63) as

$$J = \int dp\,dx\left(\frac{p^2}{2m\omega} + \frac{m\omega}{2}x^2\right) = \int du\,dv\,\kappa(u^2 + v^2) = 2\pi\kappa, \tag{5.68}$$

which is the same result as obtained in the $\{x, p\}$ frame, as expected[11].

Notice, that in the new coordinate system $\{w, w^*\}$, the Hamiltonian is represented in phase space by a (unit) circle, instead of by an ellipse (see figure 5.3(b)), and that it only depends on $|w|$ and not on $\mathbf{arg}(w)$.

We can, at this level, quantise the Hamiltonian of the harmonic oscillator by replacing the position and momentum variables with their respective operators in equations (5.66), and introduce the commutation relations

$$[\hat{x}, \hat{x}'] = 0 = [\hat{p}, \hat{p}'], \tag{5.69a}$$

$$[\hat{x}, \hat{p}] = i\hbar. \tag{5.69b}$$

Contextually, when substituting $x \to \hat{x}$, $p \to \hat{p}$ in equations (5.66), we also need to promote the variables w and w^* to operators, and we do so as $w \to \hat{a}$ and $w^* \to \hat{a}^\dagger$, i.e., the coordinates that map the harmonic oscillator on the circle become the creation and annihilation operators. The commutation relations for these operators can be derived from those for the position and momentum operators using equations (5.66), to obtain

$$[\hat{a}, \hat{a}] = 0 = [\hat{a}^\dagger, \hat{a}^\dagger], \tag{5.70a}$$

$$[\hat{a}, \hat{a}^\dagger] = 1. \tag{5.70b}$$

---

[10] These coordinates are essentially the action-angle coordinates. To see that, introduce polar coordinates on the complex plane $\{I, \phi\}$, so that $I = |w|$ and $\phi = \arg(w)$, where I is the action of the oscillator, and ϕ its canonically conjugated variable, i.e., the angle [10].

[11] To calculate the integral in u and v, first, use the fact that $u^2 + v^2 = 1$ from the definition of the w-variable, then switch to polar coordinates, remembering that the radius is clamped at $r = 1$ from the constraint $|w|^2 = 1$. This only leaves the angular integral, which simply gives 2π.

Because now we are dealing with non-commuting operators, we need to stop at the first equality of equation (5.70a) when calculating the Hamiltonian, and we need to use the commutation relation for the creation and annihilation operators, if we want to move to the second equality in equation (5.70a). This, however, gives a different result than the classical Hamiltonian, i.e.,

$$\hat{H} = \frac{1}{2}\hbar\omega(\hat{a}\hat{a}^\dagger + \hat{a}^\dagger\hat{a}) = \hbar\omega\left(\hat{a}^\dagger\hat{a} + \frac{1}{2}\right), \tag{5.71}$$

where the zero point energy of the oscillator $\hbar\omega/2$, embodying the Heisenberg uncertainty principle, makes its appearance because of the commutation relation between $\hat{a}$ and $\hat{a}^\dagger$.

Taking this route for the quantisation of the harmonic oscillator allows one to gain more insight on the physical origin of the creation and annihilation operators. Equations (5.66) establish a coordinate transformation (or, more rigorously, a contact, or canonical transformation [10]) between $\{x, p\}$, where the harmonic oscillator traces an ellipse in phase space, and $\{w, w^*\}$, where the harmonic oscillator is simply represented by a (unit) circle of constant radius. Upon quantisation, then, the creation and annihilation operators are the operator counterparts of the canonically conjugated (complex) variables w and w^*. However, since quantum mechanics only allows observables to be connected to real quantities [26], while $\{x, p\} \in \mathbb{R}$ generate operators associated with observable quantities (the position and momentum of the system), the operators generated by $\{w, w^*\} \in \mathbb{C}$ do not correspond to any observable of the system, since they are associated with complex, rather than real, variables.

Appendix B: Gaussian integrals of fields

To understand how to perform the Gaussian integral appearing in the expression of the partition function $Z_0(J)$ because of the quadratic form in the field ξ, i.e., the first term of (5.48), it is instructive to first have a look at how the standard Gaussian integral

$$I_1 = \int dx \exp(-ax^2 + bx) = \sqrt{\frac{\pi}{a}} \exp\left(\frac{b^4}{4a}\right), \tag{5.72}$$

generalises in the case of n dimensions. Consider then vectors $\mathbf{x}, \mathbf{b} \in \mathbb{R}^n$, and a complex, non-singular, symmetric (i.e., $A_{ij} = A_{ji}$) $n \times n$ matrix $A \in \mathbb{C}$. We can build a quadratic form using $\mathbf{x}$ and A by using the scalar product defined on $\mathbb{R}^n$, i.e.,

$$(\mathbf{x}, A\mathbf{x}) = \sum_{i,j} x_i^T A_{ij} x_j. \tag{5.73}$$

Notice, that this quadratic form contains only terms that are either proportional to $A_{ii} x_i^2$ (when $i = j$) or to $A_{ij} x_i x_j$ (when $i \neq j$). When generalising I_1 to any number of dimensions, then the term ax can be replaced with $(\mathbf{x}, A\mathbf{x})$, while the term bx can be replaced by $(\mathbf{b}, \mathbf{x}$, where $\mathbf{b}$ is just a constant vector. The n-dimensional version of I_1 then becomes

$$I_n = \int d^n x \, \exp\left[-\frac{1}{2}(\mathbf{x}, A\mathbf{x}) + (\mathbf{b}, \mathbf{x})\right], \tag{5.74}$$

where $d^n x = dx_1 dx_2 \cdots dx_n$. We can solve this integral by expanding the exponent and solving each one-dimensional integral separately. By doing so, we note that thanks to the form of $(\mathbf{x}, A\mathbf{x})$, the single integrals are concatenated with each other, and one should start calculating then from x_n all the way back to x_1. To understand how this works, and get an idea on how to generalise this to arbitrary n, let us make explicitly the calculation for $n = 2$, which results in the following expression:

$$
\begin{aligned}
I_2 &= \int dx_1 dx_2 \, \exp\left[-\left(\frac{A_{11}}{2}x_1^2 + A_{12}x_1 x_2 + \frac{A_{22}}{2}x_2^2\right) + b_1 x_1 + b_2 x_2\right] \\
&= \int dx_1 \, \exp\left[-\frac{A_{11}}{2}x_1^2 + b_1 x_1\right] \int dx_2 \, \exp\left[-\frac{A_{22}}{2}x_2^2 + (b_2 - A_{12}x_1)x_2\right] \\
&= \int dx_1 \, \exp\left[-\left(\frac{A_{11}}{2}x_1^2 + b_1 x_1\right)\right]\sqrt{\frac{2\pi}{A_{22}}}\exp\left[\frac{(b_2 - A_{12}x_1)^2}{2A_{22}}\right] \\
&= \sqrt{\frac{2\pi}{A_{22}}}\int dx_1 \, \exp\left[-\frac{\det A}{2A_{22}}x_1^2 - \left(b_1 - \frac{A_{12}}{A_{22}b_2}\right)x_1 + \frac{b_2^2}{2A_{22}}\right] \\
&= \sqrt{\frac{(2\pi)^2}{\det A}}\exp\left[-\frac{1}{2\det A}\left(b_1^2 A_{22} - 2A_{12}b_1 b_2 + A_{11}b_2^2\right)\right],
\end{aligned}
\tag{5.75}
$$

where we have made use several times of the symmetry property of the matrix A, i.e., $A_{12} = A_{21}$ to make the determinant of A appear in the calculations. This result can be then readily generalised to any dimension as

$$I_n = \int d^n x \, \exp\left[-\frac{1}{2}(\mathbf{x}, A\mathbf{x}) - (\mathbf{b}, \mathbf{x})\right] = \sqrt{\frac{(2\pi)^n}{\det A}}\exp\left[\frac{1}{2}(\mathbf{b}, A^{-1}\mathbf{b})\right]. \tag{5.76}$$

We can use this result to calculate the Gaussian integral with respect to the field ξ, that defines equation (5.49), i.e.,

$$\mathcal{I}_0 = \int \mathscr{D}\xi \, \exp\left\{-\frac{1}{2}\int \frac{d^4 k}{(2\pi)^4}[\xi(k)(|\mathbf{k}|^2 - \omega^2 + m^2)\xi(-k)]\right\}. \tag{5.77}$$

Notice, however, that while I_n is an integral of a function of n variables $f(x_1, \ldots, x_n) \in \mathbb{R}^n$, the integral above runs over all possible configurations of the field $\xi(k) \in \mathscr{H}$, where $\mathscr{H}$ is some properly defined infinite-dimensional Hilbert space (for example, the space of square integrable functions). The point is, that $\mathcal{I}_0$ is an integral over an infinite-dimensional (rather than finite-dimensional as for I_n) space. If we then want to apply the formula valid for I_n, we need to first reduce this infinitely dimensional space down to a finite-dimensional one. We can do that as follows: if we discretise k so that we only have a finite amount of k-vectors available, then $\xi(k)$ will become a field defined over a discrete set of k-values, and the number of configurations that it can acquire will also become finite. We can indicate the discretised version of $\xi(k)$ by $\xi(k_n) \equiv \xi_n$, where $n = 1, ..., N$ and N is finite. We can then restore the infinite dimensionality of the space where ξ_n lives by taking the limit of $N \to \infty$. By doing so we can reduce $\mathscr{I}_0$ to the following expression

$$\mathscr{I}_0 = \lim_{n \to \infty} \int d\xi_1 \cdots d\xi_n \exp\left[-\frac{1}{2}(\xi_n, \hat{A}\xi_n)\right], \tag{5.78}$$

where $\hat{A}$ is the (finite-dimensional) matrix representation of the Klein–Gordon operator $\nabla^2 - \partial_t^2 + m^2$, and $(\xi, \hat{A}\xi)$ is the scalar product defined in the Hilbert space where the field $\xi(k)$ is also defined, i.e.,

$$(f(k), g(k)) = \int \frac{d^4k}{(2\pi)^4} f^*(k) g(k). \tag{5.79}$$

We can now use the result in equation (5.76) (with $\mathbf{b} = 0$) to calculate the integrals with respect to the set of (finitely many) discretised fields ξ_n, and then take the limit to obtain the following result

$$\mathscr{I}_0 = \sqrt{\frac{2\pi}{\det \hat{A}(k)}}. \tag{5.80}$$

This term, however, is independent on the field ξ and can be understood as a normalisation constant, and as such, it is convenient to absorb it into the definition of the integration measure $\mathscr{D}\xi$, to avoid calculating the determinant of an operator, which could be quite a tedious, and complex, calculation. More details on how to calculate the determinant of an operator, and on the explicit expression of this normalisation constant are given in reference [3–5, 9, 19].

In general, however, the Gaussian path integral for the field ξ might also include some sources. In this case, the integral above gets modified, for example, as follows

$$\mathscr{I} = \int \mathscr{D}\xi \exp\left\{-\frac{1}{2}\int \frac{d^4k}{(2\pi)^4}[\xi(k)(|\mathbf{k}|^2 - \omega^2 + m^2)\xi(-k) + J(k)\xi(-k)]\right\}, \tag{5.81}$$

and the general solution given by equation (5.76) applies here, with $J(k)$ playing the role of the vector $\mathbf{b}$, to obtain, apart from

$$\mathscr{I} = \mathscr{I}_0 \exp\left[\frac{1}{2}(J(k), \hat{A}^{-1}J(k))\right], \tag{5.82}$$

where the scalar product is not to be understood as the scalar product in the Hilbert space where the field $J(k)$ lives, i.e., equation (5.79) and the inverse operator $\hat{A}^{-1}$ is the Green's function associated with the differential operator $\hat{A}(k)$. For the case of a scalar field, $\hat{A}(k) = |\mathbf{k}|^2 - \omega^2 + m^2$, and the corresponding Green's function is the Feynman propagator defined in equation (5.40).

References

[1] Maggiore M 2005 *A Modern Introduction to Quantum Field Theory* (Oxford: Oxford University Press)
[2] Srednicki M 2007 *Quantum Field Theory* (Cambridge: Cambridge University Press)

[3] Brown L S 1992 *Quantum Field Theory* (Cambridge: Cambridge University Press)

[4] Das A 2006 *Field Theory: A Path Integral Approach* 2nd edn (Singapore: World Scientific)

[5] Feynman R P and Hibbs A R 2010 *Quantum Mechanics and Path Integrals* amended edn (New York: Dover)

[6] Peskin M E and Schroeder D V 2019 *An Introduction to Quantum Field Theory* (Boca Raton, FL: CRC Press)

[7] Weinberg S 2005 *The Quantum Theory of Fields* (Cambridge: Cambridge University Press)

[8] Kleinert H 2009 *Path Integrals in Quantum Mechanics, Statistics, Polymer Physics, and Financial Markets* (Singapore: World Scientific)

[9] Rivers R J 1998 *Path Integral Methods in Quantum Field Theory* (Cambridge: Cambridge University Press)

[10] Arnold V I 1989 *Mathematical Methods of Classical Mechanics* (Berlin: Springer)

[11] Groenewold H J 1946 On the principles of elementary quantum mechanics *Physica* **12** 405

[12] Hall B C 2013 *Quantum Theory for Mathematicians* (Berlin: Springer)

[13] Hillery M and Drummond P D 2014 *The Quantum Theory of Nonlinear Optics* (Cambridge: Cambridge University Press)

[14] Boyd R W 2008 *Nonlinear Optics* 3rd edn (Amsterdam: Elsevier)

[15] Messiah A 2014 *Quantum Mechanics* (New York: Dover)

[16] Mukhanov V and Winitzki S 2007 *Introduction to Quantum Effects in Gravity* (Cambridge: Cambridge University Press)

[17] Itzykson C and Zuber J B 1980 *Quantum Field Theory* (New York: Dover)

[18] Landau L D and Lifshitz E M 1980 *Quantum Electrodynamics* 4th edn (Oxford: Butterworth-Heinemann)

[19] Shulman L S 2005 *Techniques and Applications of Path Integration* (Mineola, NY: Dover)

[20] Albeverio S, Høegh-Krohn R and Mazzucchi S 2008 *Mathematical Theory of Feynman Path Integrals: An Introduction* (Berlin: Springer)

[21] Dirac P A M 1933 The Lagrangian quantum mechanics *Phys. Z. Sowjetunion* **3** 64

[22] Brown L M (ed) 2005 *Feynmanas Thesis: A New Approach to Quantum Theory* (Singapore: World Scientific)

[23] Feynman R P 1948 Space-time approach to non-relativistic quantum mechanics *Rev. Mod. Phys.* **20** 367

[24] Calzetta E A and Hu B-L B 2023 *Nonequilibrium Quantum Field Theory* (Cambridge: Cambridge University Press)

[25] Ryder L H 1996 *Quantum Field Theory* (Cambridge: Cambridge University Press)

[26] Hatfield B 1992 *Quantum Field Theory of Point Particles and Strings* (Boston, MA: Addison-Wesley)

[27] Bacciagaluppi G 2013 *Quantum Theory at the Crossroads Reconsidering The 1927 Solvay Conference* (Cambridge: Cambridge University Press)

[28] Shinzinger R and Laura P A A 2003 *Conformal Mapping: Methods and Applications* (New York: Dover)

IOP Publishing

A Field Theory Approach to Photonics

Marco Ornigotti

Chapter 6

Quantum theory of the electromagnetic field

This chapter applies the Hamiltonian framework developed in the previous chapter to the electromagnetic field, providing quantisation both in free space and in dispersive media, introducing quantum nonlinear optics, and it is divided into three parts.

Part I comprises the linear and nonlinear quantum theory of the electromagnetic field within the canonical (i.e., Hamiltonian) formalism. After presenting the canonical quantisation of the free electromagnetic field by introducing a quantisation box and representing the field Hamiltonian as a collection of harmonic oscillators linked to the cavity modes, section 6.1 delves into describing the properties of Fock (section 6.1.1) and coherent (section 6.1.2) states of the electromagnetic field. Section 6.1.3 concludes part I and presents a quantum theory of optical beams, which represents the quantum counterpart of the Gaussian beams introduced in chapter 1.

Part II (section 6.2) deals with the nonlinear interactions of the quantum field, namely $\chi^{(2)}$ and $\chi^{(3)}$ processes, giving an explicit example of how the Hamiltonian formalism developed in chapter 4 can be used to describe such nonlinear interactions. Explicit calculations for both the transition probability (section 6.2.2) and the quantum state of the field after the interaction (section 6.2.3) are given for second-order processes, and their third-order process counterparts are also briefly discussed in section 6.2.4. Throughout part II, we assume we are considering the full quantum problem, i.e., we do not apply the conventional *undepleted* pump approximation typical of experimental nonlinear optics, but we discuss the case of a full quantum pump (i.e., a single-photon state). In chapter 7 we will present, using path integrals, a comparison between the case of a full quantum pump and an undepleted pump.

Part III starts with section 6.3 and introduces the reader to the quantisation of the electromagnetic field in terms of path integrals. This section starts by presenting the problems arising when quantising in path integral formalism due to gauge freedom,

doi:10.1088/978-0-7503-5789-0ch6

and presents the Faddeev Popov quantisation method as a solution to this issue. This method is presented both in a traditional, i.e., QFT, manner, and in a more intuitive manner utilising only elementary concepts like Legendre transformations.

Finally, section 6.4 concludes part III and presents a calculation of the propagator for the electromagnetic field in both the Hamiltonian and Lagrangian frameworks, which complements the calculations presented in chapter 2 and allows the reader to compare the two formalisms.

The reader familiar with quantum optics will find many familiar concepts explained in this chapter, although some of them are recast from a slightly different perspective, with the aim to provide new insight on nonlinear processes in the quantum domain, and to give an explicit example of how techniques and methods of field theory can be applied to quantum optics.

The discussion in section 6.1 is pretty standard and follows essentially the traditional way the topic is introduced in quantum optics books (see, e.g., references [1–3]). The path integral quantisation presented in section 6.3 is taken from standard quantum field theory and adapted to the scope of this book. The details of it can be found in any quantum field theory (QFT) book treating path integrals, amongst such are the book by Srednicki [4], Maggiore [5], or Ryder [6]. For the nonlinear optics part, the interested reader can check the book of Boyd [7] for the classical part of nonlinear optics, and the book of Drummond [8] for a field theoretical approach to nonlinear optics.

The counterpart of part II of this chapter, i.e., quantum nonlinear optics in path integral formalism, will be discussed in chapter 7, after the path integral for the dressed electromagnetic field will be introduced as a fundamental building block to describe quantum fields inside materials.

6.1 Part I: canonical quantisation of the electromagnetic field

As discussed in section 5.2, the starting point for canonical quantisation is the Hamiltonian of the field. For a free electromagnetic field, this is given by equation (4.20), i.e.,

$$H = \int d^3x \, \mathscr{H} = \frac{1}{2} \int d^3x \, [|\mathbf{E}|^2 + |\mathbf{B}|^2]. \tag{6.1}$$

The canonically conjugated momentum to the vector potential $\mathbf{A}$, taken as the main field variable for the electromagnetic field (see the discussion in section 4.1), can be calculated from the electromagnetic Lagrangian (4.5) to be (see the second of equations (4.16))

$$\pi^i = \frac{\partial \mathscr{L}}{\partial(\partial_0 A_i)} = E^i. \tag{6.2}$$

Then, the set of canonically conjugated variables we can use for the quantisation of the electromagnetic field in the Hamiltonian formalism (see chapter 5) are $\{\mathbf{A}, \mathbf{E}\}$, with the vector potential playing the role of the 'position' (or primary field) of the field, and the electric field playing the role of the 'momentum' of the field. Using equations (5.9) we can then write

$$\mathbf{A} = \int d^3\tilde{k}[\mathscr{A}(\mathbf{k})\hat{\mathbf{e}}(\mathbf{k})\exp(ikr)+\text{c.c.}], \tag{6.3a}$$

$$\mathbf{E} = \int d^3\tilde{k}[-i\omega(\mathbf{k})][\mathscr{A}(\mathbf{k})\hat{\mathbf{e}}(\mathbf{k})\exp(ikr)+\text{c.c.}], \tag{6.3b}$$

where $d^3\tilde{k} = d^3k/[2\omega(\mathbf{k})(2\pi)^3]$, and $\omega(\mathbf{k}) = |\mathbf{k}| \equiv k$ since the electromagnetic field is massless, and $\hat{\mathbf{e}}(\mathbf{k})$ is a local unit vector that accounts for the vector nature of the field. The explicit expression of the local unit vectors is different for different contexts. In special relativity, where the fields are treated as 4-vectors, these can be a set of four unit vectors defying the polarisations of the electromagnetic field (including, as they are called, longitudinal and scalar photons) and leading to the so-called Gupta–Bleuer quantisation [9–11]. In the nonrelativistic regime, instead, we can define a set of three vectors, accounting for both the transverse and longitudinal components of the field (i.e., its polarisations). In the paraxial limit, the unit vector corresponding to longitudinal polarisation simply accounts for the propagation direction ($\hat{\mathbf{e}}_3(\mathbf{k}) = \hat{\mathbf{z}}$), while the other two form an orthogonal basis in the transverse plane and are defined as $\hat{\mathbf{e}}_2(\mathbf{k}) = \hat{\mathbf{z}} \times \mathbf{k}/|\hat{\mathbf{z}} \times \mathbf{k}|$, and $\hat{\mathbf{e}}_1(\mathbf{k}) = \hat{\mathbf{e}}_2(\mathbf{k}) \times \mathbf{k}/|\hat{\mathbf{e}}_2(\mathbf{k}) \times \mathbf{k}|$, where $\mathbf{k}$ is the wave vector of the single plane wave component appearing in equations (6.3).

In principle, the above relations are all that one needs in order to canonically quantise the electromagnetic field by simply repeating the procedure highlighted in chapter 5. In quantum optics, however, it is customary to assume that the electromagnetic field is localised within an optical cavity with perfectly reflecting boundaries. This is done to avoid problems mainly due to the appearance of divergent quantities, such as an infinite zero-point energy for the quantised electromagnetic field, due to the infinite nature of the quantisation volume[1]. Constraining the electromagnetic field into a cavity (frequently also called *quantisation box*) renders the volume finite, and solves the problem of infinite zero-point energy.

Another important aspect to take into account when quantising the electromagnetic field is its gauge invariance. Since the field variables are $\{\mathbf{A}, \mathbf{E}\}$, the quantisation procedure in not gauge invariance, because of the presence of the vector potential. Therefore, fixing a gauge is needed, and here we choose, as typical in quantum optics, the Coulomb gauge, i.e., $\nabla \cdot \mathbf{A} = 0$. A more general discussion on the general quantisation of the electromagnetic field without specifying a gauge, and without the necessity to introduce a quantisation cavity is presented in section 6.4, and discussed within the path integral framework. Here, we go through the usual, Hamiltonian quantisation box procedure, to obtain results similar to standard quantum optics literature.

---

[1] This is not an issue in QFT, since there are techniques (for example, renormalisation) that allow one to safely disregard these terms in any physically meaningful calculation. In quantum optics, however, renormalisation techniques are not that commonly used, and the easiest method of using a fictitious optical cavity with finite volume is preferred, as a more intuitive and less convoluted method to achieve the same goal.

To set up the problem of quantising the (nonrelativistic) electromagnetic field using a quantisation box, we make the following assumptions:

(1) We consider a finite quantisation volume and, without loss of generality, we assume this to be represented by a cube of side L, i.e., volume $V = L^3$. We furthermore require that the electromagnetic field is nonzero only inside the cube, and zero everywhere else.

(2) We think of the cube as a perfectly closed cavity, so that the allowed wave vectors $\mathbf{k} = k_x\hat{\mathbf{x}} + k_y\hat{\mathbf{y}} + k_z\hat{\mathbf{z}}$ to propagate inside the cavity are quantised according to $k_j = 2\pi n_j/L$, where $j = \{x, y, z\}$, and $n_j \in \mathbb{Z}$.

(3) Since we assumed working in the Coulomb gauge, the electromagnetic field is purely transverse, and can be then represented by two orthogonal states of polarisation, that we label, without loss of generality, with the index $\lambda = \{1, 2\}$ (this corresponds, for example, to TE and TM polarisation [12]). This allows us to replace $\hat{\mathbf{e}}(\mathbf{k}) \rightarrow \{\hat{\mathbf{e}}_\lambda(\mathbf{k})\}$, with $\hat{\mathbf{e}}_\mu(\mathbf{k}) \cdot \hat{\mathbf{e}}_\lambda(\mathbf{k}) = \delta_{\mu\lambda}$. Moreover, the transversality condition of the field, imposed by the Coulomb gauge, becomes $\hat{\mathbf{e}}_\lambda(\mathbf{k}) \cdot \mathbf{k} = 0$.

(4) We consider monochromatic electromagnetic fields, so that $d^3\tilde{k} \rightarrow d^3k$.

Some comments to these assumptions might be in order. Choosing a finite quantisation volume allows us to circumvent the infinite energy problem, but, on the other hand, we need to be careful that the final result does not depend on the particular choice of the quantisation volume, as this is only a mathematical tool used to make calculations easier, and bears no physical meaning. Also, the assumption (4) of monochromaticity is not really necessary, but we included it to make the calculations easier.

The more general approach of canonical quantisation of polychromatic fields is described in the book by Loudon [3], for example, and essentially follows the steps described below. The choice of a finite quantisation volume also implies assumption (2), which, ultimately, transforms the integrals $\int d^3k$ in equations (6.3) into sums over all possible allowed states inside the cavity, namely

$$\int d^3k \rightarrow \sum_{k_x}\sum_{k_y}\sum_{k_z} \equiv \sum_{\mathbf{k}}. \tag{6.4}$$

As a side note, we also mention that, due to the introduction of the cavity, the integral representation of the Dirac delta function reduces to that of the Kronecker delta, i.e.,

$$\int_{\text{cavity}} d^3x \, \exp\left[\pm i(\mathbf{k} - \mathbf{k}') \cdot \mathbf{r}\right] = V\delta_{\mathbf{k},\mathbf{k}'}, \tag{6.5}$$

where V is the volume of the quantisation cavity.

Moreover, since we have defined a cavity, we can use the (complete, orthonormal) set of cavity modes to represent the vector potential and electric field appearing in equations (6.3). In the simple case of a cubic cavity, the cavity modes are simply plane waves, with k-vector assuming only discrete values according to assumption (2). We can therefore rewrite equations (6.3) as follows

$$\mathbf{A} = \sum_{\mathbf{k}}\sum_{\lambda=1}^{2}[\mathscr{A}(\mathbf{k})\hat{\mathbf{e}}_{\lambda}(\mathbf{k})\exp[i(\mathbf{k}\cdot\mathbf{r} - \omega_{k}t)]+\text{c.c.}], \tag{6.6a}$$

$$\mathbf{E} = \sum_{\mathbf{k}}\sum_{\lambda=1}^{2}[i\omega_{k}\mathscr{A}(\mathbf{k})\hat{\mathbf{e}}_{\lambda}(\mathbf{k})\exp[i(\mathbf{k}\cdot\mathbf{r} - \omega_{k}t)]+\text{c.c.}], \tag{6.6b}$$

where $\omega_{k} = k$ is the dispersion relation of each single mode of the cavity, and the summation over λ accounts for the two orthogonal states of polarisation of the electromagnetic field.

We can now promote the field amplitudes to operators, i.e., $\mathscr{A}(\mathbf{k}) \rightarrow \hat{a}_{\lambda}(\mathbf{k})$, and $\mathscr{A}^{*}(\mathbf{k}) \rightarrow \hat{a}_{\lambda}^{\dagger}(\mathbf{k})$, obeying the commutation rules

$$[\hat{a}_{\lambda}(\mathbf{k}), a_{\mu}(\mathbf{k}')] = 0 = [\hat{a}_{\lambda}^{\dagger}(\mathbf{k}), \hat{a}_{\mu}^{\dagger}(\mathbf{k}')], \tag{6.7a}$$

$$[\hat{a}_{\lambda}(\mathbf{k}), \hat{a}_{\mu}^{\dagger}(\mathbf{k}')] = \delta_{\lambda\mu}\delta_{\mathbf{k},\mathbf{k}'}. \tag{6.7b}$$

Notice, that the creation and annihilation operators for the electromagnetic field contain also a dependence on the parameter λ, i.e., the field polarisation. Upon quantisation, in fact, this degree of freedom becomes a valid quantum number, essentially signifying that each cavity mode is intrinsically degenerate in polarisation, and that this degeneracy is lifted, once the polarisation of the mode is specified [1]. The quantised electromagnetic field Hamiltonian then reads, according to equation (5.15)

$$\hat{H} = \sum_{\mathbf{k}}\sum_{\lambda=1}^{2} \hbar\omega_{k}\left[\hat{a}_{\lambda}^{\dagger}(\mathbf{k})\hat{a}_{\lambda}(\mathbf{k}) + \frac{1}{2}\right]. \tag{6.8}$$

One can then introduce the Fock states for the electromagnetic fields as the eigenstates of the electromagnetic number operator $\hat{N}_{\lambda}(\mathbf{k}) = \hat{a}_{\lambda}^{\dagger}(\mathbf{k})\hat{a}_{\lambda}(\mathbf{k})$, so that

$$\hat{N}_{\lambda}(\mathbf{k})|n_{\lambda}(\mathbf{k})\rangle = n_{\lambda}(\mathbf{k})|n_{\lambda}(\mathbf{k})\rangle, \tag{6.9a}$$

$$\langle n_{\lambda}(\mathbf{k})|n_{\mu}(\mathbf{k}')\rangle = \delta_{\lambda\mu}\delta_{\mathbf{k},\mathbf{k}'}, \tag{6.9b}$$

$$\hat{a}_{\lambda}(\mathbf{k})|n_{\lambda}(\mathbf{k})\rangle = \sqrt{n_{\lambda}(\mathbf{k})}\,|n_{\lambda}(\mathbf{k})\rangle, \tag{6.9c}$$

$$\hat{a}_{\lambda}^{\dagger}(\mathbf{k})|n_{\lambda}(\mathbf{k})\rangle = \sqrt{n_{\lambda}(\mathbf{k}) + 1}\,|n_{\lambda}(\mathbf{k})\rangle, \tag{6.9d}$$

and, analogously to section 5.2.1, we can introduce the multimode Fock states as

$$|n_{\lambda_{1}}(\mathbf{k}_{1}), n_{\lambda_{2}}(\mathbf{k}_{2}), ..., n_{\lambda_{N}}(\mathbf{k}_{N})\rangle \equiv |n_{\lambda_{1}}(\mathbf{k}_{1})\rangle|n_{\lambda_{2}}(\mathbf{k}_{2})\rangle \cdots |n_{\lambda_{N}}(\mathbf{k}_{N})\rangle. \tag{6.10}$$

Finally, we can write the quantum representation of the vector potential, electric and magnetic field as follows

$$\hat{\mathbf{A}}(\mathbf{r},\,t) = \sum_{\mathbf{k}}\sum_{\lambda=1}^{2}\sqrt{\frac{\hbar}{2\varepsilon_0 V\omega_k}}\,\hat{\mathbf{e}}_\lambda(\mathbf{k})[\hat{a}_\lambda(\mathbf{k})\exp\left[i(\mathbf{k}\cdot\mathbf{r}-\omega_k t)\right]+\text{h.c.}],$$

$$\equiv \hat{\mathbf{A}}^+(\mathbf{r},\,t) + \hat{\mathbf{A}}^-(\mathbf{r},\,t),$$

(6.11a)

$$\hat{\mathbf{E}}(\mathbf{r},\,t) = i\sum_{\mathbf{k}}\sum_{\lambda=1}^{2}\sqrt{\frac{\hbar\omega_k}{2\varepsilon_0 V}}\,\hat{\mathbf{e}}_\lambda(\mathbf{k})[\hat{a}_\lambda(\mathbf{k})\exp\left[i(\mathbf{k}\cdot\mathbf{r}-\omega_k t)\right]+\text{h.c.}]$$

$$\equiv \hat{\mathbf{E}}^+(\mathbf{r},\,t) + \hat{\mathbf{E}}^-(\mathbf{r},\,t),$$

(6.11b)

$$\hat{\mathbf{B}}(\mathbf{r},\,t) = i\sum_{\mathbf{k}}\sum_{\lambda=1}^{2}\sqrt{\frac{\hbar}{2\varepsilon_0 V\omega_k}}\,[\mathbf{k}\times\hat{\mathbf{e}}_\lambda(\mathbf{k})][\hat{a}_\lambda(\mathbf{k})\exp\left[i(\mathbf{k}\cdot\mathbf{r}-\omega_k t)\right]+\text{h.c.}]$$

$$\equiv \hat{\mathbf{B}}^+(\mathbf{r},\,t) + \hat{\mathbf{B}}^-(\mathbf{r},\,t),$$

(6.11c)

where $\hat{\mathbf{E}}^\pm(\mathbf{r},\,t)$ $(\hat{\mathbf{B}}^\pm(\mathbf{r},\,t))$ are the positive (+) and negative (−) frequency parts of the electric (magnetic) fields, and h.c. stands for Hermitian conjugate [13]. The same is valid for the vector potential as well. Notice, moreover, that the constant $\sqrt{\hbar/2\varepsilon_0 V\omega_k}$ appearing in the definition of the vector potential operator and, consequently, in the electric and magnetic fields, has been inserted, in analogy with the harmonic oscillator (see appendix A of chapter 5), to guarantee that $[\hat{\mathbf{A}}^+(\mathbf{r},\,t),\hat{\mathbf{A}}^-(\mathbf{r}',\,t)] = \delta_{\mathbf{r},\mathbf{r}'}$. Finally, the expression of the magnetic field has been derived using $\mathbf{B} = \nabla\times\mathbf{A}$.

The electromagnetic field operators defined above are written in terms of a superposition of all the cavity modes and the correspondent electromagnetic field state $|n_{\lambda_1}(\mathbf{k}_1),\, n_{\lambda_2}(\mathbf{k}_2),\, \dots,\, n_{\lambda_N}(\mathbf{k}_N)\rangle$ is referred to as a multimode state (or multimode field), since the total field is characterised by multiple quantum numbers (i.e., as a superposition of different cavity modes with different polarisation). Since these modes are mutually independent (i.e., the multimode Hilbert space is just the tensor product of single-mode Hilbert spaces), analysing the properties of a single mode of the field is sufficient to grasp the physics behind it. The results obtained for a single mode can then be easily be generalised to multimode fields. For this reason, in the rest of this chapter, for the sake of simplicity, we will limit our description to single-mode fields, i.e., electromagnetic fields characterised only by a single quantum number (one polarisation state, one particular cavity mode, etc).

To represent the electric and magnetic field operators of a single-mode field, we fix a polarisation, say $\hat{\mathbf{e}}_\lambda(\mathbf{k}) = \hat{\mathbf{x}}\,\delta_{\lambda 1}$ for the electric field, so that $(\mathbf{k}\times\hat{\mathbf{e}}_\lambda(\mathbf{k})) = \hat{\mathbf{y}}$ for the magnetic field, if we choose the electromagnetic field to propagate along the $\hat{\mathbf{z}}$-direction, and we concentrate on a single cavity mode with $\mathbf{k} = \bar{\mathbf{k}}$, so that $\omega_k \equiv \omega$. Because the vectorial character of the field is contained in its polarisation, and this is fixed by definition, we drop the vector nature of the field, since in the single-mode approximation the polarisation dynamics does not influence the quantum properties of the field. We then get

$$\hat{E}(z,\,t) = \mathscr{E}_0[\hat{a}\exp(-i\omega t) - \hat{a}^\dagger\exp(i\omega t)]\sin(kz) \equiv \hat{E}^+(z,\,t) + \hat{E}^-(z,\,t),\quad(6.12a)$$

$$\hat{B}(z, t) = \mathscr{E}_0[\hat{a}\exp(-i\omega t) - \hat{a}^\dagger\exp(i\omega t)]\cos(kz) \equiv \hat{B}^+(z, t) + \hat{B}^-(z, t), \quad (6.12b)$$

where $\mathscr{E}_0$ is the electric field amplitude[2]. In addition to this, we will also use the notaiton $|n\rangle \equiv |0, 0, ..., n_{\lambda=1}(\mathbf{k} = \bar{\mathbf{k}}), 0, 0, ...\rangle$ to indicate the Fock state of a single-mode field, characterised by the choice of polarisation ($\lambda = 1 \rightarrow \hat{\mathbf{x}}$) and k-vector $\bar{\mathbf{k}}$.

6.1.1 Single-mode electromagnetic Fock states

We now want to look at the general properties of (electromagnetic) Fock states. In the quantum optics community, these states are commonly referred to as number states, because essentially they convey the information about the number of photons in the electromagnetic field [1, 3]. Let us start by calculating the expectation value of the electric field operator over number states, as a measure of the classical electric field

$$\langle n|\hat{E}(z, t)|n\rangle = \mathscr{E}_0\sin(kz)[\langle n|\hat{a}|n\rangle\exp(-i\omega t)+\text{h.c.}] = 0, \quad (6.13)$$

since both terms give as a result an expectation value between orthogonal states, i.e., $\langle n|\hat{a}|n\rangle = \sqrt{n}\langle n|n - 1\rangle = 0$ (and similarly for $\hat{a}^\dagger$). This result is essentially telling us that there is no classical counterpart to number states, since the classical electric field associated with them is zero. This means, that number states are quantum states of the electromagnetic field, whose electric field is not well defined, since the average electric field of a number state is zero[3].

This result, however, does not lead to a paradox, since the energy stored in the electromagnetic field is not lost, but simply shifted to a different observable. If we, in fact, calculate the fluctuations of the field, i.e., the expectation value of the variance of the field operator, we get

$$\langle n|\Delta\hat{E}^2(z, t)|n\rangle = \langle n|\hat{E}^2(z, t)|n\rangle - [\langle n|\hat{E}(z, t)|n\rangle]^2$$
$$= 2\mathscr{E}_0^2\sin^2(kz)\left(n + \frac{1}{2}\right), \quad (6.14)$$

which is nonzero, as expected. We can moreover notice, that the result above is nonzero also for $n = 0$, i.e., when there are no photons in the electromagnetic field. This nonzero result with zero excitations of the field (i.e., photons) reflects the zero-point energy of all the oscillators associated with all the cavity modes of the quantised electromagnetic field. For this reason, the fluctuations with $n = 0$ are called *vacuum field fluctuations*.

According to their definition, number states are the eigenstates of the number operator $\hat{N} = \hat{a}^\dagger\hat{a}$, This means, that the energy stored in a number state is a

---

[2] Notice, that normally one would have $\mathscr{E}_0$ for the electric field amplitude and $\mathscr{B}_0 = \mathscr{E}_0/c$ for the magnetic field amplitude. Here, however, since we have implicitly chosen natural units, i.e., $c = 1$, we get $\mathscr{B}_0 = \mathscr{E}_0$.

[3] This is one of the reasons, why number states are often referred to as the most quantum amongst the states of the electromagnetic field.

well-defined quantity. In fact, the total energy of a single-mode electromagnetic field can be written, using equation (6.8), as

$$\hat{H} = \hbar\omega\left(\hat{N} + \frac{1}{2}\right),$$ (6.15)

and therefore the expectation value of the energy contained in a number state is given by

$$\langle n|\hat{H}|n\rangle = \langle n|\hbar\omega\left(\hat{N} + \frac{1}{2}\right)|n\rangle = \hbar\omega\left(n + \frac{1}{2}\right),$$ (6.16)

since $\hat{N}|n\rangle = n|n\rangle$. This also allows us to say that the number of photons is a well-defined quantity for number states. This can be seen explicitly by calculating the probability of finding n photons in a number state $|m\rangle$, which is calculated as follows

$$P_n = |\langle n|m\rangle|^2 = \delta_{nm}.$$ (6.17)

This is a consequence of the fact that the number of photons in a number state is determined with absolute certainty, as can be seen from the fluctuations of the photon number operator, i.e.,

$$\langle \Delta\hat{N}^2 \rangle = 0.$$ (6.18)

This photon distribution, called sub-Poissonian, is what characterises number states and it constitutes the main reason why these states are often called the *most quantum* amongst the states of the electromagnetic field. On the opposite side of the spectrum, the *most classical* states of the electromagnetic fields, i.e., those that are reproducing very closely the behaviour of the classical electromagnetic field, are the coherent states, which are discussed in the next section.

6.1.2 Single-mode electromagnetic coherent states

The canonical quantisation procedure gives us the simple picture of a quantum field being represented as a collection of quantum harmonic oscillators, allowing us to use many of the results of simple harmonic oscillators to construct the properties and different states of the field. Coherent states are no exception to this, and are essentially defined as the field counterpart of the states of the harmonic oscillator possessing minimal uncertainty in both the position and momentum degrees of freedom of the field [14–16]. We can use this as guidance to introduce coherent states for the electromagnetic field as well. Since we are working in the single-mode regime, this task is made easier, since essentially coherent states of a single-mode field are the coherent states of a single quantum harmonic oscillator. This is done in Appendix 6.4.2. Here, we use the final result in appendix 6.4.2 to define coherent states as the (left) eigenstates of the annihilation operator as

$$\hat{a}|\alpha\rangle = \alpha|\alpha\rangle,$$ (6.19)

where $\alpha \in \mathbb{C}$, whose physical meaning, as discussed below, is such that $|\alpha|^2$ represents the average number of photons in the electromagnetic field. Since number states constitute a complete set, coherent states can also be represented in terms of number states as follows [1]

$$|\alpha\rangle = \exp\left(-\frac{|\alpha|^2}{2}\right)\sum_{n=0}^{\infty}\frac{\alpha^n}{\sqrt{n!}}|n\rangle. \tag{6.20}$$

Since coherent states contain all possible number states, the expectation value of the electric field operator over them can now be, in general, different from zero. We have in fact[4]

$$\langle\alpha|\hat{E}(z, t)|\alpha\rangle = 2|\alpha|\mathscr{E}_0 \sin(kz - \omega t + \theta), \tag{6.21}$$

where $\theta = \arg(\alpha)$. This corresponds to the expression of a monochromatic, scalar, *classical* electromagnetic field propagating along the z-direction. Coherent states, therefore, are those states of the electromagnetic field that are closer to the classical field, in the sense that the expectation value of the electromagnetic field over them reproduces the classical electromagnetic field. The fluctuations of the field can be then calculated as

$$\langle\alpha|\Delta\hat{E}^2(z, t)|\alpha\rangle = \mathscr{E}_0^2, \tag{6.22}$$

which are identical to those of the vacuum state $|0\rangle$. This is again an interesting result, that needs a bit of discussion. While on one hand coherent states reproduce, in the average limit, the classical field, on the other hand they retain their quantumness by displaying field fluctuations, which have the same magnitude of the vacuum field fluctuations. This fact leads to the conclusion that coherent states $|\alpha\rangle$ and the vacuum state $|0\rangle$ are linked somehow. This is becoming clear by introducung the displacement operator

$$\hat{\mathscr{D}}(\alpha) = \exp\left[\alpha\hat{a}^\dagger - \alpha^*\hat{a}\right], \tag{6.23}$$

and defining the coherent states as displaced vacuum states through the relation [1, 15]

$$|\alpha\rangle = \hat{\mathscr{D}}|0\rangle, \tag{6.24}$$

which shows explicitly why the fluctuation properties of coherent states and the vacuum state are the same. A detailed discussion on displacement operators and their properties is given, for example, in reference [15].

Let us now discuss the physical meaning of the complex parameter α characterising the coherent states. Contrary to number states, the number of photons in a coherent state is not constant but fluctuates according to Poissonian statistics[5]. If we

---

[4] In order to obtain the result below, we have also used the fact that coherent states are (right) eigenstates of the creation operator, i.e., $\langle\alpha|\hat{a}^\dagger = \langle\alpha|\alpha^*$.

[5] A Poissonian process is a statistical process characterised by having mean equal to its variance.

calculate the average number of photons in a coherent state and the fluctuations of the photon number, in fact, we see that

$$\langle \hat{N} \rangle = \langle \alpha | \hat{N} | \alpha \rangle = |\alpha|^2, \tag{6.25a}$$

$$\langle \Delta \hat{N} \rangle = |\alpha|^2. \tag{6.25b}$$

Consequently, the probabilty of detecting n photons in a coherent state $|\alpha\rangle$ follows Poissonian statistics, and is given explicitly by

$$P_n = |\langle n | \alpha \rangle|^2 = \frac{|\alpha|^{2n}}{n!} \exp(-|\alpha|^2). \tag{6.26}$$

6.1.3 Quantisation and optical beams

The quantisation scheme presented in section 6.1 is based on defining a complete set of cavity modes and using them to represent the electromagnetic field inside the quantisation volume. Then, the amplitude of each cavity mode is promoted to be an operator, and quantisation is carried out. Although this procedure is fairly general, the particular choice of cavity modes is dictated by the geometry of the quantisation (plane waves for cubic cavities, spherical waves for spherically symmetric cavities, cylindrical waves for cylindrical cavities, and so on [13]). Furthermore, as the choice of the quantisation cavity is totally arbitrary, there is no preferential choice, and typically one chooses the framework that best suits their needs.

Introducing a cavity, however, automatically limits the volume of space in which the quantisation takes place, and naturally introduces a discrete set of modes, which are a good representation of the quantum field inside the cavity, but fail to represent it in the rest of space, where continuous modes constitute a better choice, as they reflect the infinite nature of space outside the cavity.

A more general approach to field quantisation is to avoid using a cavity and quantise the electromagnetic field in the whole space. In this case, a suitable set of continuous modes[6] needs to be introduced to carry on the quantisation procedure. Moreover, the commutation relations must reflect the continuous nature of the eigenvalue spectrum of the field, opposite to the case of the closed cavity, where the spectrum is discrete. So if the commutation relation for discrete cavity modes is something like $[\hat{a}_k, \hat{a}_{k'}^\dagger] = \delta_{k,k'}$, for a continuous set of modes they need to be properly generalised to something like $[\hat{a}(k), \hat{a}^\dagger(k')] = \delta(k - k')$ [3]. In addition to all this, one needs to pay particular attention to the various infinities that arise, as a consequence of the quantisation volume being infinite. In particular, one should learn to deal with them and how to recognise those infinities that can be regularised, and those that cannot. This is essentially the subject of the field of renormalisation, which constitutes an important part of QFT.

---

[6] These modes can be seen as the limit of cavity modes, when the volume of the cavity goes to infinity, thus encompassing the whole space.

The standard approach of QFT is then that of using the same plane wave expansion used above to carry on the quantisation in free space, without any cavity. This, of course, is possible, since plane waves are a complete set for $\mathbb{R}^3$ as well, and not only inside a cubic-shaped cavity.

In photonics, however, there might be the need to use a different mode set than plane waves, especially when dealing with optical beams solution of the paraxial (equation (2.39)) or Helmholtz (equation (2.5)) equation, as those introduced in chapter 1. This section then presents the quantisation procedure to quantise the electromagnetic field, when represented as a paraxial, monochromatic optical beam. The same line of reasoning can be then applied to the Helmholtz and the wave equation.

To start with, let us recall the paraxial equation (2.39)

$$i\frac{\partial E(\mathbf{R}, z)}{\partial z} = -\frac{1}{2k_0}\nabla_\perp^2 E(\mathbf{R}, z), \qquad (6.27)$$

where $\mathbf{R} = \{x, y\} = \{R, \theta\}$ are the transverse coordinates, z is the propagation direction, $\nabla_\perp^2 = \partial_x^2 + \partial_y^2$ is the Laplacian in the transverse coordinates, and $E(\mathbf{R}, z)$ is the scalar part of the electromagnetic field, i.e., polarisation is fixed and doesn't change upon propagation.

To apply the canonical quantisation procedure highlighted in the previous section, we need to derive the Hamiltonian associated with the paraxial field $E(\mathbf{R}, z)$. To this aim, we let us first define the canonically conjugated momentum to the paraxial field $E(\mathbf{R}, z)$ using the Lagrangian in equation (3.14)

$$\mathscr{L}_p = \frac{1}{2}(\nabla E) \cdot (\nabla E) + i\,k\,E\,\partial_z E, \qquad (6.28)$$

as the derivative of $\mathscr{L}_p$ with respect to $\partial_z E$, i.e.,

$$\Pi(\mathbf{R}, z) = \frac{\partial \mathscr{L}_p}{\partial(\partial_z E)} = i\,k\,E(\mathbf{R}, z), \qquad (6.29)$$

so that the Hamiltonian can be defined through a Legendre transformation as

$$\mathscr{H}_p = \Pi\,\partial_z E - \mathscr{L}_p = -\frac{1}{2}(\partial_i E) \cdot (\partial^i E). \qquad (6.30)$$

Notice, that if we represent the paraxial Hamiltonian above in Fourier space, then it reduces to $\mathscr{H}_p = \Pi^2/2$, since the canonically conjugated momentum $\Pi(\mathbf{R}, z)$ defined in equation (6.105) is nothing other than the Fourier transform of $\partial_i E$ [14]. The experienced reader might find a similarity between the paraxial Hamiltonian in Fourier space and the Hamiltonian for a free quantum particle, i.e., $H = p^2/2m$. This is not a coincidence, since the evolution of a paraxial, monochromatic electromagnetic field can be isomorphically mapped into the evolution of a free quantum particle described by the Schrödinger equation [17]. The paraxial equation (6.27) can be then recovered from the Hamilton equations (3.21), in particular from the equation for $\partial_z \Pi$.

The paraxial modes solution of the paraxial equation (6.27) form a complete orthonormal set, which can be then used to represent an arbitrary electromagnetic field as a continuous superposition of such modes in Fourier space as

$$E(\mathbf{R}, z) = \sum_{p,q \in \mathbb{M}} \int d^2k \, [a(\mathbf{k})u_{pq}(\mathbf{k}, z)\exp(i\, \mathbf{k} \cdot \mathbf{R}) + \text{h.c.}], \qquad (6.31)$$

where $\mathbb{M}$ is the appropriate set for $\{p, q\}$ and depends on the particular choice of modes, $\mathbf{k} = \{k_x, k_y\}$ is the transverse wave vector, and $u_{pq}(\mathbf{k}, z)$ is the Fourier transform of the paraxial mode $u_{pq}(\mathbf{R}, z)$, solution of equation (6.27).

Next, we can promote the spectral coefficients $a(\mathbf{k})$ and $a^*(\mathbf{k})$ to creation annihilation operators by associating with them the usual commutation rules

$$[\hat{a}(\mathbf{k}), \hat{a}^\dagger(\mathbf{k}')] = \delta(\mathbf{k} - \mathbf{k}'), \qquad (6.32a)$$

$$[\hat{a}(\mathbf{k}), \hat{a}(\mathbf{k}')] = 0 = [\hat{a}^\dagger(\mathbf{k}), \hat{a}^\dagger(\mathbf{k}')]. \qquad (6.32b)$$

Because of the fact that the set of paraxial modes $u_{pq}(\mathbf{R}, z)$ forms a complete and orthogonal set, one has a choice of representation for the quantum field. In fact, one could either choose to represent the field in terms of momentum space states, i.e., the traditional plane wave representation, modified by the presence of the paraxial modes as in equation (6.31), or one could represent the quantum field directly in the Fock space spanned by the mode functions themselves, whose basis states can be uniquely addressed by the set of quantum numbers $\{p, q\}$. This can be done by introducing the mode operators

$$\hat{a}_{pq} = \frac{1}{(2\pi)^2} \int d^2k \, u_{pq}^*(\mathbf{k}, z)\hat{a}(\mathbf{k}), \qquad (6.33a)$$

$$\hat{a}_{pq}^\dagger = \frac{1}{(2\pi)^2} \int d^2k \, u_{pq}(\mathbf{k}, z)\hat{a}^\dagger(\mathbf{k}), \qquad (6.33b)$$

with their associated commutation relations

$$[\hat{a}_{pq}, \hat{a}_{nm}^\dagger] = \delta_{pn}\,\delta_{qm}, \qquad (6.34a)$$

$$[\hat{a}_{pq}, \hat{a}_{nm}] = 0 = [\hat{a}_{pq}^\dagger, \hat{a}_{nm}^\dagger], \qquad (6.34b)$$

and the prescription that $u_{pq}(\mathbf{k}, z) = \langle \mathbf{k}, z | p, q \rangle$. For the case of a single-photon state, for example, the two representations are related as follows

$$|p, q\rangle \equiv \hat{a}_{pq}^\dagger|0\rangle = \int d^2k \, u(\mathbf{k}, z)\hat{a}^\dagger(\mathbf{k})|0\rangle = \int d^2k \, u(\mathbf{k}, z)|\mathbf{k}, z\rangle, \qquad (6.35)$$

where the first vacuum refers to the vacuum in Fock space spanned by $|p, q\rangle$, while the second vacuum, under the integral sign, refers to the vacuum state in momentum space. Notice, moreover, that

$$\langle p, q | p, q \rangle = 1 = \int d^2k \, |u_{pq}(\mathbf{k}, z)|^2, \qquad (6.36)$$

which provides the correct normalisation condition for Fock states in both representations.

A formal description of paraxial quantisation can be found in the book of Garrison and Chiao [18], or, in more accessible form, in references [19, 20], the latter of which is specifically focused on the quantisation of Hermite–Gaussian (HG) and Laguerre–Gaussian (LG) beams.

6.2 Part II: quantum nonlinear optics in canonical formalism

The discussion on the field quantisation presented in the previous sections only accounts for the free field, i.e., an electromagnetic field propagating in vacuum, without interacting with any other field, or even with itself. In terms of photonics, this just corresponds to an optical beam (or, more generally, an electromagnetic field) propagating in free space. With few tweaks, the theory above can also be adapted to optical beams propagating in a linear medium, where the interaction of light with matter is essentially described by the refractive index of the material [3, 12].

Many of the relevant photonics applications, however, involve the study of the nonlinear interaction between light and matter, enabling the generation of new frequencies for the electromagnetic field, and many other effects. This is the domain of nonlinear optics, which essentially deals with the self-interaction of the electro-magnetic field, mediated by the material, that is typically referred to as the nonlinear medium. Developing a canonical quantisation formalism to describe such inter-actions is then very useful, since it allows one to tackle a large number of problems.

Usually, nonlinearities are introduced in photonics by means of a nonlinear relation between the matter polarisation (i.e., the mesoscopic response of the material to the electromagnetic field interacting with it) and the electromagnetic field, which is represented as a series expansion of the matter polarisation in powers of the electric field in frequency domain as [7]

$$P_i(\omega) = \varepsilon_0 \left[\chi_{ij}^{(1)}(\omega)E_j(\omega) + \chi_{ijk}^{(2)}(\omega)E_j(\omega)E_k(\omega) + \chi_{ijk\ell}^{(3)}(\omega)E_j(\omega)E_k(\omega)E_\ell(\omega) + \cdots \right], \quad (6.37)$$

where the nth order optical susceptibility $\chi^{(n)}(\omega) \equiv \chi^{(n)}(\omega; \omega_1, \omega_2, \ldots, \omega_n)$ is a rank $(n + 1)$ tensor describing the interaction of n fields, each oscillating at its own frequency ω_i, to produce a contribution to the matter polarisation at frequency $\omega = \omega_1 + \omega_2 + \cdots + \omega_n$. The magnitude of this term, i.e., $|\chi^{(n)}|$, can be taken as a measure of the strength of the nonlinear interaction. An exception to this rule is the case $n = 1$, which corresponds to the linear optical response of the material, for which $\chi_{ij}^{(1)}(\omega) = \varepsilon_{ij}(\omega) - 1$, where $\varepsilon_{ij}(\omega)$ is the permittivity tensor, describing the linear optical properties of the medium [7, 12]. In the linear regime, moreover, an electric field oscillating at frequency ω generates a matter polarisation oscillating at the same frequency.

In conventional nonlinear materials $|\chi^{(n+1)}(\omega)| \ll |\chi^{(n)}(\omega)|$ [7, 21]. This has two practical consequences: first, nth order nonlinarities are dominating over higher-order ones, and the latter can be, in general, neglected, if working at low enough electric field amplitudes, which is the case for most of the experimental setups in photonics nowadays. On the other hand, this also opens for the possibility of treating the optical nonlinearity in a perturbative manner, therefore allowing the use of the tools developed in chapter 5. In present photonics applications and laboratory setups, only the second- and third-order nonlinearities are typically accessible experimentally with the range of powers available to laser pulses nowadays. Fifth order nonlinearities are also accounted for, normally when studying the propagation and stability of optical solitons and solitary waves [22].

The usual approach to tackle the nonlinear effects described by equation (6.37) is to substitute this equation in the wave equation (2.4), to obtain a set of nonlinear coupled partial differential equations, one for each electric field oscillating at a different frequency ω_i.

To describe such nonlinear processes inside a canonical formalism framework, we need instead to have access to a nonlinear interaction Hamiltonian that accounts for such processes and that, at the equations of motion level, reproduces the set of nonlinear coupled mode equations generated by using equations (6.37) and (2.4). The form of equation (6.37) suggests choosing an interaction Hamiltonian that is polynomial in the electric field $E(\omega)$, whose explicit form, taking into account all possible orders of nonlinearities, can be written as

$$\hat{H}_{int} = \sum_{n=2}^{\infty} \int [d\omega]_{n+1} \, \chi_{i_1 i_2 \cdots i_{n+1}}^{(n)}(\omega_1; \omega_2, ..., \omega_n) E_{i_1}(\omega_1) E_{i_2}(\omega_2) \cdots E_{i_n}(\omega_n) E_{i_{n+1}}(\omega_{n+1}), \quad (6.38)$$

where $[d\omega]_{n+1} = d\omega_1 d\omega_2 \cdots d\omega_n d\omega_{n+1}$, and $\chi_{i_1 i_2 \cdots i_n}^{(n)}(\omega_1; \omega_2, ...\omega_n)$ is the nth order nonlinear optical susceptibility [7, 21], playing here the role of frequency-dependent coupling constant[7]. One could easily verify that, together with the free Hamiltonian, this interaction Hamiltonian generates the correct equations of motion, i.e., equation (2.4).

Before delving into the examples for second- and third-order nonlinearities, it is worth fixing the notation, in order to avoid confusion. Traditionally, nonlinear optics refers to the order of a nonlinear interaction at the equations of motion level, i.e., the order of the nonlinearity expresses the number of electric fields appearing in equation (6.37). In field theory, however, the order of the nonlinear interaction refers to the Lagrangian (Hamiltonian) level. For example, nonlinear optics would call a third-order nonlinearity the processes coming from the term proportional to $\chi^{(3)}(\omega)$ in equation (6.37), and encompassing the interaction between three fields, while field theory would label this as a four-field interaction, since it corresponds to a term in the Hamiltonian above containing four fields. This mismatch of jargon can cause some confusion. Here, we adopt the photonics convention of labelling the order of the nonlinearity with the order of the susceptibility tensor, to remain close to the traditional jargon of nonlinear optics.

Below, we discuss second-order nonlinearities in detail, deriving the transition probability and the explicit form of the quantum state of the field after the nonlinear

---

[7] The frequency dependence of the nth order susceptibility is written in a way to emphasise energy conservation amongst the varoius frequency components, i.e., $\omega_1 = \omega_2 + \omega_3 + \cdots + \omega_n$.

interaction, then briefly mentioning how these concepts can be generalised to third-order nonlinearities. The results presented here emphasise the quantum treatment of these processes. For a comprehensive discussion on the classical aspects of second- and third-order nonlinear optical processes, the reader is addressed to the books by Boyd [7], Butcher and Cotter [21] and Yariv [23], and those by Weiner [24] and Wegener [25] as a reference for nonlinear optics of pulsed electromagnetic fields.

6.2.1 Second-order nonlinearities

Second-order optical nonlinearities are regulated by the second-order susceptibility $\chi_{ijk}^{(2)}(\omega; \omega', \omega'')$, which contains all the information about the nonlinear material enabling the self-interaction of the electromagnetic field [7, 21, 23], involves the interaction of three fields, conventionally called *pump* (p), *signal* (s), and *idler* (i), linked together by the energy ($\omega_p = \omega_s + \omega_i$) and momentum [$\mathbf{k}(\omega_p) = \mathbf{k}(\omega_s) + \mathbf{k}(\omega_i)$] conservation laws [7]. Without loss of generality, we assume that the nonlinear material providing the interaction is modelled by a crystal infinitely extended along the x- and y-directions, and has length L along the z-direction. A sketch of the geometry of the nonlinear crystal and the correspondent electromagnetic interaction is reported in figure 6.1.

The interaction Hamiltonian for second-order optical nonlinearities can be derived from equation (6.38) with $n = 2$ and reads

$$\hat{H}_{\text{int}}^{(2)} = \frac{g}{3!} \int d\omega d\omega' d\omega'' \, \chi_{jk\ell}^{(2)}(\omega; \omega', \omega'') \hat{E}_j(\omega) \hat{E}_k(\omega') \hat{E}_\ell(\omega''), \tag{6.39}$$

where $g \ll 1$ is the coupling constant[8], and $\hat{E}(\omega)$ is the Fourier transform in time of the electric field operator, defined as

$$\hat{\mathbf{E}}(z, t) = i \, \hat{\mathbf{f}} \int d\omega \sqrt{\frac{\omega}{4\pi \, n(\omega)}} [\hat{a}(\omega) \exp\{i[k(\omega)z - \omega t]\} - \text{h.c.}]$$
$$\equiv \hat{\mathbf{E}}^+(z, t) + \hat{\mathbf{E}}^-(z, t), \tag{6.40}$$

where $n(\omega)$ is the refractive index of the nonlinear medium, $k(\omega) = k_0 n(\omega)$, and the creation and annihilation operators in frequency domain obey the usual bosonic commutation relations[9]

---

[8] Normally, the strength of the interaction would be regulated by the susceptibility. Here, instead, we have decided to make g explicilty appear, to be able later on to make a power series expansion with respect to the coupling constant. Practically, one could think of g as being related to the magnitude of the nonlinear interaction, i.e., $g \simeq |\chi^{(2)}|$.

[9] The expression of the Fourier transform of the field operator in equations (6.42) differs from equation (2.9.2) of reference [3] by a missing factor of $1/A$ inside the square root, which accounts for the transverse area of the field. Here, we don't have this factor, since we are assuming that the nonlinear crystal is infinitely extended in the transverse dimension. This area-dependent factor, therefore, is absorbed in the definition of the continuous-spectrum creation and annihilation operators $\hat{a}(\omega)$ and $\hat{a}^\dagger(\omega)$, in accordance with equations (6.2.3) and (6.2.4) of reference [3].

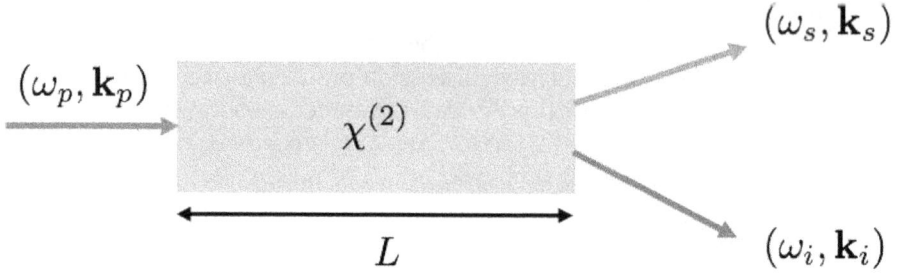

Figure 6.1. Pictorial representation of a possible second-order nonlinear interaction (parametric down-conversion) of the electromagnetic field in a $\chi^{(2)}$ medium. The medium is represented by a light blue box of length L (and it is assumed to be infinitely extended in the transverse direction, for simplicity). The pump beam, characterised by frequency and momentum $(\omega_p, \mathbf{k}_p)$ impinges on the nonlinear crystal (red arrow). As a result of the nonlinear interaction, two new beams (signal, in green, and idler, in blue) are generated, with frequency and momentum $(\omega_s, \mathbf{k}_s)$ for the signal beam, and $(\omega_i, \mathbf{k}_i)$ for the idler beam. The nonlinear process conserves both energy and momentum, forcing the relations $\omega_p = \omega_s + \omega_i$ and $\mathbf{k}_p = \mathbf{k}_s + \mathbf{k}_i$ to hold between the pump, signal and idler electromagnetic modes.

$$[\hat{a}(\omega), \hat{a}^\dagger(\Omega)] = \delta(\omega - \Omega), \tag{6.41a}$$

$$[\hat{a}(\omega), \hat{a}(\Omega)] = 0 = [\hat{a}^\dagger(\omega), \hat{a}^\dagger(\Omega)]. \tag{6.41b}$$

For simplicity, we have assumed the field to be scalar and paraxial, so that its polarisation can be described by the transverse unit vector $\hat{\mathbf{f}} = f_x \hat{\mathbf{x}} + f_y \hat{\mathbf{y}}$. Inverting this relation gives the frequency-dependent field operator as

$$\hat{E}^+(\omega) = i\sqrt{\frac{\omega}{2\,n(\omega)}}\,\hat{a}(\omega) \equiv i\mathscr{E}(\omega)\hat{a}(\omega), \tag{6.42a}$$

$$\hat{E}^-(\omega) = -i\sqrt{\frac{\omega}{2\,n(\omega)}}\,\hat{a}^\dagger(\omega) \equiv -i\mathscr{E}(\omega)\hat{a}^\dagger(\omega). \tag{6.42b}$$

Notice, that the electric field operator in the frequency domain is z-independent. Substituting these expressions into equation (6.39) leads to[10]

$$\hat{H}_{\text{int}}^{(2)}(z) = -\frac{ig}{3!}\int d\omega d\omega' d\omega'' \, [\bar{\chi}^{(2)}(z, \omega; \omega', \omega'')\hat{a}^\dagger(\omega')\hat{a}^\dagger(\omega'')\hat{a}(\omega) + \text{h.c.}], \tag{6.43}$$

where

$$\bar{\chi}^{(2)}(z, \omega; \omega', \omega'') = \frac{\delta(z)}{2}\sqrt{\frac{\omega\omega'\omega''}{2\,n(\omega)\,n(\omega')\,n(\omega'')}}\,\chi_{jk\ell}^{(2)}(\omega; \omega', \omega'')\hat{\mathbf{f}}_j\hat{\mathbf{f}}_k\hat{\mathbf{f}}_\ell, \tag{6.44}$$

---

[10] In deriving this expression for the interaction Hamiltonian, we have only retained energy-conserving terms, i.e., those terms obeying $\omega = \omega' + \omega''$, and we have implicitly made use of normal ordering.

and

$$\delta(z) = \exp\{i[k(\omega) - k(\omega') - k(\omega'')]z\}, \tag{6.45}$$

is the phase-matching function [7, 21, 23]. The two terms appearing in the expression of the interaction Hamiltonian above (together with their permutations) describe all possible second-order phenomena. The first term, for example, describes transfer of energy from the pump mode (where a photon is annihilated through the operator $\hat{a}(\omega)$) to both the signal and idler modes, where one photon is created in each mode thanks to the operators $\hat{a}^\dagger(\omega')$ and $\hat{a}^\dagger(\omega'')$. This corresponds to the process of *down-conversion*, where a photon of high energy (the pump photon) gets converted into two photons (the signal and idler) of lower energy.

The second term, instead, describes the opposite process, i.e., the combination of two low energy photons (signal and idler) to create a higher energy one (pump) and is known as *sum frequency generation* (SFG) if $\omega' \neq \omega''$, i.e., if the two photons come from different modes and *second-harmonic generation* (SHG) if $\omega' = \omega''$, i.e., if the two photons come from the same mode [7, 21]. The Feynman diagrams corresponding to the relevant (i.e., up to permutations) second-order nonlinear optical processes are depicted in figure 6.2.

An important point to make here is that the amplitude probability of all these processes is the same, since they are generated by the same interaction Hamiltonian. What distinguishes between the different processes (i.e., the different diagrams) is the boundary conditions, i.e., the choice of the initial state that the electromagnetic field finds itself in at the beginning of the interaction. For example, the diagram in figure 6.2(a), describing down-conversion is equivalent to the diagram in figure 6.2(b), describing parametric amplification of the signal mode. However, one can select which process to activate by suitably setting the initial conditions: by choosing an initial state of the form $|\psi_1\rangle = |1_{\omega_p}, 0_{\omega_s}, 0_{\omega_i}\rangle$, signifying that a single pump photon enters the nonlinear crystal (represented by the vertex dot in figure 6.2), while the signal and idler modes are in their vacuum states (panel (a)), the only possible process will be that of down-conversion. By instead taking the state $|\psi_2\rangle = |1_{\omega_p}, 0_{\omega_s}, 1_{\omega_i}\rangle$ as initial state (panel (b)), the presence of both a pump and idler photons at the input facet of the crystal allows the onset of parametric amplification instead of down-conversion.

(a)   (b)

Figure 6.2. Relevant Feynman diagrams for second-order processes. Both these diagrams obey the energy conservation rule $\omega_p = \omega_s + \omega_i$. The pump photon, with frequency ω_p, is represented in red, while the signal (ω_s) and idler (ω_i) are represented in blue and teal, respectively. (a) A pump photon gets converted into a signal and idler photon in a down-conversion process. (b) A pump and idler photon mix to create a signal photon. This process (called difference frequency generation (DFG), since $\omega_p - \omega_i = \omega_s$) can also be seen as parametric amplification of the signal photon, mediated by the idler photon. Permutations of both (a) and (b) give all the other possible Feynman diagrams for second-order processes.

6.2.2 Transition probability

We now have all the ingredients to calculate the transition probability for a $\chi^{(2)}$-process. This can be done, according to section 5.3, by calculating the n-point Green's function. Since $\chi^{(2)}$ nonlinearities involve three photons, the 3-point Green's function will suffice for this calculation. This can be calculated using equation (5.34) with $n = 3$, which gives the following result

$$G_3(\omega_p, \omega_s, \omega_i) = \mathcal{N}\langle 0|\hat{T}\left\{\hat{E}(\omega_p)\hat{E}(\omega_s)\hat{E}(\omega_i)\exp\left[-ik_0\int_0^L dz\ \hat{H}_{int}^{(2)}(z)\right]\right\}|0\rangle, \quad (6.46)$$

where $\mathcal{N}$ is a normalisation constant (the denominator in equation (5.34)) and now $\hat{T}$ plays the role of z-ordering operator, instead of time-ordering operator, since the field is evolving along the z-direction[11].

To find the explicit expression for $G_3(\omega_p, \omega_s, \omega_i)$ we first expand the exponential above in a power series with respect to the interaction parameter g, up to first order, to obtain

$$\exp\left[-ik_0\int_0^L dz\ \hat{H}_{int}^{(2)}(z)\right] = 1 - ik_0\int_0^L dz\ \hat{H}_{int}^{(2)}(z) + \mathcal{O}(g^2)$$

$$= 1 - \frac{g\Delta}{3!}\int d\omega d\omega' d\omega''\ X^{(2)}(\omega; \omega', \omega'')[\hat{a}^\dagger(\omega')\hat{a}^\dagger(\omega'')\hat{a}(\omega) \quad (6.47)$$

$$+ \text{h.c.}] + \mathcal{O}(g^2),$$

where $X^{(2)}(\omega; \omega', \omega'') = \bar{\chi}^{(2)}(z, \omega; \omega', \omega'')/\delta(z)$ and

$$\Delta = k_0\int_0^L dz\ \delta(z) = k_0 L \exp\left[i\left(\frac{\Delta k\ L}{2}\right)\right]\text{sinc}\left(\frac{\Delta k\ L}{2}\right), \quad (6.48)$$

and $\Delta k = k(\omega) - k(\omega') - k(\omega'')$ is the phase-matching condition [7, 23], i.e., the conservation of momentum during the interaction. Substituting the expansion (6.47) into equation (6.46), and expanding the electric field operators in terms of creation and annihilation operators gives the following (normal ordered[12]) result

$$G_3(\omega_p, \omega_s, \omega_i) = G_3^{(free)}(\omega_p, \omega_s, \omega_i) - \frac{\mathcal{N}\ g\Delta}{3!}\int d\omega d\omega' d\omega''\ X^{(2)}(\omega; \omega', \omega'')$$

$$\times \langle 0|\hat{E}(\omega_p)\hat{E}(\omega_s)\hat{E}(\omega_i)\hat{a}^\dagger(\omega')\hat{a}^\dagger(\omega'')\hat{a}(\omega)|0\rangle, \quad (6.49)$$

---

[11] This is consistent with the assumption that the nonlinear medium is infinitely extended along the transverse direction, and has length L along the propagation direction. This assumption, in fact, implies that the electric field varies with z-inside the crystal, as the nonlinear interaction can happen at any point along the crystal length.

[12] Normal ordering simply means, that one arranges products of operators in such a way that all creation operators are on the left of all annihilation operators. All other terms containing non-normal ordered products of operators are then set to zero. This ordering is quite useful for representing quantum fields, as it amounts to ignoring the vacuum field fluctuations, and allows one to write the field Hamiltonian simply as $\hbar\omega\hat{a}^\dagger\hat{a}$. For more information about the various kinds of ordering schemes in quantum field theory, the reader is referred to reference [26], or [27].

where $G_3^{(free)}(\omega_p, \omega_s, \omega_i) = 0$ is the 3-point Green's function of the free field, which is zero since it is the (normal ordered) expectation value of an odd number of operators [4, 5][13].

To calculate the expectation value $\langle 0|\hat{E}\hat{E}\hat{E}\hat{a}^\dagger\hat{a}^\dagger\hat{a}|0\rangle$ appearing in the second line above, we substitute equations (6.42) for the electric field operator in equation (6.49) and then make use of Wick's theorem to calculate the expectation value of the product of six operators. In doing so, we need the frequency domain representation of the Feynman propagator, which, according to equation (5.38), can be written as follows

$$D_F(\omega - \Omega) = [\hat{a}(\omega), \hat{a}^\dagger(\Omega)] = \delta(\omega - \Omega). \tag{6.50}$$

This result reflects the fact that the frequency plays the role of an active degree of freedom for the electric field, meaning that the nonlinear process can combine fields of different frequencies to generate new ones. Moreover, since we have assumed that we disregard the transverse spatial structure of the beam, and we are considering only the dynamics along the propagation direction where, by the transversality condition imposed by the Coulomb gauge, we can effectively set $\mathbf{k} = 0$ in equation (5.40) and integrate with respect to k_0 to obtain the expression above for the Feynman propagation in frequency domain.

Using this result, and applying Wick's theorem to equation (6.49) gives following result

$$\begin{aligned} \mathscr{G}_3 &= \langle 0|\hat{E}(\omega_p)\hat{E}(\omega_s)\hat{E}(\omega_i)\hat{a}^\dagger(\omega')\hat{a}^\dagger(\omega'')\hat{a}(\omega) \\ &= \mathscr{E}(\omega_p)\mathscr{E}(\omega_s)\mathscr{E}(\omega_i)\{\text{terms containing 1 propagator} \\ &\quad + \text{terms containing 2 propagators} \\ &\quad + \text{terms containing 3 propagators}\}. \end{aligned} \tag{6.51}$$

The 3-point correlation function defined in equation (6.49) contains, after having used equations (6.42) for the electric field operator, a total of eight terms. To better understand how Wick's theorem works, let us use equation (5.41) to calculate two representatives of these eight terms, namely $\langle 0|\hat{a}(\omega_p)\hat{a}^\dagger(\omega_s)\hat{a}^\dagger(\omega_i)\hat{a}^\dagger(\omega')\hat{a}^\dagger(\omega'')\hat{a}(\omega)|0\rangle$ and $\langle 0|\hat{a}(\omega_p)\hat{a}(\omega_s)\hat{a}^\dagger(\omega_i)\hat{a}^\dagger(\omega')\hat{a}^\dagger(\omega'')\hat{a}(\omega)|0\rangle$, and comment the result. According to equation (5.41), the first of these terms can be written as

$$\begin{aligned} &\langle 0|\hat{a}(\omega_p)\hat{a}^\dagger(\omega_s)\hat{a}^\dagger(\omega_i)\hat{a}^\dagger(\omega')\hat{a}^\dagger(\omega'')\hat{a}(\omega)|0\rangle \\ &= \langle 0|[: \hat{a}(\omega_p)\hat{a}^\dagger(\omega_s)\hat{a}^\dagger(\omega_i)\hat{a}^\dagger(\omega')\hat{a}^\dagger(\omega'')\hat{a}(\omega): \\ &\quad + D_F(\omega_p - \omega_s): \hat{a}^\dagger(\omega_i)\hat{a}^\dagger(\omega')\hat{a}^\dagger(\omega'')\hat{a}(\omega): \\ &\quad + \text{seven more similar terms containing one propagator} \\ &\quad + D_F(\omega_p - \omega_s)D_F(\omega_i, \omega): \hat{a}^\dagger(\omega')\hat{a}^\dagger(\omega''): \\ &\quad + \text{three more similar terms containing two propagators}]|0\rangle = 0, \end{aligned} \tag{6.52}$$

---

[13] This result simply follows from the fact, that the zero order expansion of equation (6.46) is $\langle 0|\hat{E}(\omega_p)\hat{E}(\omega_s)\hat{E}(\omega_i)|0\rangle \propto \langle 0|(\hat{a}_p - \hat{a}_p^\dagger)(\hat{a}_s - \hat{a}_s^\dagger)(\hat{a}_i - \hat{a}_i^\dagger)|0\rangle = 0$.

where the notation $:\hat{A}\hat{B}:$, again, denotes normal ordering (see chapter 5). Ultimately, this term evaluates to zero because the number of creation and annihilation operators inside the expectation value is not the same. Because of this, in fact, the terms proportional to one propagator contain the expectation value of a normal ordered product containing at least one annihilaiton operator (which evaluates to zero, by definition of normal ordering), while the terms containing two propagators are proportional to the expectation value of a product containing either all creation operators (as in this case) or all annihilation operators, which, for both cases, evaluates to zero.

If we apply equation (5.41) to the second expectation value chosen, namely $\langle 0|\hat{a}(\omega_p)\hat{a}(\omega_s)\hat{a}^\dagger(\omega_i)\hat{a}^\dagger(\omega')\hat{a}^\dagger(\omega'')\hat{a}(\omega)|0\rangle$, we get instead the following result

$$
\begin{aligned}
&\langle 0|\hat{a}(\omega_p)\hat{a}(\omega_s)\hat{a}^\dagger(\omega_i)\hat{a}^\dagger(\omega')\hat{a}^\dagger(\omega'')\hat{a}(\omega)|0\rangle \\
&= \langle 0|[D_F(\omega_p - \omega_i): \hat{a}(\omega_s)\hat{a}^\dagger(\omega')\hat{a}^\dagger(\omega'')\hat{a}(\omega): \\
&\quad + \text{ eight more similar terms containing one propagator} \\
&\quad + D_F(\omega_p - \omega_i)D_F(\omega_s - \omega'): \hat{a}^\dagger(\omega'')\hat{a}(\omega): \\
&\quad + \text{ three other terms containing two propagators} \\
&\quad + D_F(\omega_p - \omega_i)D_F(\omega_s - \omega')D_F(\omega'' - \omega)]|0\rangle \\
&= D_F(\omega_p - \omega_i)D_F(\omega_s - \omega')D_F(\omega'' - \omega).
\end{aligned}
$$

(6.53)

In this case, since the number of creation and annihilation operators appearing in the expectation value is the same, we are able to pair all creation and annihilation operators together in such a way that we can obtain a nonzero result, corresponding to the term that sees all pairs of creation and annihilation operators paired together, i.e., a three-propagator term.

To calculate the other terms, one can follow the same line of reasoning. In general, it is useful to remember that the only nonzero expectation value containing four operators reduces to the product of two Feynman propagators, i.e.,

$$
\begin{aligned}
\langle 0|\hat{a}^\dagger(\omega_1)\hat{a}^\dagger(\omega_2)\hat{a}(\omega_3)\hat{a}(\omega_4)|0\rangle &= D_F(\omega_1 - \omega_3)D_F(\omega_2 - \omega_4) + \text{permutations} \\
&= \delta(\omega_1 - \omega_3)\delta(\omega_2 - \omega_4) + \text{ permutations}.
\end{aligned}
$$

(6.54)

From this result, it is easy to see that any term in equation (6.51) containing one single Feynman propagator gives a nonzero result only if is of the form

$$
\begin{aligned}
P_1 &= \langle 0|\hat{a}^\dagger(\omega_i)\hat{a}^\dagger(\omega')\hat{a}^\dagger(\omega'')\hat{a}(\omega)\hat{a}(\omega_p)\hat{a}(\omega_s)|0\rangle \\
&= D_F(\omega_i - \omega)\langle 0|\hat{a}^\dagger(\omega')\hat{a}^\dagger(\omega'')\hat{a}(\omega_p)\hat{a}(\omega_s)|0\rangle,
\end{aligned}
$$

(6.55)

which reduces to a term containing three propagators. The terms containing two propagators in equation (6.54), on the other hand, are all zero, since they do not contain the same number of creation and annihilation operators (they do, in fact, contain either four $\hat{a}^\dagger$ terms and two $\hat{a}$-terms or some other unequal combination), which, once contracted, give something proportional to two propagators times with $\langle \hat{a}^\dagger\hat{a}^\dagger\rangle$ or $\langle \hat{a}\hat{a}\rangle$, which both evaluate to zero.

Therefore, the only surviving terms in equation (6.51) are those with three propagators (including those of the form (6.54)), and the expectation value evaluates to

$$\mathcal{G}_3 = 3!\delta(\omega_p - \omega)\delta(\omega_s - \omega')\delta(\omega_i - \omega''), \tag{6.56}$$

where the factor 3! accounts for the fact that all the permutations, once inserted in the integral in equation (6.49) give the same result, up to renaming of dummy variables. Substituting this result into equation (6.51) gives the final result

$$G_3(\omega_p, \omega_s, \omega_i) = -\mathcal{N}g\Delta X^{(2)}(\omega_s + \omega_i; \omega_s, \omega_i)\mathcal{E}(\omega_p)\mathcal{E}(\omega_s)\mathcal{E}(\omega_i). \tag{6.57}$$

This represents the transition amplitude for a general $\chi^{(2)}$ process. The probability for this process to happen is given by the modulus square of the above relation, i.e.,

$$\begin{aligned} P(\omega_s, \omega_i) &= |G_3(\omega_s, \omega_i)|^2 \\ &= \mathcal{N}^2 g^2 \, |X^{(2)}(\omega_s + \omega_i; \omega_s, \omega_i) \, \Delta|^2 \mathcal{E}^2(\omega_s + \omega_i)\mathcal{E}^2(\omega_s)^2\mathcal{E}^2(\omega_i). \end{aligned} \tag{6.58}$$

6.2.3 Quantum state of the field after the interaction

We now tackle the problem of how to calculate the quantum state of the field after the nonlinear interaction (for a $\chi^{(2)}$-process). As mentioned in the discussion below equation (6.45), to do this we need to specify the boundary conditions, i.e., what is the state of the field at the beginning of the interaction (i.e., at the input facet of the nonlinear crystal). Doing so will automatically select the appropriate Feynman diagram amongst those represented in figure 6.2 and the interaction will produce a suitable output state, compatible with the input state and the given interaction. However, as discussed above, the transitional probability is independent of the particular process, but it is the same for all the second-order processes.

As an example, we discuss the case of parametric down-conversion. whose Feynman diagram is given in figure 6.2(a), and we assume that at $z = 0$ the electromagnetic field is in the state $|\psi_{\text{initial}}\rangle = |1_p, 0_s, 0_i\rangle = \hat{\psi}(0, \omega)|0\rangle^{14}$, where $\hat{\psi}(0, \omega) = \hat{a}^\dagger(\omega_p)$ is the *state operator*, i.e., the operator defining the initial state of the system. According to the theory developed in chapter 5, we can look at the evolution of a quantum system using the Heisenberg equation of motion (equation (5.16)), which evolves operators in time (or, as in this case, propagation direction), rather than evolving the state itself[15], meaning that to find the expression of the quantum state of the electromagnetic field after the interaction we first need to

---

[14] Please note that here we only mention the relevant electromagnetic modes, i.e., pump, signal and idler. We also assume that all the other modes at all the other (infinitely many) frequencies are also in the vacuum state, but omit their presence for convenience.

[15] We are implicitly using the Heisenberg representation of quantum mechanics, where operators are evolving, while states remain constant in time, as opposed to the Schrödinger equation, where the state evolves in time and operators are time independent [14].

determine how the state operator at $z = 0$ evolves under the nonlinear Hamiltonian using equation (5.16)[16], i.e.,

$$\frac{\partial \hat{\psi}(z, \omega)}{\partial z} = \frac{k_0}{i} \left[\hat{\psi}(z, \omega), \hat{H}_{\text{int}}^{(2)}(z) \right], \qquad (6.59)$$

where $k_0 = 1/\bar{\lambda}$ plays the role of $\hbar$, since evolution is taking place in propagation direction, instead of time [17]. We can then write the final state as $|\psi_{\text{final}}\rangle = \hat{\psi}(z, \omega)|0\rangle$. The general solution of the equation above reads [14], for an electromagnetic field propagating in a crystal of length L,

$$\hat{\psi}(L, \omega) = \exp\left[ik_0 \int_0^L dz \, \hat{H}_{\text{int}}^{(2)}(z) \right] \hat{\psi}(0, \omega) \exp\left[-ik_0 \int_0^L dz \, \hat{H}_{\text{int}}^{(2)}(z) \right]. \quad (6.60)$$

To find the explicit expression for $\hat{\psi}(z, \omega)$ we can again use the fact, that $g \ll 1$ in the expression of the interaction Hamiltonian (see equation (6.39)) holds, and expand both exponentials using equation (6.47) up to first order in g, to obtain,

$$\begin{aligned}
\hat{\psi}(L, \omega) &= \hat{\psi}(0, \omega) + ik_0 \int_0^L dz \, \left[\hat{\psi}(0, \omega), \hat{H}_{\text{int}}^{(2)}(z) \right] + \mathcal{O}(g^2) \\
&= 2\hat{\psi}(0, \omega) - \frac{g\Delta}{3!} \int d\omega' \, d\omega'' \left[X^{(2)}(\omega_p; \omega', \omega'') \hat{a}^\dagger(\omega') \hat{a}^\dagger(\omega'') \right. \\
&\quad \left. + (2X^{(2)}(\omega'; \omega_p, \omega''))^* \hat{a}(\omega'') \hat{a}^\dagger(\omega') \right],
\end{aligned} \qquad (6.61)$$

where in deriving the above expression we have assumed that $X^{(2)}(\omega; , \omega', \omega'') = X^{(2)}(\omega; , \omega'', \omega')$ [7]. Notice, that the expression of $\hat{\psi}(z, \omega)$ contains two terms: the first one, describing the diagram in figure 6.2(a) and corresponding to parametric down-conversion, sees the conversion of one pump photon into a signal (ω'') and idler (ω') photon pair. The second term, instead, corresponds to the Feynman diagram in figure 6.2 and represents DFG, where one pump photon mixes with an idler (ω') photon to give a signal photon (ω'').

The state of the electromagnetic field after the nonlinear interaction can then be calculated explicitly by applying the state operator $\hat{\psi}(L, \omega)$ to the vacuum state, obtaining

$$|\psi_{\text{final}}\rangle = \hat{\psi}(L, \omega)|0\rangle = |\psi_{\text{initial}}\rangle - \frac{g\Delta}{3!} \int d\omega' \, d\omega'' \, X^{(2)}(\omega_p; \omega', \omega'')|0_p, 1_{\omega'}, 1_{\omega''}\rangle. \quad (6.62)$$

Notice how the second term in equation (6.61) evaluates to zero when applied to the vacuum state, due to the presence of the annihilation operator $\hat{a}(\omega'')$. This, once again, reflects the fact that, although the state operator $\hat{\psi}(L, \omega)$ contains information on the states that every second-order process can generate, only that state compatible with the initial condition will actually evaluate to something that is nonzero. By changing the initial state $|\psi_{\text{initial}}\rangle$ one can calculate all the possible states

---

[16] Notice that we have neglected a term proportional to $[\hat{\psi}(z, \omega), \hat{H}_0]$, i.e., the free evolution of the state operator, since this only amounts to a global phase factor, and can therefore be neglected [3, 14].

that can be generated by a second-order nonlinear process by adapting equation (6.60) to the initial state operator $\hat{\psi}(0, \omega)$ corresponding to the particular choice of the initial state $|\psi_{\text{initial}}\rangle$. This process, moreover, can also be adapted to higher-order nonlinear processes, by replacing the second-order Hamiltonian in equation (6.60) with the appropriate interaction Hamiltonian describing the nonlinear process at hand.

6.2.4 Third-order nonlinearities

Similarly to second-order nonlinearities, third-order nonlinearities are characterised by a rank-4, frequency-dependent nonlinear susceptibility $\chi^{(3)}_{jk\ell m}(\omega_p; \omega_1, \omega_2, \omega_3)$ and involve the interaction of four fields, so that the interaction Hamiltonian describing third-order nonlinearities is given by

$$\hat{H}^{(3)}_{\text{int}} = \frac{g}{4!} \int d\Omega \; \chi^{(3)}_{jk\ell m}(\omega; , \omega', \omega'', \omega''') \; \hat{E}_j(\omega)\hat{E}_k(\omega')\hat{E}_\ell(\omega'')\hat{E}_m(\omega'''), \quad (6.63)$$

where $d\Omega$ is shorthand for $d\omega \, d\omega' \, d\omega'' \, d\omega'''$. The Feynman diagrams for third-order nonlinear processes are depicted in figure 6.3. The calculations of the transition probability follow the same line of reasoning of section 6.2.1, except for the fact that one needs to calculate $G_4(\omega)$ instead of $G_3(\omega)$ to get the transition probability, as the nonlinearity contains four, instead of three, fields. By repeating the same calculations as in section 6.2.1 using the interaction Hamiltonian (6.63) instead of (6.39) one arrives at the following expression for the transition probability for third-order nonlinear processes

$$\begin{aligned} P(\omega_1, \omega_2, \omega_3) &= |G_4(\omega_1, \omega_2, \omega_3)|^2 \\ &= \mathcal{N}^2 g^2 |X^{(3)}(\omega_1 + \omega_2 + \omega_3; \omega_1, \omega_2, \omega_3)\Delta|^2 \\ &\times \mathscr{E}^2(\omega_1)\mathscr{E}^2(\omega_2)^2\mathscr{E}^2(\omega_3)\mathscr{E}^2(\omega_1 + \omega_2 + \omega_3), \end{aligned} \quad (6.64)$$

where now Δ contains the third-order phase-matching condition, i.e., $\Delta k = k(\omega_1) - k(\omega_2) - k(\omega_3) - k(\omega_4)$.

Figure 6.3. Relevant Feynman diagrams for third-order processes. Both these diagrams obey the energy conservation rule $\omega_4 = \omega_1 + \omega_2 + \omega_3$. The four fields involved in the interaction are represented in red (ω_4), blue (ω_1), orange (ω_2), and teal (ω_3). (a) A photon with frequency ω_4 (red) gets converted into a blue (ω_1), orange (ω_2), and teal (ω_3) photon triplet, in a down-conversion process. (b) A photon with frequency ω_4 (red) and one of frequency ω_3 (teal) mix to create a blue (ω_1) and orange (ω_2) photon pair. This process is called four-wave mixing (FWM). Permutations of both (a) and (b) give all the other possible Feynman diagrams for third-order processes.

6.3 Part III: path integrals quantisation of the electromagnetic field

We now return to the problem of quantising the free electromagnetic field. Here, however, instead of pursuing the canonical (i.e., Hamiltonian) way presented in part I, we approach the problem from a Lagrangian perspective, i.e., through path integrals. Unlike section 6.1, where we choose to work in the Coulomb gauge, due to the nonrelativistic nature of most of the experiments in photonics, in this section we consider the fully relativistic situation, since this is the natural framework of path integrals anyway. In this way, we can remain within the traditional framework of QFT and directly use some of the techniques developed there (see, e.g., references [4–6]) to deal with field quantisation.

As discussed in section 5.5.1, the starting point for describing the properties of the electromagnetic field is the partition function

$$Z[J] = \int \mathcal{D}A_\mu \, \exp\left\{-i \int d^4x \left[-\frac{1}{4}F_{\mu\nu}F^{\mu\nu} + A_\mu J^\mu\right]\right\}. \tag{6.65}$$

The quantisation of the field, in this case, is taken care by the path integration with respect to the vector potential A_μ. Once the explicit expression of $Z[J]$ is known, we can generate any correlation function of the free and interacting electromagnetic field by taking derivatives of $Z[J]$ with respect to the current J, as illustrated in chapter 5.

Since we are dealing with a free field, the integrand in the exponent of equation (6.65) is a quadratic form in the vector potential A_μ[17], it would be tempting to solve the path integral using Gaussian integrals, as done in chapter 5. However, contrary to the case of a scalar field discussed in chapter 5, here we also need to take into account the gauge invariance of the partition function, since $Z[J]$ is written as a path integral in terms of the vector potential, instead of the electric or magnetic fields (which are, instead, gauge invariant quantities).

If we apply the gauge transformation $A'_\mu = A_\mu + \partial_\mu\Lambda$ to the vector potential appearing in equation (6.65), the free Lagrangian term $F_{\mu\nu}F^{\mu\nu}$ remains invariant under such transformation, while the current term becomes $A_\mu J^\mu + \partial_\mu\Lambda J^\mu$. This is not a problem, since the extra term $\partial_\mu\Lambda J^\mu$ can be neglected. To see this, just integrate by parts, and use the fact that $\partial_\mu J^\mu = 0$, i.e., J^μ obeys the continuity equation [12]. Up to this point, everything seems fine. However, if we think about the path integration appearing in equation (6.65), we quickly realise that the integration is extended to *all* possible configurations A_μ of the vector potential, regardless if they are connected by a gauge transformation or not. As a result of this, when we integrate over A_μ, all those vector potentials that are linked by a gauge transformation are essentially counted twice, once as the configuration A_μ and once as the configuration $A'_\mu = A_\mu + \partial_\mu\Lambda$. Since there are infinitely many gauge functions Λ

---

[17] This can be easily checked by integrating by part the term $F_{\mu\nu}F^{\mu\nu}$ and write the result as $A^\mu R_{\mu\nu} A_\nu$, where $R_{\mu\nu} = \delta_{\mu\nu}\partial^\alpha\partial_\alpha - \partial_\mu\partial_\nu$. This calculation is explicitly made in chapter 7 for the dressed electromagnetic field (see equation (7.106)).

that connect A_μ and A'_μ, this results in overcounting infinitely many configurations, thus making the path integration to always give an infinite result. Of course, the easy solution to this problem would be to fix a gauge *a priori* (e.g., the Coulomb gauge), for example by following the gauge-fixing procedure highlighted in section 4.2.1, which results in avoiding the overcounting problem because, practically, eliminates all those gauge-connected configurations of the vector potential from the domain of the path integration. Although this way of fixing the overcounting problem by specifying a gauge works nicely for the electromagnetic field for all practical problems in photonics, it cannot be easily generalised to other gauge fields, or to situations in which one would like to postpone the choice of gauge until the very end of the quantisation procedure. In both these cases, a more general approach, that doesn't rely on choosing a gauge *a priori*, is needed[18].

The general method of quantising a gauge field through path integrals, without specifying *a priori* a choice of gauge is known in literature with the name of *Faddeev–Popov quantisation* [28, 29], and will be the focus of this last part of the chapter. However, given the complexity of the topic, here we only present the Faddeev–Popov quantisation for the electromagnetic field, as the application of the method simplifies noticeably. For a more general discussion on the application of this procedure to the full non-Abelian case of a Yang–Mills gauge theory, we refer the interested reader to the books by Ryder [6] or Faddeev [28], for a traditional derivation, or the book by Srednicki [4] for a more modern approach to the topic[19].

6.3.1 Faddeev–Popov quantisation

The main problem of the partition function appearing in equation (6.65) is that the path integration counts all possible A_μ, thus overcounting those connected by a gauge transformation. In its present form, however, equation (6.65) offers little insight into this problem, as we can't really see how this problem manifests. Therefore, we need to rewrite the path integral in equation (6.65) in a way that it displays explicitly the overcounting problem. To do so, it is instructive to look at a gauge transformation from a slightly different perspective than usual, introducing the concept of equivalence relation and equivalence class.

Mathematically, an equivalence relation, indicated by the symbol $\sim$, is a binary relation between elements of a given set X, which has three fundamental properties: it is reflexive (given $a \in X$, then $a \sim a$), symmetric (if $a, b \in X$, then $a \sim b$ if and only if $b \sim a$) and transitive (given $a, b, c \in X$, if $a \sim b$ and $b \sim c$, then $a \sim c$). An equivalence relation also defines an *equivalence class* $[a] = \{x \in X : x \sim a\}$. The equivalance class $[a]$ contains all elements that are *the same* as a under the equivalence relation $\sim$.

---

[18] The necessity to have a general method of quantisation for gauge fields arises prevalently in QFT, where the Yang–Mills field, i.e., the quark field, is a non-Abelian gauge field. In this case, a more general gauge-fixing procedure, the Faddeev–Popov method, is needed in order to solve the problem.

[19] This topic, however, is deeply rooted in the geometrical nature of gauge fields, and finds an elegant and intuitive explanation in differential geometry. The interested reader is referred to references [30–32], for a more geometry-focused explanation of the Faddeev–Popov quantisation method.

A gauge transformation can be then seen as an equivalence relation between vector potentials, i.e., $A_\mu \sim A'_\mu$ if they are related by a gauge transformation, i.e., if $A'_\mu = A_\mu + \partial_\mu\Lambda$. To better understand this, think of it in terms of electric and magnetic fields: two vector potentials A_μ and A'_μ linked by a gauge transformation give rise to the same electric and magnetic fields. In this sense, then, A_μ and A'_μ are equivalent to each other. We can then use this equivalence relation to define an equivalence class for the vector potential, i.e.,

$$[A_\mu] = \{\bar{A}_\mu \in \mathcal{M} : \bar{A}_\mu \sim A_\mu\}, \tag{6.66}$$

where $\mathcal{M}$ is the space of all configurations of the vector potential. Practically, we have grouped all the vector potentials linked by a gauge transformation in a single set. Moreover, since all the $\bar{A}_\mu \in [A_\mu]$ are equivalent under $\sim$, we can take $\bar{A}_\mu$ as a representative of this equivalence class. This is illustrated schematically in figure 6.4. The set of all equivalence classes of $\mathcal{M}$ by the equivalence class $\sim$ (i.e., the gauge transformations) is denoted $\mathcal{M}/\sim$ and it is called the quotient set of $\mathcal{M}$ [33].

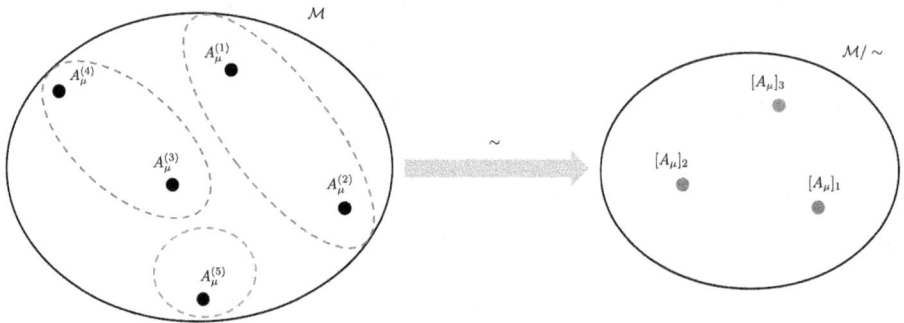

Figure 6.4. Schematic representation of the concept of equivalence relation, equivalence class, and quotient set. The set $\mathcal{M}$ on the left contains all possible configurations of the vector potential, here represented by five points, corresponding to the five configurations $A_\mu^{(1, 2, 3, 4, 5)}$. The vector potentials $A_\mu^{(1)}$ and $A_\mu^{(2)}$ are related by a gauge transformation, represented by the red, dashed ellipse encircling them. Both of these fields are then identified, under the equivalence relation $\sim$ generated by this (red) gauge transformation, to the red point $[A_\mu]_1$ in the quotient space $\mathcal{M}/\sim$. Similarly, the vector potentials $A_\mu^{(3)}$ and $A_\mu^{(4)}$ are linked by a different gauge transformation (represented by the blue, dashed ellipse) and they are identified, under the same equivalence relation $\sim$ (this time generated by the blue gauge transformation), to the blue point $[A_\mu]_2$ in the quotient space $\mathcal{M}/\sim$. Finally, the vector potential $A_\mu^{(5)}$ is linked only to itself by yet another gauge transformation (represented by the green, dashed ellipse) and is identified, under the same equivalence relation $\sim$ (generated by the green gauge transformation, which corresponds, in this case, to the identity) to the green point $[A_\mu]_3$ in the quotient space $\mathcal{M}/\sim$. Notice, that while $\mathcal{M}$ contains five different configurations of the vector potential, $\mathcal{M}/\sim$ only contains three (i.e., the three equivalence classes). The difference in number of elements between $\mathcal{M}$ and $\mathcal{M}/\sim$, i.e., 2, corresponds to the number of different gauge transformations (the identity, i.e., the green gauge transformation, does not count as one) needed to group up the vector potentials in equivalence classes. Introducing equivalence classes then broke down the set $\mathcal{M} = \{A_\mu^{(1)}, A_\mu^{(2)}, A_\mu^{(3)}, A_\mu^{(4)}, A_\mu^{(5)}\}$ into the product of the quotient set $\mathcal{M}/\sim=\{[A_\mu]_1, [A_\mu]_2, [A_\mu]_3\}$ times the set containing the gauge functions necessary to define the equivalence classes, i.e., $\mathcal{M}_\Lambda = \{\Lambda_1(\text{red}), \Lambda_2(\text{blue})\}$, i.e., $\mathcal{M} = (\mathcal{M}/\sim) \times \mathcal{M}_\Lambda$. Therefore, the integral measure $\mathscr{D}A_\mu$ can be written as $\mathscr{D}\bar{A}_\mu\mathscr{D}\Lambda$, as in equation (6.67).

Intuitively, this procedure allows us to *factor out* the gauge transformations from the space $\mathcal{M}$. This allows us to reduce the domain of integration of Z from $\mathcal{M} = \{A_\mu\}$ (i.e., all possible configurations of the vector potential) to $\mathcal{M}/\sim \equiv \{\bar{A}_\mu\} \otimes \{\Lambda\}$ (i.e., to the space of the equivalence classes $\bar{A}_\mu$ defined by the gauge functions Λ), meaning that now the integration measure $\mathcal{D}A_\mu$ can be changed into $\mathcal{D}\bar{A}_\mu \mathcal{D}\Lambda$, where the integration over $\bar{A}_\mu$ now runs over all the equivalence classes, and the integration over Λ over all possible gauge transformations defining such classes.

This modifies the path integral in equation (6.65) as follows

$$Z = \int \mathcal{D}\bar{A}_\mu \exp\left[i \int d^4x(\mathcal{L} + \bar{A}_\mu J^\mu)\right]\int \mathcal{D}\Lambda, \qquad (6.67)$$

which highlights the problematic part, i.e., the path integration over Λ. This is the origin of the overcounting: while performing the path integration over A_μ, in fact, we not only integrate over all the equivalence classes (i.e., $\bar{A}_\mu$), but also over all possible gauge transformations. These, however, are already accounted for by $\bar{A}_\mu$ and result into counting the same term multiple times, i.e, in $\int \mathcal{D}\Lambda \to \infty$.

Before tackling this problem, let us neglect, for the sake of simplicity, the source term in equation (6.94), as it doesn't add anything relevant to the discussion, but only slightly complicates the formalism. Then, the path integral reduces to

$$Z = \int \mathcal{D}\bar{A}_\mu \exp\left[i \int d^4x \mathcal{L}\right]\int \mathcal{D}\Lambda. \qquad (6.68)$$

As discussed above, the overcounting problem stems from the fact that the integral in $\mathcal{D}\Lambda$ diverges. One possible solution to this would be to insert a regularisation factor in the Λ-integral, so that the integral would now converge to a finite result. In principle, the choice of the regularisation factor is arbitrary. Here, however, we want to choose a regularisation factor that leaves the overall path integral a quadratic function of the vector potential, so that Gaussian integration can still be performed. Therefore, inspired by the Gaussian regularisation of the Dirac delta function [13]

$$\delta(x) = \lim_{\varepsilon \to 0}\left[\frac{1}{\varepsilon\sqrt{\pi}}\exp\left(-\frac{x^2}{2\varepsilon^2}\right)\right], \qquad (6.69)$$

we can choose the regularisation factor as follows

$$\int \mathcal{D}\Lambda \to \int \mathcal{D}\Lambda \exp\left(-\frac{i}{2\xi}G^2(A_\mu)\right), \qquad (6.70)$$

where $G(A_\mu)$ is an arbitrary, but well-behaved, functional of the vector potential[20], and $\xi \in \mathbb{R}$.

If we add this regularisation term in equation (6.67), the integral in Z now converges, but the partition function becomes dependent on the explicit form of the functional G. This is not optimal, since *a priori* we don't have any control or prior knowledge of the explicit form of $G(A_\mu)$. This issue, however, is easily solved by using the fact that $G(A_\mu) = G(A_\mu(\Lambda))$ by virtue of gauge transformations. This allows us to practically perform a change of coordinates from the gauge function Λ to the functional G, i.e.,

$$ \int \mathscr{D}\Lambda \;\to\; \int \mathscr{D}G \, \exp\left(-\frac{i}{2\xi}G(A_\mu)\right) = \int \mathscr{D}\Lambda \, \det\left(\frac{\partial G}{\partial \Lambda}\right) \exp\left(-\frac{i}{2\xi}G(A_\mu)\right), \quad (6.71) $$

where $\det(\partial G/\partial \Lambda)$ is the Jacobian of the coordinate transformation that links G and Λ [13]. The partition function then assumes the following form[21]

$$ Z = \int \mathscr{D}A_\mu \, \exp\left[i \int d^4x \left(\mathscr{L} - \frac{1}{2\xi}G^2(A_\mu)\right)\right] \int \mathscr{D}\Lambda \, \det\left(\frac{\partial G}{\partial \Lambda}\right). \quad (6.72) $$

Before proceeding any further, let us reflect a moment on the physical meaning of the term $\partial G/\partial \Lambda$. This, as stated above, is the Jacobian of the coordinate transformation linking the functional G to the gauge function Λ. This transformation of coordinates is nothing other than a gauge transformation. Since $G(A_\mu)$ is a function of the vector potential, under a gauge transformation it will transform as

$$ G(A'_\mu) = G(A_\mu + \partial_\mu\Lambda) \simeq G(A_\mu) + M\,\Lambda + \cdots, \quad (6.73) $$

where M will in general be a differential operator. If we, on the other hand, see $G(A'_\mu)$ as an implicit function of Λ (because of the gauge transformation), we have that

$$ G(A'_\mu) = G(A_\mu + \partial_\mu\Lambda) \simeq G + \left(\frac{\partial G}{\partial \Lambda}\right)\Lambda + \cdots, \quad (6.74) $$

and we can therefore identify $\partial G/\partial \Lambda$ with the differential operator M, i.e.,

$$ \frac{\partial G}{\partial \Lambda} = M. \quad (6.75) $$

The partition function for the electromagnetic field then becomes

$$ Z = \int \mathscr{D}A_\mu \int \mathscr{D}\Lambda \, \det M \, \exp\left[i \int d^4x \left(\mathscr{L} - \frac{1}{2\xi}G^2(A_\mu)\right)\right]. \quad (6.76) $$

---

[20] Notice, that $G(A_\mu)$ is implicitly a function of Λ since $A'_\mu = A_{mu} + \partial_\mu\Lambda$ holds. Therefore, $G(A_\mu) = G(A_\mu(\Lambda))$, which also justifies equation (6.71).

[21] We have dropped the $\tilde{\ }$ symbol from the vector potential, for simplicity.

Conceptually, equations (6.72) and (6.76) are exactly the same. Practically, now that we have established that $\det(\partial G/\partial \Lambda)$ is the determinant of a differential operator, we can give an explicit expression to it and find the final form of the partition function. This can be done by introducing the so-called *Grassmann fields* (see appendix 6.4.2), for which the following identity holds

$$\int \mathscr{D}\psi \mathscr{D}\psi^\dagger \exp\left(i \int \psi^\dagger M \psi\right) = \det M. \tag{6.77}$$

Substituting this equality into equation (6.76) leads to the following result

$$Z = \int \mathscr{D}A_\mu \mathscr{D}\psi \mathscr{D}\psi^\dagger \exp\left(i \int d^4x \, \mathscr{L}_{\text{eff}}\right), \tag{6.78}$$

where the effective Lagrangian density $\mathscr{L}_{\text{eff}}$ is defined as

$$\mathscr{L}_{\text{eff}} = -\frac{1}{4}F_{\mu\nu}F^{\mu\nu} - \frac{1}{2\xi}G^2(A_\mu) - \psi^\dagger M \psi. \tag{6.79}$$

The first term in the expression above is the usual electromagnetic Lagrangian density. The second term can be interpreted as a gauge-fixing term; for example, if we choose $G(A_\mu) = \partial_\mu A_\mu$, the gauge-fixing term reduces to that of equation (4.15), with ξ being the gauge parameter. The third term is known in literature as the Faddeev–Popov ghost Lagrangian density, and the field ψ, a scalar field with fermionic statistics (i.e., a Grassman field), is known as a ghost field. The appearance of ghost fields in the effective Lagrangian is the *price we have to pay* for quantising the electromagnetic field without choosing the gauge *a priori*.

A closer inspection of equation (6.79), however, reveals that there is no coupling between the vector potential A_μ and the ghost field ψ, meaning that the two path integrations can be done independently. This leaves the contrigution of the ghost field as an inessential multiplicative constant in the expression of Z, which can therefore be neglected by choosing a proper normalisation for Z. For this reason, in QFT people often say that the electromagnetic field is a ghost-free gauge field. This, however, is not always the case. For non-Abelian gauge fields, such as the Yang–Mills field, for example, this is not the case. If one would in fact repeat this quantisation procedure for a non-Abelian gauge field, one would find out that there is a coupling between the gauge field and the ghost field, and therefore ghost fields cannot be factored out of the partition function anymore[22].

If we then neglect the contribution of the ghost field and reinstate the source term, the Faddeev–Popov-quantised expression of the partition function for the electro-magnetic reads

$$Z[J] = \int \mathscr{D}A_\mu \exp\left[i \int d^4x \left(-\frac{1}{4}F_{\mu\nu}F^{\mu\nu} - \frac{1}{2\xi}G^2(A_\mu) + A_\mu J^\mu\right)\right]. \tag{6.80}$$

---

[22] See for example the book by Ryder [6] for a detailed discussion about the role of ghost fields in non-Abelian gauge fields.

This integral can now be safely computed, and the calculations are presented in section 6.3.3. Before doing that, we present an alternative interpretation of the Faddeev–Popov quantisation method, which only relies on the more familiar concept of Legendre transformation.

6.3.2 A more intuitive approach to Faddeev–Popov quantisation

The derivation of the Faddeev–Popov quantisation presented in the previous section is what Ryder in his book calls *heuristic* derivation [6]. The formal derivation, however, involves a deeper knowledge of group theory and functional theory, and it is outside the scope of this book. For the reader not familiar with QFT, however, even the heuristic derivation presented above might seem difficult to approach, and offering little intuition. For this reason, this section aims at presenting the Faddeev–Popov quantisation method under a different perspective, one that is more insightful and relies only on simple concepts known to most of the readers familiar with basic mathematics and physics, i.e., the Dirac delta function and the Legendre transform. This discussion follows closely that of reference [34].

The starting point of our discussion is the definition of the integral of the Dirac delta function [13]

$$\int dx\ \delta(x) = 1. \tag{6.81}$$

This expression also holds when the argument of the Dirac delta function is not simply x, but an arbitrary (well-behaved) function $g(x)$, that we assume, for the sake of this discussion, to have a single zero at x_0, i.e., $g(x_0) = 0$. Using then the well-known relation [13]

$$\delta(g(x)) = \left[\frac{\partial g(x)}{\partial x} \Big|_{x=x_0} \right]^{-1} \delta(x - x_0) \equiv \Delta^{-1}(x_0)\delta(x - x_0), \tag{6.82}$$

we can rewrite equation (6.81) as

$$\int dx\ \Delta(x_0)\delta(g(x)) = 1. \tag{6.83}$$

Notice, how the expression above gives us a resolution of the identity based on the Dirac delta function, i.e., it allows us to write 1 in a complicated manner as a function of a Dirac delta function.

We can use this result directly in equation (6.65), instead of going through the definition of equivalence classes and so on, by multiplying the integrand by 1 and then represent 1 as an integral over the gauge functions Λ using the result above, i.e.,

$$1 = \int \mathscr{D}\Lambda\ \Delta(A_\mu)\delta(G(A_\mu)), \tag{6.84}$$

where now the functional $G(A_\mu)$ plays the role of the function $g(x)$, and A_μ plays the role of x_0, and

$$\Delta(A_\mu) = \det\left(\frac{\partial G}{\partial \Lambda}\right)\bigg|_{\Lambda=0} \equiv \det(M).\qquad(6.85)$$

This *hand-wavy* way of applying the Faddeev–Popov quantisation condition is known in QFT literature as the Faddeev–Popov trick. At this point, a natural question arises: if we have the possibility of understanding Faddeev–Popov quantisation through the trick above, instead of the formal derivation (or the heuristic one presented in the previous section), can we also understand how and why it works, without making use of higher mathematics, that might not be, in general, familiar to us? The answer is rather simple: the Faddeev–Popov trick is nothing other than a Legendre transform of the partition function Z, from its non-gauge-fixed definition in equation (6.65), to the gauge-fixed definition of the partition function given by equation (6.80).

To prove our statement, we first need to give a suitable definition of Legendre transform. We have encountered this transform before, in chapter 2, where it was used to link the Lagrangian and Hamiltonian formulation of a given field theory (see equation (3.17)). Here, however, we give a slightly more general definition of it, in a form that is best suited to be used in conjunction with equation (6.83) [35]: given a convex function[23] $f(x)$ and two functions $F(x, p) = px - f(x)$ and $G(x, p) = \partial F(x, p)/\partial x$, we define the function $g(p)$ as the Legendre transform of $f(x)$ via the following relation

$$g(p) = F(x, p)|_{G(x, p)=0},\qquad(6.86)$$

where $G(x, p) = 0$ is intended to be solved with respect to x.

To fix ideas on this definition, let us first consider the following example. Take $f(x) = \mathscr{L}(y, x)$ to be the Lagrangian of a mechanical system characterised by generalised position y and generalised velocity x. Accordingly, we have that $F(x, p) = px - \mathscr{L}(y, x)$ and $G(x, p) = \partial F(x, p)/\partial x = p - \partial\mathscr{L}/\partial x$. The constraining condition $G(x, p) = 0$, solved with respect to x now reads $p = \partial\mathscr{L}/\partial x$, which is the definition of the canonical momentum. The Legendre transform then reads

$$g(p) = F(x, p)|_{G(x, p)=0} = [px - \mathscr{L}]_{p=\partial\mathscr{L}/\partial x} = \frac{\partial\mathscr{L}}{\partial x}x - \mathscr{L} \equiv \mathscr{H},\qquad(6.87)$$

which is the Hamiltonian of the same mechanical system, written as a function of the canonically conjugated position y and momentum p.

What is now left to do is to prove that the Faddeev–Popov trick is equivalent to taking the Legendre transform of the partition function. Rather than proving that equations (6.65) and (6.80) are related by a Legendre transform (which would involve lifting the definition of Legendre transfrom given above to functionals and deal with functional derivatives and functional distributions), we instead prove that applying the Faddeev–Popov trick to the expression

---

[23] A convex function is one, whose second derivative is always positive, i.e., $f(x)$ is convex, if $d^2f(x)/dx^2 \geqslant 0$.

$$\mathscr{L}(p) = \int dx \; F(x, p) \, \Delta(p, x_0) \, \delta(G(x, p)), \tag{6.88}$$

where $\Delta(x_0, p)$ is a function to be determined, gives exactly the Legendre transform in equation (6.87). This can be seen as the 'normal function' version of the partition function (6.65). In fact, upon identifying p with the vector potential A_μ, x with the gauge function Λ, and $F(x, p)$ with the exponential of the action, the integral above is then fully analogous to equation (6.76). The advantage of working on equation (6.88) instead of equation (6.65) is that in equation (6.88) we only have to deal with normal functions and apply the normal rules of calculus, without invoking any functional analysis. Moreover, if we prove that applying the Faddeev–Popov trick to equation (6.88) gives indeed the Legendre transform of the function $f(x)$ used to define $F(x, p)$, then proving that the same is true for functionals follows straightforwardly.

To apply the Faddeev–Popov trick to equation (6.88), let us first assume that $G(x, p) = 0$ occurs at $x = x_0$, and that the function $G(x, p)$ does not admit any other zeros. Then, we can use equation (6.82) to rewrtie the Dirac delta, obtaining

$$\mathscr{L}(p) = \int dx \; F(x, p)\Delta(x_0, p)\frac{\delta(x - x_0)}{|(\partial G(x, p)/\partial x)_{x=x_0}|}. \tag{6.89}$$

The quantiy $\Delta(x_0, p)$ is independent on the integration variable x and amounts to an arbitrary constant that, without loss of generality, we can choose to be $\Delta(x_0, p) = |(\partial G(x, p)/\partial x)_{x=x_0}|$, so that we get rid of the scaling factor appearing when we go from $\delta(G(x, p))$ to $\delta(x - x_0)$. We can then carry on the integration, which leads to

$$\mathscr{L}(p) = \int dx \; F(x, p)\delta(x - x_0) = F(x_0, p). \tag{6.90}$$

Since x_0 is the solution of $G(x, p) = 0$, we can rewrite the result above in the following, equivalent form

$$\mathscr{L}(p) = \int dx \; F(x, p)\delta(x - x_0) = F(x_0, p) = F(x, p)\Big|_{G(x, p)=0} = g(p), \tag{6.91}$$

which is precisely the Legendre transformation defined in equation (6.86).

With the same argument, but extended to the case of functionals instead of functions, we can immediately go from the partition function in equation (6.65) to the form in equation (6.80) by simply taking the Legendre transform of the (non-gauge-fixed) integrand of the partition function (i.e., the term inside the path integral in equation (6.65)), using the gauge condition as the constraining function.

6.3.3 Partition function for the electromagnetic field

We conclude this chapter by calculating the explicit form of the partition function for the quantised electromagnetic field. After the Faddeev–Popov method has been applied to the electromagnetic functional, in fact, the path integration can now be safely performed using Gaussian integration, as discussed in chapter 5.

The effective Lagrangian emerging from the Faddeev–Popov quantisation is given by equation (6.79). As discussed in section 6.3.1, for the electromagnetic field the Fadeev–Popov ghosts only amount to a normalisation factor, so we can neglect the last term in equation (6.79), and add a source term instead, of the form $J^\mu(x)A_\mu(x)$ to account for possible field sources, obtaining

$$\mathscr{L}_{eff} = -\frac{1}{4}F_{\mu\nu}F^{\mu\nu} - \frac{1}{2\xi}(\partial^\mu A_\mu)^2 + J^\mu A_\mu, \tag{6.92}$$

where, for simplicity, we have chosen $G(A_\mu) = \partial^\mu A_\mu$ as the explicit expression for the gauge-fixing function.

Calculating the partition function is actually simpler in Fourier space, so we introduce the Fourier transform of the vector potential as

$$A_\mu(x) = \int \frac{d^4k}{(2\pi)^4}\tilde{A}_\mu(k)\exp(ikx). \tag{6.93}$$

Compared to the Lagrangian (4.5) for the free electromagnetic field, the Lagrangian above has the advantage to be naturally gauge-fixed, i.e., the gauge condition appears explicitly in the Lagrangian through the term $(\partial^\mu A_\mu)^2$. This is equivalent to adding the gauge-fixing term in equation (4.15) to the free Lagrangian (4.5).

Substituting equation (6.93) (and its equivalent for the current $J^\mu(x)$) into equation (6.92) we then get

$$\int d^4x\,\mathscr{L}_{eff} = \frac{1}{2}\int \frac{d^4k}{(2\pi)^4}\left\{-\tilde{A}_\mu(k)\left[k^2\delta^{\mu\nu} + \left(1 - \frac{1}{\xi}\right)k^\mu k^\nu\right]\tilde{A}_\nu(-k) \right.$$
$$\left. + \tilde{J}^\mu(k)\tilde{A}_\mu(-k) + \tilde{J}^\mu(-k)\tilde{A}_\mu(k)\right\}, \tag{6.94}$$

where $\tilde{J}^\mu(k)$ is the Fourier transform of the source term $J^\mu(x)$. Notice that in the second term within the square brackets the term proportional to –1 comes from the free term $F_{\mu\nu}F^{\mu\nu}$, while the term proportional to ξ comes from the gauge-fixing condition. Notice, moreover, that if we operate the substitution $\delta^{\mu\nu} \to g^{\mu\nu}$ in the expression above, we can immediately generalise the result of this calculation to the case of curved spacetime [36].

To calculate the path integral we can essentially repeat the same calculation as done for the scalar field in section 5.5.1, i.e., make a change of variables similar to equation (5.47), then perform the Gaussian integration to obtain the partition function. The assumption for this procedure to be viable, however, is that the operator appearing in the square bracket in the equation above is invertible, since performing the Gaussian integration implies calculating the determinant of said operator.

For the case of the electromagnetic field, however, the operator

$$k^2 g^{\mu\nu} + \left(1 - \frac{1}{\xi}\right)k^\mu k^\nu \equiv k^2 P^{\mu\nu}(k), \tag{6.95}$$

is not invertible, since it has a zero eigenvalue. This can be readily checked by recognising that $P^{\mu\nu}(k)$ is a projector operator and, as such, is idempotent, i.e.,

$$P^{\alpha\beta}(k)P^{\lambda}_{\beta}(k) = P^{\alpha\lambda}(k), \qquad (6.96)$$

which is the tensor equivalent of $P^2 = P$ [14]. Therefore, as a projector, $P^{\mu\nu}(k)$ can only have zero or one eigenvalues. Notice, moreover, that at least one of the eigenvalues of $P^{\mu\nu}(k)$ will always be zero if we choose $\xi = 0$, since

$$P^{\mu\nu}(k)k_{\nu} = g^{\mu\nu}k_{\nu} - \left(1 - \frac{1}{\xi}\right)\frac{k^{\mu}k^{\nu}k_{\nu}}{k^2} = k^{\mu} - k^{\mu} = 0. \qquad (6.97)$$

Physically, this condition is equivalent to the Landau (Lorentz) gauge $\partial^{\mu}A_{\mu} = 0$ and endows $P^{\mu\nu}(k)$ with the physical meaning of projecting the electromagnetic field in the subspace orthogonal to k_{ν} [24]. However, since it is also true that [4]

$$g_{\mu\nu}P^{\mu\nu}(k) = g_{\mu\nu}g^{\mu\nu} - g_{\mu\nu}\frac{k^{\mu}k^{\nu}}{k^2} = 4 - \frac{k_{\mu}k^{\mu}}{k^2} = 4 - 1 = 3, \qquad (6.98)$$

the other three eigenvalues of $P^{\mu\nu}(k)$ must be all equal to one, in order for the equality above to be true.

If we then assume that we work within the premises of the observations above, i.e., we restrict ourselves to the Lorentz gauge, we can compute the path integral (6.65), using equation (6.94) as the action of the free electromagnetic field in free space. In doing so, let us notice that the quadratic term in $\tilde{A}_{\mu}(k)$ appearing in equation (6.94) involves the projector $P_{\mu\nu}(k)$, for which $P^{\mu\nu}k_{\nu} = 0$ holds. If we then imagine decomposing $A_{\mu}(k)$ into a basis formed by a set of linearly-independent k_{μ}, by virtue of equation (6.97) the component of the vector potential along k_{ν} does not contribute to this term. Moreover, it does not contribute to the linear term either, since $\tilde{J}_{\mu}(k)$ must obey the continuity equation, i.e., $\partial_{\nu}J^{\nu} = 0 \rightarrow k_{\nu}\tilde{J}^{\nu} = 0$. This means, that the path integration $\int \mathcal{D}A_{\mu}$ appearing in equation (6.65) should be performed only on the other three components of the vector potential, for which $k^{\mu}\tilde{A}_{\mu}(k) = 0$, i.e., the Coulomb gauge condition $\nabla \cdot \mathbf{A} = 0$ holds [25]. Then, we can perform the path integral as a Gaussian integral in the vector potential $\tilde{A}_{\mu}(k)$ and obtain the following result, for the partition function for the free electromagnetic field

$$Z_0(J) = \exp\left[\frac{i}{2}\int\frac{d^4k}{(2\pi)^4}\,\tilde{J}_{\mu}(k)\frac{P^{\mu\nu}(k)}{k^2 + i\varepsilon}\tilde{J}_{\nu}(-k)\right], \qquad (6.99)$$

---

[24] On the other hand, if we choose $\xi = 1$ we get $P^{\mu\nu}(k)k_{\nu} = k^{\mu}$, which corresponds to the Feynman gauge.

[25] In general, this would correspond to the Lorentz gauge $\partial^{\mu}A_{\mu} = 0$, but we have the freedom to define A_0 as the non-contributing component of the vector potential, and reduce the Lorentz gauge condition to the Coulomb one. This is done mostly to deal with a quantisation procedure that reduces the quantised electromagnetic field to the usual quantum optical quantised field, rather than the more general, Lorentz covariant one.

where we have used the fact, that

$$[k^2 \, P^{\mu\nu}(k)] = -\frac{1}{k^2}\left[g_{\mu\nu} + (\xi - 1)\frac{k_\mu k_\nu}{k^2}\right] \equiv -\frac{\tilde{P}_{\mu\nu}}{k^2}, \qquad (6.100)$$

since $P^{\mu\nu}(k)$ essentially projects any 4-vector into a subspace perpendicular to a given k_ν (see equation (6.97)), and in that subspace, $P^{\mu\nu}(k)$ behaves as an identity matrix [4]. We can also obtain the position–space representation of the partition function by taking the inverse Fourier transform of the current, following equation (6.93), to obtain

$$Z_0(J) = \exp\left[\frac{i}{2}\int d^4x \, d^4y \, J_\mu(x)G_F^{\mu\nu}(x - y)J_\nu(y)\right], \qquad (6.101)$$

where

$$G_F^{\mu\nu}(x - y) = \int \frac{d^4k}{(2\pi)^4}\frac{\tilde{P}^{\mu\nu}(k)}{k^2 + i\varepsilon}\exp\left[ik(x - y)\right], \qquad (6.102)$$

is the Feynman propagator of the free electromagnetic field.

6.4 Feynman propagator for the electromagnetic field

We conclude this chapter by deriving the explicit expression of the Feynman propagator for the electromagnetic field, using both the canonical and path integral formalism. In section 2.2, equation (2.15), we have written the Green's function for the Helmholtz equation in terms of its Fourier representation as

$$G(\mathbf{r} - \mathbf{r}') = \left(\frac{1}{2\pi}\right)^3 \int d^3k \, \frac{\exp[i\mathbf{k}\cdot(\mathbf{r} - \mathbf{r}')]}{k_0^2 - k^2}. \qquad (6.103)$$

Here, we want to repeat this calculation, using the tools developed in chapter 5, as an example of application of them to the case of the electromagnetic field.

6.4.1 Feynman propagator in canonical formalism

The Feynman propagator can be calculated adapting equation (5.35) to the electromagnetic field as follows

$$G_F^{\mu\nu}(x - y) = \langle 0|\hat{T}\hat{A}^\mu(x)\hat{A}^\nu(y)|0\rangle. \qquad (6.104)$$

As can be seen, since we are now dealing with vector, rather than scalar, fields, the Feynman propagator is a rank-2 tensor. We can, however, repeat the same line of reasoning used in section 5.4.1 to calculate the quantity above. First, we expand the time-ordering operator to obtain the equivalent of equation (5.36)

$$G_{\mu\nu}(x - y) = \Theta(x^0 - y^0)[\hat{A}_\mu(x), \hat{A}_\nu(y)] + \Theta(y^0 - x^0)[\hat{A}_\mu(y), \hat{A}_\nu(x)]. \qquad (6.105)$$

Then, we calculate the commutator using the Fourier representation (6.3) of the vector potential to obtain

$$[\hat{A}_\mu^+(x), \hat{A}_\nu^-(y)] = \int d^3\tilde{k} \, \Lambda_{\mu\nu} \exp[-i\mathbf{k} \cdot (\mathbf{x} - \mathbf{y})]\exp[-ik(x^0 - y^0)], \quad (6.106)$$

where $\Lambda_{\mu\nu} = [\hat{e}(\mathbf{k})]_\mu[\hat{e}(\mathbf{k})]_\nu = [\hat{e}(\mathbf{k}) \cdot \hat{e}(\mathbf{k})]_{\mu\nu}$. Next, we use equation (5.38) to transform the three-dimensional k-integral above into a four dimensional one and substitute equations (5.38) and (6.106) into equation (6.105) to obtain, after some straightforward algebra

$$G_{\mu\nu}(x - y) = \int \frac{d^4k}{(2\pi)^4} \frac{i \, \Lambda_{\mu\nu}}{k^2 + i\varepsilon} \exp[-ik(x - y)]. \quad (6.107)$$

The expression above is very similar to that of a scalar field (see equation (5.40)), where now $\omega^2(\mathbf{k}) = k_0^2$ is the dispersion relation of the free electromagnetic field, and the term $\Lambda_{\mu\nu}$ takes into account the vector nature of the field. In flat, Minkowski, spacetime, the unit vectors $\hat{e}(\mathbf{k})$ are mutually orthogonal, since, as we discussed in section 6.1, they represent the two orthogonal polarisation states of the field. Then, in flat spacetime, $\Lambda_{\mu\nu} = \eta_{\mu\nu}$, where $\eta_{\mu\nu} = \text{diag}(-1, 1, 1, 1)$ is the Minkowski metric. The result above also allows us to write the propagator for the electromagnetic field in curved spacetime, by just replacing the Minkowski metric with the appropriate metric describing the curved spacetime at hand, so that $\Lambda_{\mu\nu} = g_{\mu\nu}$ [36].

Notice, however, that the expression for the Feynman propagator for the electromagnetic field is not, in general, gauge invariant, because equation (6.104) contains the vector potential. So, to properly define the Feynman propagator for the electromagnetic field, one should also specify the gauge in which this quantity is defined. A more general expression for the Feynman propagator in any arbitrary gauge can be obtained by using the following form for $\Lambda_{\mu\nu}$ [36]

$$\Lambda_{\mu\nu} = g_{\mu\nu} + k_\mu c_\nu(k) + k_\nu c_\mu(k), \quad (6.108)$$

where the explicit form of the 4-vector $c_\mu(k)$ depends on the particular choice of gauge. Using the same approach to gauge fixing as we introduced in section 4.2.1, we can rewrite the term above in a more convenient way as

$$\Lambda_{\mu\nu} = g_{\mu\nu} + (\xi - 1)\frac{k_\mu k_\nu}{k^2 + i\varepsilon}, \quad (6.109)$$

so that for $\xi = 1$ we retrieve the Landau gauge $\partial_\mu A^\mu = 0$, and for $\xi = 0$ we retrieve the Feynman gauge. The Coulomb gauge is then a special case of the Landau gauge with $A^0 = 0$.

Lastly, it is not difficult to see, that the propagator in equation (6.107) reduces to that of the Helmholtz equation defined in equation (2.15). If we, in fact, assume that the field is monochromatic, then the frequency of the field needs to be fixed, so that $k^0 = \omega = k_0$, and we can also assume $x^0 = y^0$, leading to $\exp[i\, k(x - y)]= \exp[i\, \mathbf{k}(\mathbf{x} - \mathbf{y}) - k_0(x^0 - y^0)] = \exp[i\, \mathbf{k} \cdot (\mathbf{x} - \mathbf{y})]$. With these assumptions, then, we can neglect the k^0-integral, which only gives a constant contribution to the propagator, and factor out the term $\Lambda_{\mu\nu}$, since it is independent of the integration variable, to rewrite equation (6.107) for a monochromatic field as

$$G_{\mu\nu}^{(monochr.)}(\mathbf{x} - \mathbf{y}) = \Lambda_{\mu\nu} \int \frac{d^3k}{(2\pi)^3} \frac{\exp[i\,\mathbf{k} \cdot (\mathbf{x} - \mathbf{y})]}{|\mathbf{k}|^2 - k_0^2 + i\varepsilon} \equiv \Lambda_{\mu\nu} G(\mathbf{x} - \mathbf{y}), \quad (6.110)$$

where $G(\mathbf{x} - \mathbf{y})$ is the Helmholtz propagator defined in equation (2.15).

6.4.2 Feynman propagator in path integral formalism

As seen in section 5.5.1, the propagator for a free field can be derived from the partition function by calculating its derivatives with respect to the external source. For the case of the electromagnetic field, then, we have, according to equation (6.101)

$$G_F^{\mu\nu}(x - y) = \frac{\partial^2 Z_0(J)}{\partial J_\mu(x) \partial J_\nu(y)} = \int \frac{d^4k}{(2\pi)^4} \frac{\tilde{P}^{\mu\nu}(k)}{k^2 + i\varepsilon} \exp\left[ik(x - y)\right], \quad (6.111)$$

which is in agreement with equation (6.102). It is worthy comparing this result to that obtained in section 6.4.1 within the canonical formalism. While in canonical formalism we had to specify the form of the tensor $\Lambda_{\mu\nu}$ in such a way that it accommodates a gauge choice (see equation (6.110)), if we derive the Feynman propagator directly from path integrals, the Faddeev–Popov quantisation procedure automatically generates the right expression for $\Lambda_{\mu\nu}$, i.e., $\tilde{P}_{\mu\nu}$, since it already contains the implicit choice of gauge (Lorentz, or, if we choose A_0 to be the superfluous component of the vector potential, Coulomb). Moreover, following the same line of reasoning of section 6.4.1 we can then rewrite the propagator above for a monochromatic field, obtaining equation (6.110), where now $\Lambda_{\mu\nu}$ is replaced by $\tilde{P}^{\mu\nu}(k) \equiv P^{\mu\nu}$, which is independent of k.

Appendix A: Coherent states of the quantum harmonic oscillator

In this appendix, we derive coherent states for the quantum harmonic oscillator based only on the noncommutativity of a suitable pair of canonically conjugated operators. First, we present the traditional calculation done using $\{\hat{x}, \hat{p}\}$, obtaining the usual Gaussian form of minimal uncertainty wave packets (i.e., coherent states) and afterwards we generalise the results for the creation and annihilation operators $\{\hat{a}, \hat{a}^\dagger\}$, deriving then equation (6.20).

To start with, let us recall some results of operator theory. Let us define $\hat{A}$ and $\hat{B}$ as a pair of self-adjoint, noncommuting operators, and let $\hat{\mathscr{A}} = \hat{A} - \langle \hat{A} \rangle$ and $\hat{\mathscr{B}} = \hat{B} - \langle \hat{B} \rangle$ measure the deviation of these operators from their average value. The variance of $\hat{A}$ and $\hat{B}$ can then be written as $\Delta A^2 = \langle \hat{\mathscr{A}}^2 \rangle$ and $\Delta B^2 = \langle \hat{\mathscr{B}}^2 \rangle$, respectively. Since these two operators do not commute, Robertson's theorem [37] applies, i.e.,

$$\Delta A \Delta B \geqslant \frac{1}{2} |\langle [\hat{A}, \hat{B}] \rangle|, \quad (6.112)$$

which is the general form of the uncertainty principle of quantum mechanics. Let us now consider a quantum system characterised by the wave function $\Psi(x)$ (assumed normalised to one), and introduce the transformation $\Phi(x) = (\hat{\mathscr{A}} + i\lambda\hat{\mathscr{B}})\Psi(x)$, where $\lambda \in \mathbb{R}$. This new wave function is handy, because its norm directly contains information on the uncertainty on the operators $\hat{A}$ and $\hat{B}$. Calculating the norm $\langle\Phi|\Phi\rangle$, in fact, gives

$$\langle\Phi|\Phi\rangle = \int dx \; \Phi^*(x)\Phi(x) = \Delta A^2 + i\lambda\langle[\hat{A},\hat{B}]\rangle + \lambda^2\Delta B^2 \equiv F(\lambda) \geqslant 0, \quad (6.113)$$

where the last inequality comes from the fact that the norm is positive definite by definition [14]. Notice, that $F(\lambda) \in \mathbb{R}$, since the commutator is in general a purely imaginary quantity, and can be written as $\langle[\hat{A},\hat{B}]\rangle \equiv i\langle\hat{C}\rangle$ [37]. Requiring that $F(\lambda) \geqslant 0$ is equivalent to equation (6.113). This can be seen by treating $F(\lambda)$ as a polynomial in λ. Then, for $F(\lambda)$ to be positive or zero, the discriminant of the associated λ-polynomial must be negative. One can easily check that this condition reduces to equation (6.112).

The function $F(\lambda)$, however, gives us direct access to the uncertainty in the operators $\hat{A}$ and $\hat{B}$, and can be used to find the states of minimal uncertainty, by looking for those values $\lambda_{\min}$ that minimise $F(\lambda)$, i.e., $dF(\lambda)/d\lambda\,|_{\lambda=\lambda_{min}} = 0$. Imposing this condition leads to

$$\lambda_{\min} = -\frac{i}{2}\frac{\langle[\hat{A},\hat{B}]\rangle}{\Delta B^2}, \quad (6.114)$$

and the states of minimal uncertainty are those that fulfill equation (6.113), with $\lambda = \lambda_{\min}$, i.e, those that are solution of the following equation

$$(\hat{\mathscr{A}} + i\lambda_{\min}\hat{\mathscr{B}})\Psi(x) = 0. \quad (6.115)$$

If we choose position and momentum of the harmonic oscillator as operators, i.e., we set $\hat{A} = \hat{x}$ and $\hat{B} = \hat{p}$, such that $\langle\hat{A}\rangle = 0$ and $\langle\hat{B}\rangle = \hbar k_0$ (with k_0 being some central wave vector), equation (6.114) becomes

$$\lambda_{\min} = \frac{\hbar}{2(\Delta p)^2} = \frac{2(\Delta x)^2}{\hbar}, \quad (6.116)$$

which leads to the following expression for equation (6.115)

$$(\hat{a} + i\lambda_{\min}\hat{b})\Psi(x) = \left[\hat{x} + i\frac{2(\Delta x)^2}{\hbar}(\hat{p} - \hbar k_0)\right]\Psi(x) = 0. \quad (6.117)$$

In position space, $\hat{p} = -i\hbar d/dx$ and the operator equation above becomes the following differential equation for $\Psi(x)$

$$\left[x - 2i(\Delta x)^2\left(-i\frac{d}{dx} + k_0\right)\right]\Psi(x) = 0, \quad (6.118)$$

whose solution is the well-known Gaussian state

$$\Psi(x) = \mathcal{N} \exp\left[-\frac{x^2}{4(\Delta x)^2} + ik_0 x\right], \tag{6.119}$$

with $\mathcal{N}$ a normalisation constant. The wave function above is that of the ground state of a quantum harmonic oscillator, which is now reinterpreted as the state of minimal uncertainty for the position–momentum pair of canonically conjugated operators. For a quantum harmonic oscillator, therefore, coherent states correspond to its own eigenstates (in position representation).

We now derive the expression of coherent states in terms of number states. To do so, we use equation (5.66) to link the position and momentum operators to the creation and annihilation operators, to then obtain

$$\hat{A} = \frac{1}{2}(\hat{a} + \hat{a}^\dagger), \tag{6.120a}$$

$$\hat{B} = \frac{1}{2i}(\hat{a} - \hat{a}^\dagger), \tag{6.120b}$$

where we have dropped the constants, for simplicity. Moreover, we choose the state $|0\rangle$ as the Fock state representation of the ground state of the oscillator (we can justify this choice in general, by saying that we take the state with the least numbers of excitations, i.e., zero, as the ground state of the quantum harmonic oscillator. This, moreover, is equivalent to say that $\Psi(x) = \langle x|0\rangle$). Combining this with the commutation rules $[\hat{a}, \hat{a}^\dagger] = 1$ we get, from equation (6.116), $\lambda_{min} = 1$, and equation (6.115) reduces to

$$\hat{a}|0\rangle = 0, \tag{6.121}$$

which is the definition of the annihilation operator, as the operator that destroys the vacuum state.

Proving that also the excited states of a harmonic oscillator are minimal uncertainty states goes along the same lines, with the difference that the starting state must be chosen to be the general oscillator state $|n\rangle$ (i.e., $\Psi_n(x) = \langle x|n\rangle$), instead of the ground state. Since the harmonic oscillator eigenstates are a complete set, we can write an arbitrary coherent state as a superposition of harmonic oscillator states as $\Psi_\alpha(x) = \langle x|\alpha\rangle = \sum_n c_n \Psi_n(x) = \sum_n c_n\langle x|n\rangle$ from which it follows $|\alpha\rangle = \sum_n c_n|n\rangle$, which is the Fock state representation of a coherent state, with $c_n \in \mathbb{C}$ to be determined. In this case, we choose

$$\hat{\mathcal{A}} = \frac{1}{2}(\hat{a} + \hat{a}^\dagger) - \left\langle \frac{1}{2}(\hat{a} + \hat{a}^\dagger) \right\rangle, \tag{6.122a}$$

$$\hat{\mathcal{B}} = \frac{1}{2i}(\hat{a} - \hat{a}^\dagger) - \left\langle \frac{1}{2i}(\hat{a} - \hat{a}^\dagger) \right\rangle, \tag{6.122b}$$

where the expectation values have to be calculated with respect to the coherent state $|\alpha\rangle$. Since $|\alpha\rangle$ is written as a superposition of minimal uncertainty states, $\lambda_{min} = 1$ still holds, and the characteristic equation (6.115) now becomes

$$\hat{a}|\alpha\rangle = \sum_{n=0}^{\infty} \frac{c_{n+1}\sqrt{n+1}}{c_n} c_n |n\rangle \equiv \alpha|\alpha\rangle, \qquad (6.123)$$

which is the definition of coherent states as eigenstates of the annihilation operator [1]. Notice, moreover, that the quantity $\alpha = c_{n+1}\sqrt{n+1}/c_n$ defines a recursion relation for the expansion coefficients of the coherent state in terms of the number states, which corresponds to

$$c_n = \frac{\alpha^n}{\sqrt{n!}} c_0, \qquad (6.124)$$

and c_0 can be determined by imposing the normalisation condition $\langle\alpha|\alpha\rangle = 1$, obtaining $c_0 = \exp(-|\alpha|^2/2)$. Putting everything together gives the expression of the coherent states $|\alpha\rangle$ of the electromagnetic field in terms of number states, which is equation (6.20), i.e.,

$$|\alpha\rangle = \exp\left(-\frac{|\alpha|^2}{2}\right) \sum_{n=0}^{\infty} \frac{\alpha^n}{\sqrt{n!}} |n\rangle. \qquad (6.125)$$

Appendix B: A primer on Grassmann variables and fields

The goal of this appendix is to introduce enough knowledge about Grassmann fields, to justify equation (6.77). Some of these concepts will be reused in chapter 8, when dealing with electrons in 2D materials. The idea of Grassmann fields essentially generalises the anti-commutativity of fermion-type particles (like electrons, for example) to QFT. In the canonical QFT approach, fermionic fields $\psi(x)$ are simply replaced with a set of anti-commuting operators, for which the (equal-time) anti-commutation relation

$$\{\psi(x), \psi(y)\} = \psi(x)\psi(y) + \psi(y)\psi(x) = 0, \qquad (6.126)$$

holds [6].

Since the path integral formalism doesn't make direct use of operators, but works instead with field configurations and their integrals, it would be convenient to have *anti-commuting* fields that describe fermions, instead of anti-commuting operators. These anti-commuting fields, called *Grassman fields* inherit their properties from a special set of anti-commuting *c-numbers*, known in mathematical literature as Grassmann numbers [38], which have the following properties

$$\{a_k, a_\ell\} = a_k a_\ell + a_\ell a_k = 0, \qquad (6.127a)$$

$$a_k^2 = 0, \qquad (6.127b)$$

where $\{k, \ell\} \in \{1, 2, ..., n\}$, with n being the dimension of the space the Grassmann numbers are considered [38]. The first expression defines their anti-commuting character, while the second expression allows one to express any function of Grassmann numbers as an, at most, linear combination of Grassman numbers. For $n = 2$, for example, there are only two Grassmann numbers, a_1 and a_2, and the most general function $f(a_1, a_2)$ an be written as follows

$$f(a_1, a_2) = c_0 + c_1 a_1 + c_2 a_2 + c_3 a_1 a_2 = c_0 + c_1 a_1 + c_2 a_2 - c_3 a_2 a_1, \quad (6.128)$$

with $c_k \in \mathbb{C}$. It is not hard to see, that for an arbitrary value of n, the second of equations (6.127) always allows representing any function $f(a_1, ..., a_n)$ with a finite number of terms, either proportional to a_k, or to $a_k a_\ell \cdots a_s$, with $k \neq \ell \neq \cdots \neq s$ [38].

Another striking property of Grassmann numbers, also deriving from the second of equations (6.127), is that integration and derivation are equivalent operations. For the function defined in equation (6.128), this translates to

$$\frac{\partial f}{\partial a_1} = \int da_1 \, f(a_1, a_2) = c_1 + c_3 a_2, \quad (6.129a)$$

$$\frac{\partial f}{\partial a_2} = \int da_2 \, f(a_1, a_2) = c_2 - c_3 a_1, \quad (6.129b)$$

To understand why is that so one would have to first define what $\int da_k$ means in terms of Grassmann numbers, and then calculate the integrals. To do this properly, however, one would need differential geometry tools, to show that if a_k are Grassmann numbers, then the differentials da_k are also Grassmann numbers, and that $\{a_k, da_k\} = 0$. Here, we do not demonstrate this, but we take it for granted, directing the interested reader to the book by Nakahara [39], for example, for a more rigorous discussion on the topic. As a result of this proper analysis, one would find out that

$$\int da_k = 0, \quad (6.130a)$$

$$\int da_k \, a_k = 1. \quad (6.130b)$$

Using these rules, one can easily figure out equations (6.129).

The same line of resoning can be applied to Grassmann-valued fields. If ψ and $\psi^\dagger$ are (complex) independent Grassmann-valued fields (shortly called from now on *Grassmann fields*), then, using the relations above, they have to obey the following set of rules

$$\psi^2 = (\psi^\dagger)^2 = 0, \quad (6.131a)$$

$$\int d\psi = \int d\psi^\dagger = 0, \quad (6.131b)$$

$$\exp\left(-\psi^\dagger\psi\right) = 1 - \psi^\dagger\psi, \tag{6.131c}$$

$$\int d\psi^\dagger d\psi \, \exp\left(-\psi^\dagger\psi\right) = 1. \tag{6.131d}$$

We can use these properties as operative definitions of exponentials and integrals of Grassmann fields, and build upon them to arrive at the result of equation (6.77). Moreover, from equation (6.131d) it follows that

$$\int d\psi^\dagger d\psi \, \exp\left(-\psi^\dagger\psi\right) = \int d\psi_1^\dagger d\psi_2^\dagger d\psi_1 d\psi_2 \, \psi_1^\dagger\psi_1\psi_2^\dagger\psi_2 = 1. \tag{6.132}$$

This is all we need to derive equation (6.77). To illustrate this, let us fix out attention to the particular case of two Grassmann fields ψ_1 and ψ_2. Generalisation of these results to higher numbers of fields is then straightforward.

To keep things clear and simple enough, we can introduce the vectors $\psi \equiv (\psi_1, \psi_2)^T$ and $\psi^\dagger \equiv (\psi_1^\dagger, \psi_2^\dagger)^T$ to describe the fields and their Hermitian conjugates in a compact manner, and write equation (6.131c) in terms of the two Grassmann fields as

$$\begin{aligned}
\exp\left(\psi^\dagger\psi\right) &= 1 - \psi^\dagger\psi = 1 - \left(\psi_1^\dagger\psi_1 + \psi_2^\dagger\psi_2\right) \\
&+ \psi_1^\dagger\psi_1\psi_2^\dagger\psi_2,
\end{aligned} \tag{6.133}$$

Now, let us perform a coordinate transformation that brings $\{\psi, \psi^\dagger\}$ to $\{\eta, \eta^\dagger\}$ according to $\psi = M\eta$, $\psi^\dagger = N\eta^\dagger$, where M, N are 2×2 matrices. The Grassmann fields then transform according to[26]

$$\begin{aligned}
\psi_1\psi_2 &= (M_{11}\eta_1 + M_{12}\eta_2)(M_{21}\eta_1 + M_{22}\eta_2) \\
&= (M_{11}M_{22} - M_{12}M_{21})\eta_1\eta_2 = \det M \, \eta_1\eta_2.
\end{aligned} \tag{6.134}$$

Since equation (6.132) must hold independently of the particular choice of Grassmann fields, it must in particular hold for the η_k fields as well. If we limit ourselves to the η_k fields[27], we then need to requite that[28]

$$d\psi_1 d\psi_2 = \frac{1}{\det M} d\eta_1 d\eta_2. \tag{6.135}$$

---

[26] For simplicity, we report only the transformation rules for the fields ψ_k. Those for the Hermitian conjugate fields $\psi_k^\dagger$ are obtained by simply replacing every entry of M with N.

[27] Again, the same result holds for the $\eta_k^\dagger$ fields, and, therefore, for any product of integrals involving η_k and $\eta_k^\dagger$ fields, i.e., equation (6.132).

[28] Notice the discrepancy with the usual change of coordinates, where the differentials transform with $\det M$, and not $1/\det M$.

repeating the same kind of calculations for the conjugated fields allows us to write

$$1 = \frac{1}{\det(MN)} \int d\eta^\dagger d\eta \, \exp\left(-\eta^\dagger N^T M \eta\right)$$

$$= \frac{1}{\det(M^T N)} \int d\eta^\dagger d\eta \, \exp\left(-\eta^\dagger N^T M \eta\right) \qquad (6.136)$$

$$\equiv \frac{1}{\det A} \int d\eta^\dagger d\eta \, \exp\left(-\eta^\dagger A \eta\right),$$

where to go from the first to the second line we have use the property of the determinant, that $\det(MN) = \det(M^T N)$, and then defined $A \equiv M^T N$ to rewrite the exponential as $\eta^\dagger N^T M \eta = \eta^\dagger M^T N \eta = \eta^\dagger A \eta$. This formula, then, can be generalised to any number of dimensions, and in the case of infinite dimensions gives exactly the result presented in equation (6.77).

References

[1] Gerry C and Knight P L 2004 *Introductory Quantum Optics* (Cambridge: Cambridge University Press)
[2] Mandel L and Wolf E 1995 *Optical Coherence and Quantum Optics* (Cambridge: Cambridge University Press)
[3] Loudon R 2000 *The Quantum Theory of Light* (Oxford: Oxford University Press)
[4] Srednicki M 2007 *Quantum Field Theory* (Cambridge: Cambridge University Press)
[5] Maggiore M 2005 *A Modern Introduction to Quantum Field Theory* (Oxford: Oxford University Press)
[6] Ryder L H 1996 *Quantum Field Theory* (Cambridge: Cambridge University Press)
[7] Boyd R W 2008 *Nonlinear Optics* 3rd edn (Amsterdam: Elsevier)
[8] Hillery M and Drummond P D 2014 *The Quantum Theory of Nonlinear Optics* (Cambridge: Cambridge University Press)
[9] Gupta S N 1950 Theory of longitudinal photons in quantum electrodynamics *Proc. Phys. Soc.* A **63** 681
[10] Bleuer K 1950 Eine neue methode zur behandlung der longitudinalen und skalaren photonen *Helv. Phys. Acta* **23** 567
[11] Itzykson C and Zuber J B 1980 *Quantum Field Theory* (New York: Dover)
[12] Jackson J D 1998 *Classical Electrodynamics* (New York: Wiley)
[13] Byron F W and Fuller R W 1992 *Mathematics of Classical and Quantum Physics* (Mineola, NY: Dover)
[14] Messiah A 2014 *Quantum Mechanics* (New York: Dover)
[15] Barnett S M and Radmore P M 2005 *Methods in Theoretical Quantum Optics* (Oxford: Oxford Science Publications)
[16] Boffi S 2010 *Da Laplace a Heisenberg. Una Introduzione alla Meccacnica Quantistica e alle Sue Applicazioni* (Biblioteca delle Scienze)
[17] Longhi S 2009 Quantum optical analogies using photonic structures *Laser Photon. Rev.* **3** 243
[18] Garrison J and Chiao R 2008 *Quantum Optics* (Oxford: Oxford University Press)
[19] Aiello A and Woerdman J P 2005 Exact quantization of a paraxial electromagnetic field *Phys. Rev.* A **72** 060101

[20] Wünsche A 2004 Quantization of Gauss-Hermite and Gauss-Laguerre beams in free space *J. Opt. B: Quantum Semiclass. Opt.* **6** S47

[21] Butcher P N and Cotter D 1990 *The Elements of Nonlinear Optics* (Cambridge: Cambridge University Press)

[22] Trillo S and Torruellas W (ed) 2001 *Spatial Solitons* (Berlin: Springer)

[23] Yariv A 1985 *Optical Electronics* (New Delhi: CBS College Publishing)

[24] Weiner A M 2009 *Ultrafast Optics* (New York: Wiley)

[25] Wegener M 2005 *Extreme Nonlinear Optics* (Berlin: Springer)

[26] Weinberg S 2005 *The Quantum Theory of Fields* (Cambridge: Cambridge University Press)

[27] Mandl F and Shaw G 1984 *Quantum Field Theory* (New York: Wiley)

[28] Faddeev L D and Slavnov A A 1991 *Gauge Fields: An Introduction to Quantum Field Theory* (Boulder, CO: Westview Press)

[29] Faddeev L D and Popov V 1967 Feynman diagrams for the Yang-Mills field *Phys. Lett.* B **25** 29

[30] Baez J and Muniain J P 1994 *Gauge Fields, Knots and Gravity* (Singapore: World Scientific)

[31] Nash C 1991 *Differential Topology and Quantum Field Theory* (New York: Academic)

[32] Göckeler M and Schücker T 1989 *Differential Geometry, Gauge Theories, and Gravity* (Cambridge: Cambridge University Press)

[33] Munkres J R 2017 *Topology* (London: Pearson Education)

[34] Ornigotti M and Aiello A 2014 The Faddeev-Popov method demystified arXiv:1407.7256

[35] Rockafellar R T 1970 *Convex Analysis* (Princeton, NJ: Princeton University Press)

[36] Birrel N D and Davies P C W 1984 *Quantum Fields in Curved Space* (Cambridge: Cambridge University Press)

[37] Robertson H P 1929 The uncertainty principle *Phys. Rev.* **34** 163

[38] DeWitt B 1984 *Supermanifolds* (Cambridge: Cambridge University Press)

[39] Nakahara M 2003 *Geometry, Topology, and Physics* (Boca Raton, FL: CRC Press)

Part III

Path integrals for classical and quantum optics applications

IOP Publishing

A Field Theory Approach to Photonics

Marco Ornigotti

Chapter 7

Applications of path integrals in photonics

The aim of this chapter is to familiarise the reader with using path integrals in photonics. This is done through a series of different examples, from ray optics to quantum nonlinear optics, that can be approached and solved using the path integral representation of the electromagnetic field. In particular, two different classes of path integrals approaches will be discussed here, namely the path integrals á la Feynman–Hibbs [1], and the relativistic (i.e., quantum field theory (QFT)) path integrals introduced in chapters 5 and 6. For both cases, the examples are worked out in a fairly detailed manner, so that the reader can familiarise themselves with the methods and techniques that come with the formalism.

The distinction between Feynman–Hibbs and relativistic path integrals naturally divides this chapter into two parts. Part I, discussed in section 7.1, presents how path integrals á la Feynman–Hibbs [1] can be be used to solve complex problems in photonics by presenting two examples, i.e., the propagation of an electromagnetic field in a graded-index medium in section 7.1.1, and the dynamics of a parametric amplifier in section 7.1.2. This part is somehow stand-alone, with respect to the concept of path integrals introduced in chapters 5 and 6, since it makes use of the more traditional, quantum mechanical definition of path integrals as the sum over all possible trajectories, instead of field configurations. Nevertheless, a lot of insight can be gained by looking at problems in photonics that can be solved with this formalism, and the examples presented in part I of this chapter are aimed at showing the potential of this method in a context that is closer to more familiar topics in optics and photonics.

The first example, discussed in section 7.1.1, uses path integrals to describe the propagation of the electromagnetic field inside a graded-index medium, in the paraxial regime. This problem has been tackled and solved by Gòmez-Reino [2] for a paraxial field. However, in this section, we take a slightly different perspective with respect to reference [2], and represent the problem in terms of a massive quantum particle (a photon dressed by the refractive index of the medium) evolving in a

doi:10.1088/978-0-7503-5789-0ch7

harmonic oscillator potential with a z-dependent characteristic frequency, following the discussion in reference [3]. Here, we will learn how path integrals can automatically account for the wave corrections to ray bundles, giving therefore access to paraxial optics from a new perspective.

The second example is then presented in section 7.1.2 and is taken from nonlinear optics. Here, following the works of Hillery and Zubairy [4, 5], we discuss how the electromagnetic field can be written in terms of path integrals over coherent states, and use this formalism to tackle a specific problem in nonlinear optics, namely that of a parametric amplifier. Contrary to section 7.1.1, where the path integrals originates from ray bundles, here the path integral originates from quantum states in phase space.

Using the same formalism to tackle two such different problems will allow the reader to draw some analogies between the two problems.

Part II of this chapter, presented in section 7.2, makes instead use of the concepts introduced in chapters 5 and 6 and constructs, building upon the work of Bechler [6], a theory of the electromagnetic field inside a lossy and dispersive medium. Here, light–matter interaction will be presented both in the linear and nonlinear case using the formalism of path integrals and Feynman diagrams.

To efficiently discuss the examples belonging to this part of the chapter, some work is needed to prepare all the necessary ingredients. For this reason, section 7.2.1 recasts the Huttner–Barnett model discussed in section 4.7 in terms of path integrals. Once this has been done, the first example of this part, namely the calculation of the dielectric function for a lossy dispersive media from the Huttner–Barnett model, is carried out in section 7.2.2. Section 7.2.3 takes this further by presenting the path integral quantisation of the electromagnetic field and the explicit calculation of the dressed electormagnetic propagator, together with the electric field operator. This concludes the discussion of the linear part of light–matter interaction.

From section 7.2.5 onwards, the chapter focuses on quantum nonlinear optics. First, the formalism of path integrals for interacting fields is adapted to the electromagnetic field (section 7.2.5 and 7.2.6), then second-order nonlinear processes are discussed extensively, with the full quantum problem being discussed in section 7.2.7, and the traditional undepleted pump approximation in section 7.2.8. Third-order nonlinearities are then briefly discussed in section 7.2.9, and the chapter concludes with a brief discussion on cascaded nonlinearities in section 7.2.10. These example are based on the work by Bechler [6], for the linear part, and on more recent works [3, 7] for the extension to the nonlinear part.

7.1 Part I: path integrals in photonics

The path integral approach presented in the next two sections deviates a bit from that presented in chapter 6. Rather than focusing on the wave properties of light, in fact, here we focus more on the geometrical optics limit of the electromagnetic field, and establish the formal connection between nonrelativistic quantum particles and optical rays. This connection is guaranteed by the fact that the paraxial wave equation (2.37), describing slowly varying optical fields (which, in first approximation, can be treated as optical rays) can be put in a one-to-one correspondence with the Schrödinger equation, describing the dynamics of non-relativistic quantum particles [8].

The path integral for a massive quantum particle, described by the Lagrangian $L(\mathbf{x}, \dot{\mathbf{x}}, t)$, is defined as [1]

$$K(b, a) = \int_a^b \mathscr{D}\mathbf{x}(t)\exp\left[\frac{i}{\hbar}\int_{t_0}^{t_1} dt\ L(\mathbf{x}, \dot{\mathbf{x}}, t)\right], \qquad (7.1)$$

where $a = (\mathbf{x}_0, t_0)$ indicates the initial position $\mathbf{x}_0$ of the particle at the initial time t_0, and $b = (\mathbf{x}_1, t_1)$ represent the final position $\mathbf{x}_1$ of the particle at the end of its evolution, occurring at time t_1. The quantity above, in fact, is understood as the probability amplitude for the system to start its evolution at point $(\mathbf{x}_0, t_0)$ and be found at point $(\mathbf{x}_1, t_1)$ at the end of its evolution. In this form, the path integral above is interpreted as a sum of all possible trajectories a quantum system takes, and the quantum effects of the system are emerging from the interference between all these possible trajectories [1]. In a sense, this version of path integrals is conceptually the same, as that introduced in chapter 6, with the difference that here, instead of looking at all possible configurations of a quantum field in spacetime, we consider all possible trajectories of a quantum particle in space. Going back to figure 5.2, the path integral above implements the idea depicted in figure 5.2(b).

To use nonrelativistic path integrals to solve problems in photonics, a few adjustments need to be done. First, we need to clarify what is the framework within which we intend to use equation (7.1) to describe the properties of the electro-magnetic field. In quantum mechanics, equation (7.1) contains the essence of the wave–particle duality, and essentially associates a wave $\exp(iS/\hbar)$ with a quantum particle, which allows one to express the evolution probability of a particle in terms of a trajectory (the classical path) *corrected* by the interference between such waves coming from all possible paths the particle can take to go from the initial to the final state [1]. For the electromagnetic field, this concept of wave corrections to a particle trajectory translates in optical rays with diffraction corrections [9]. Optical rays (i.e., plane waves), in fact, move along well-defined trajectories (solution of the equation of motion for rays, which is called the eikonal equation [10]), and are characterised by a momentum $\mathbf{k}$ and a mass given by the refractive index n_0 of the medium in which they propagate (with $n_0 = 1$ for vacuum).

In this picture, the Lagrangian $L(\mathbf{x}, \dot{\mathbf{x}}, t)$ for an optical ray propagating in a general medium with refractive index $n(\mathbf{R})$ has the following form [1, 3]

$$L(\mathbf{R}, \dot{\mathbf{R}}, z) = n(\mathbf{R}, z)\sqrt{1 + \dot{\mathbf{R}}}, \qquad (7.2)$$

where $\mathbf{R} = \mathbf{R}(z) = x(z)\hat{\mathbf{x}} + y(z)\hat{\mathbf{y}}$ is the trajectory followed by the ray, with z taken as the propagation direction, and the dot implies derivation with respect to z. Without loss of generality, and to keep the discussion as simple as possible, let us consider a $1 + 1$-dimensional geometry, i.e., we consider only one transverse dimension $\mathbf{R}(z) = x(z)\hat{\mathbf{x}}$, and drop the vector nature of the transverse coordinate, since in 1D it is redundant. The extension to higher dimensions can then be made in a trivial manner from the results below. Within this set of assumptions, the Lagrangian above then simplifies to

$$L(x, \dot{x}, z) = n(x, z)\sqrt{1 + \dot{x}}, \qquad (7.3)$$

and the path integral (7.1) becomes

$$G(x, y; z) = \int_a^b \mathscr{D}\mathbf{x}(z)\exp\left[\frac{i}{\lambdabar} \int_{z_a}^{z_b} dz\, L(\mathbf{x}, \dot{\mathbf{x}}, z)\right], \qquad (7.4)$$

where now $a = (x, z_a)$ and $b = (y, z_b)$. Notice, that in writing the expression above we have renamed the kernel $K(b, a)$ as $G(x, y; z)$, to highlight its role as Green's function (or propagator) of the electromagnetic field, to remain consistent with the correspondent definition of propagator of a quantum field given in chapter 5. More importantly, in going from equation (7.1) for a quantum particle to the equation above for an optical ray, we have also changed $\hbar$, the typical scale of quantum mechanics, with $\lambdabar = 1/k$, which is the natural length scale of the electromagnetic field, i.e., its wave vector. In fact, as $\hbar$ defines in quantum mechanics the scale at which quantum effects (i.e., the wave nature of matter) cannot be neglected, and the formal limit $\hbar \to 0$ represents the classical lmit to quantum mechanics, for photonics, $\lambdabar$ represents the scale at which diffraction effects cannot be neglected, and the limit $\lambdabar \to 0$ corresponds to the limit of ray optics.

7.1.1 Light propagation in inhomogeneous media

Now that we have the explicit expression for the path integral of the electromagnetic field, represented as a bundle of rays, i.e., equation (7.4), we can introduce our first example: derive the explicit expression of the propagator (7.4) for an electromagnetic field propagating in a graded-index (GRIN) medium, a special class of inhomogeneous materials (i.e., materials where the refractive index is not constant throughout the material of the volume, but depends on position) characterised by a refractive index profile defined by the function

$$n(x, z) = n_0\sqrt{1 - f^2(z)\, x^2}, \qquad (7.5)$$

where $n_0 = n(0, z)$ is the background refractive index of the medium, and $f(z)$ is a smooth function (continuous and differentiable infinitely many times) that accounts for the changes of the refractive index along the propagation direction. The Lagrangian for this problem is then obtained by substituting the expression above into equation (7.3), i.e.,

$$L(x, \dot{x}, z) = n_0\sqrt{1 - f^2(z)x^2}\sqrt{1 + \dot{x}}, \qquad (7.6)$$

which generates the following equation of motion for the ray trajectory $x = x(z)$

$$\frac{d}{dz}\left[\frac{\dot{x}\sqrt{1 + f^2(z)x^2}}{\sqrt{1 + \dot{x}^2}}\right] - \frac{xf^2(z)\sqrt{1 + \dot{x}^2}}{\sqrt{1 + f^2(z)x^2}} = 0. \qquad (7.7)$$

Solving this equation of motion or, analogously, the path integral in equation (7.4) is in general not possible analytically, so one must revert to numerical methods in order to find a solution to this problem

Rather than doing this, however, we make some reasonable assumptions (while maintaining $f(z)$ as general as possible, thus keeping the result of these calculations valid for a broad class of materials), which allow for an elegant analytical solution, and give us the possibility to showcase the power of the path integral approach. The assumptions we make for solving this problem are then the following

(1) We assume paraxial propagation, i.e., the optical rays travel only very close to the z-direction. This corresponds to $\dot{x} \ll 1$ and allows us to Taylor expand the $\dot{x}$-dependent square root in equation (7.3);

(2) We assume that the medium is weakly inhomogeneous, i.e., that $\Delta n \equiv n(x, z) - n_0 \ll 1$. This allows us to Taylor expand the $f(z)$-dependent square root in equation (7.3).

Using these two assumptions allows us to expand equation (7.3) up to second order in x and $\dot{x}$, and gives

$$L(x, \dot{x}, z) = n_0 \left[1 + \frac{\dot{x}^2}{2} - f^2(z) \frac{x^2}{2} \right] + \mathcal{O}(x^4, \dot{x}^4), \qquad (7.8)$$

for the Lagrangian in equation (7.6) and

$$\frac{d^2}{dz^2} x(z) - f^2(z)x(z) = 0, \qquad (7.9)$$

for its correspondent equation of motion (equation (7.7)). Within this set of approximations, the Lagrangian (7.8) is that of a harmonic oscillator with a z-dependent frequency $\omega(z) = f(z)$, as it can be clearly seen from the equation of motion above. Solving the problem of the propagation of the electromagnetic field in a weakly inhomogeneous medium is then fully equivalent to solving the problem of a harmonic oscillator with z-dependent frequency.

The propagator for the Lagrangian in equation (7.8) then reads

$$G(x, y; z) = \int_a^b \mathcal{D}x(z) \exp \left[\frac{in_0}{\lambda} \int_{z_a}^{z_b} dz \left(1 + \frac{\dot{x}^2}{2} - \frac{f^2(z)x^2}{2} \right) \right]. \qquad (7.10)$$

As a first step towards obtaining the solution to the path integral above, let us recall that for $f(z) = \omega = $ constant, the path integral above admits the following analytical expression for the propagator of a simple harmonic oscillator[1]

$$G_{h.o.}(x, y; z) = \sqrt{\frac{n_0 \omega}{i\lambda \sin \omega z}} \exp \left\{ -\frac{in_0 \omega}{2\lambda \sin \omega z} [(x^2 + y^2)\cos \omega z - 2xy] \right\}, \qquad (7.11)$$

where we have set $z_a = 0$ and $z_b = z$, $x(0) = x$, and $x(z) = y$ for convenience.

---

[1] The derivation can be found in the Feynman–Hibbs book [1], or in reference [3], for example, or in any textbook discussing path integrals, both from a quantum mechanics or QFT perspective.

Since equation (7.9) is that of a harmonic oscillator, it is reasonable to expect that the explicit form of the propagator in equation (7.10) would be similar to the expression (7.11). However, the z-dependence of the frequency through the term $f(z)$ complicates the derivation of the propagator. If we want to use the expression above as guidance to solve equation (7.10), we need to find a way to cast our problem in terms of simple harmonic oscillators, i.e., make the frequency independent on z.

A possible strategy to tackle this problem is to convert the Lagrangian of a harmonic oscillator with z-dependent frequency to a series of coupled simple harmonic oscillators. This can be achieved by discretising the action integral $S = \int_{z_a}^{z_b} dz\, L(x, \dot{x}, z)$ by dividing the propagation interval $\zeta = z_b - z_a$ into N discrete steps of length $d\zeta = z_{k+1} - z_k$, so that

$$S(X, Y; z) = \int_{z_a}^{z_b} dz\, L(x, \dot{x}, z) = \sum_{k=0}^{N} S_k(x_k, x_{k-1})$$

$$= \sum_{k=0}^{N} \int_{z_{k-1}}^{z_k} dz\, L(x, \dot{x}, z) = \frac{n_0}{2d\zeta}(x_{k+1} - x_k)^2 - \frac{n_0 d\zeta}{2} f^2(z_k) x_k^2, \tag{7.12}$$

where $X = x(z_a) = x_0$, $Y = x(z_b) = x_{N+1}$ and $\dot{x} = dx/dz = (x_{k+1} - x_k)/d\zeta$, with $x_k \equiv x(z_k)$. At each step $d\zeta$, then, the action is that of a simple harmonic oscillator with constant frequency $f(z_k)$.

Discretising the propagation interval into N steps also has an impact on the number of possible paths the ray can take to go from $x(z_a)$ to $x(z_b)$, reducing their number to N as well. Mathematically, this means that the path integral $\int \mathscr{D} x(z)$ is discretised into a series of N Riemannian integrals over the finite set of paths $\{x_1(z), ..., x_N(z)\}$, i.e.,

$$\int \mathscr{D}x(z) \rightarrow \prod_{k=1}^{N} \int dx_k(z) \equiv \prod_{k=1}^{N} \int dx_k. \tag{7.13}$$

The equality between the two relations above is then restored once the limit $N \rightarrow \infty$ is taken. Implementing these discretisation steps and taking the limit $N \rightarrow \infty$ let us then write the propagator (7.4) as

$$G(X, Y; z) = \lim_{N \to \infty} \prod_{k=1}^{N} \int dx_k, \exp\left[\frac{i}{\hbar} \sum_{k=0}^{N} S(x_k, x_{k-1}) \right]. \tag{7.14}$$

Let us discuss the form of this expression, to gain more insight on what we achieved and how this can help us in our calculations. First, notice from equation (7.12) that each action term $S(x_k, x_{k-1})$ in the summation above is quadratic in x_k and x_{k-1}. This means, that the integrals $\int dx_k$ are all Gaussians and can be computed analytically using the results presented in appendix B of chapter 5. In particular, notice that these Gaussian integrals are pairwise nested, since the action contains a term proportional to $x_k x_{k-1}$, which couples all these integrals and suggests to solve them in reverse order, i.e., starting from $k = N$ and going back all the way to $k = 1$. This coupling of different Gaussian integral is interesting, because it has a neat physical meaning, which will be discussed below.

Practically, equation (7.14) allows us to first evaluate all the N Gaussian integrals independently and then obtain the final result by taking the limit $N \to \infty$. In this way, we go around the problem of solving a difficult problem (calculating the path integral of a harmonic oscillator with z-dependent frequency) by solving instead a simpler problem (N coupled simple harmonic oscillators) and then take the limit to the continuum to get back to the real solution.

Before performing these calculations explicitly, it is convenient to introduce the following scaled mass (i.e., refractive index) and frequency

$$\beta = \frac{n_0}{\hbar d\zeta}, \tag{7.15a}$$

$$\alpha_k = \beta\left[1 - \frac{(d\zeta)^2}{2}f^2(z_k)\right], \tag{7.15b}$$

and, as stated above, perform the Gaussian integrals in reverse order, i.e., first integrating with respect to x_N (which will result in a function $g_1(Y, x_{N-1})$), then with respect to x_{N-1} (which will result in a function $g_2(Y, x_{N-2})$), all the way to the last integral with respect to x_1, which will result in a function $g_N(Y, X)$, corresponding to the discretised version of the propagator. We can implement this by using equation (5.72) repeatedly to calculate the various nested integrals appearing in equation (7.14), which results in

$$G(X, Y; z) = \lim_{N \to \infty} \sqrt{\frac{c_N}{2\pi}} \exp\left[i(a_N X^2 + b_N Y^2) - c_N XY\right], \tag{7.16}$$

where $a_N = a_N(\beta, \alpha_k)$, $b_N = b_N(\beta, \alpha_k)$, $c_N = c_N(\beta, \alpha_k)$ are constants emerging from the nested Gaussian integration, whose explicit expression is given as follows [11]

$$a_N = \frac{\beta}{2} - \sum_{k=1}^{N-1}\frac{\rho_k^2}{4\sigma_k}, \tag{7.17a}$$

$$b_N = \frac{\beta}{2} - \frac{\beta^2}{4\sigma_{N-1}}, \tag{7.17b}$$

$$c_N = \beta\sum_{k=1}^{N-1}\frac{\beta}{2\sigma_k}, \tag{7.17c}$$

and the quantities ρ_k and σ_k are defined by the following recursive relations

$$\rho_k = \beta\prod_{\ell=1}^{k-1}\frac{\beta}{2\sigma_\ell}, \tag{7.18a}$$

$$\sigma_k = \alpha_k - \frac{\beta^2}{4\sigma_{k-1}}, \tag{7.18b}$$

with $\rho_1 = \beta$, and $\sigma_1 = \alpha_1$. Before going any further into the calculations, let us take a moment to discuss the physical meaning of taking all those nested Gaussian integrals. In the first slice $d\zeta_1 = z_{N+1} - z_N$, the system behaves like a simple harmonic oscillator, with scaled mass β_1 and constant frequency α_1. When the integration takes place, the information on the mass and frequency of this oscillator is passed to the one in the next slice $d\zeta_2 = z_N - z_{N-1}$. This happens by modifying the mass and frequency of the new oscillator, but not its overall physics, which keeps on being described by a simple harmonic oscillator, this time, however, with mass β_2 and frequency α_2, both functions of β_1 and α_1. This process continues until all N oscillators have been taken into account and the final integration generates equation (7.16), which is reminiscent of all the harmonic oscillators in the systems, coupled to each other because of the discretisation. Taking the limit $N \to \infty$ on equation (7.16) then corresponds to finding the solution for an infinite distribution of coupled oscillators, which ultimately corresponds to the solution of equation (7.9).

Taking the limit $N \to \infty$ of equation (7.16) amounts to taking the limits of the coefficients a_N, b_N, and c_N. Since we know that this limiting procedure must produce the propagator associated to equation (7.9) (since that is the equation of motion corresponding to $G(X, Y; z)$), we can express the limiting forms a_∞, b_∞, and c_∞ as a function of the complex function $\xi(z) = R(z)\exp[i\varphi(z)]$, which we assume being a solution of equation (7.9), i.e.,

$$\frac{d^2}{dz^2}\xi(z) + f^2(z)\xi(z) = 0. \tag{7.19}$$

This serves two aims: first, it allows us to write the limiting forms of the coefficients a_N, b_N and c_N in an easy and compact form, and second it gives more insight on their connection with the actual trajectory of the harmonic oscillator with z-dependent frequency. Performing the limit and using the function $\xi(z)$ gives us then the following result

$$a_\infty = \lim_{N\to\infty} a_N = \frac{n_0}{2\hbar}\left[-\frac{d}{dz}\log R(z)|_{z=z_a} + \dot\varphi(z_a)\cot \Phi(\zeta)\right], \tag{7.20a}$$

$$b_\infty = \lim_{N\to\infty} b_N = \frac{n_0}{2\hbar}\left[\frac{d}{dz}\log R(z)|_{z=z_b} + \dot\varphi(z_b)\cot \Phi(\zeta)\right], \tag{7.20b}$$

$$c_\infty = \lim_{N\to\infty} c_N = \frac{n_0}{\hbar}\frac{\sqrt{\dot\varphi(z_a)\dot\varphi(z_b)}}{\sin \Phi(\zeta)}, \tag{7.20c}$$

where $\Phi(\zeta) = \varphi(z_b) - \varphi(z_a)$. The explicit calculation of these quantities, originally derived in reference [11], is presented in appendix A, for reference[2]. With these

---

[2] Notice, that in [11], equation (19), corresponding to equation (7.20c) above, is wrong, since it features sin Φ inside the square root, when it shouldn't be. Its derivation in equation (A16) of the same reference, however, is correct.

results, we can now write down the Green's function for the electromagnetic field propagating in a weakly inhomogeneous medium as follows

$$G(x, y; z) = \sqrt{\frac{c_\infty}{2\pi}} \exp\left\{\frac{in_0}{2\lambdabar}\left[x^2(z)\frac{d}{dz}\log R(z)|_{z_a}^{z_b}\right] + A\left[\dot\varphi(z_b)x_b^2 + \dot\varphi(z_a)x_a^2\right] - B\,x_a x_b\right\}, \quad (7.21)$$

where $A = (in_0/2\lambdabar)\cot\Phi(\zeta)$ and $B = 2\sqrt{\dot\varphi(z_a)\dot\varphi(z_b)}$. This is the final expression for the Green's function of an optical ray propagating in a weakly inhomogeneous, 1D medium.

Some comments are in order. First, as a sanity check, equation (7.21) reduces to (7.4) for $f(z) = \omega$, for which $\xi(z) = \sqrt{n_0/(2\omega)}\exp(-i\omega z)$ holds, leading to $d\log R(z)/dz = 0$, $\dot\varphi(z_a) = \dot\varphi(z_b) = \omega$ and $\Phi(\zeta) = \omega\,\zeta$. This tells us that we have done the calculations correctly, since we are able to get back to the expresison for a simple harmonic oscillator if we set the frequency to be constant.

Second, the expression of the propagator in equation (7.21) is quite complicated and not so easy to read. We can, with the help of a little mathematics of special functions, rewrite it in a more elegant and physically meaningful form. To do so, we use Mehler's formula [12]

$$\frac{1}{\sqrt{1 - \alpha^2}}\exp\left(-\frac{\xi^2 + \eta^2 - 2\xi\eta\alpha}{1 - \alpha^2}\right) = \exp[-(\xi^2 + \eta^2)]\sum_{n=0}^{\infty}\frac{\alpha^n}{2^n n!}H_n(\xi)H_n(\eta), \quad (7.22)$$

to rewrite the propagator in equation (7.21) in terms of 1D Hermite–Gaussian (HG) beams $HG_n(x, z)$ as

$$G(x, y; z) = \sum_{n=0}^{\infty} HG_n^*(x, 0)HG_n(y, z). \quad (7.23)$$

The HG beams in one dimension are defined using Mehler's formula with the substitutions $\xi = \sqrt{n_0\dot\varphi(z_a)/\lambdabar}\,x$, $\eta = \xi = \sqrt{n_0\dot\varphi(z_a)/\lambdabar}\,y$, and $\alpha = \exp[-i\Phi(\zeta)]$ and read

$$HG(x, z) = \mathcal{N}\exp\left[i\left(n + \frac{1}{2}\right)\varphi(z)\right]\exp\left[\frac{in_0}{2\lambdabar}\left(\frac{d}{dz}\log R(z) + i\dot\varphi(z)\right)x^2\right]H_n\left(\sqrt{\frac{n_0\dot\varphi(z)}{\lambdabar}}x\right). \quad (7.24)$$

To understand what this result means in terms of the electromagnetic field propagating in a (weakly) inhomogeneous medium, let us first introduce the following quantities

$$\rho(z) = \frac{1}{2n_0}\left[\frac{d}{dz}\log R(z)\right]^{-1}, \quad (7.25a)$$

$$w_0^2(z) = \frac{\lambdabar}{n_0}\left[\frac{d}{dz}\varphi(z)\right]^{-1}, \quad (7.25b)$$

$$\psi(z) = \left(n + \frac{1}{2}\right)\varphi(z), \quad (7.25c)$$

so that we can rewrite equation (7.24) as

$$\mathrm{HG}(x,z) = \mathcal{N} \exp\left[i\psi(z)\right] \exp\left[\frac{ix^2}{4\rho(z)}\right] \exp\left[-\frac{x^2}{w^2(z)}\right] H_n\left(\frac{x}{w(z)}\right). \quad (7.26)$$

Equation (7.26) has the form of a (one-dimensional) HG beam, with beam waist $w(z)$, wavefront curvature $\rho(z)$, and Gouy phase $\psi(z)$. In photonics, HG beams are solution of the paraxial wave equation in Cartesian coordinates [13]. As such, they are electromagnetic waves that propagated along the z-direction exhibiting diffraction. This fact should raise a question for the reader.

Why are *waves* (i.e., HG beams) suddenly making an appearance as a solution of a problem that was cast in terms of *rays* propagating in a medium?

The answer is surprisingly insightful. In path integrals quantum mechanics, the propagator from the initial to the final state is computed as the interference of all possible paths the system can take to connect these two states, where each path is weighted by the quantum phase $\exp(iS/\hbar)$. Paths closer to the classical path contribute constructively to this interference, since their action differs from the classical action S_{cl} by a small quantity, i.e., $S = S_{cl} + \delta S$ (with δS small), which essentially allows constructive interference between their quantum phases. In contrast, paths far away from the classical paths have much larger actions, and their quantum phases interfere destructively. As a result of this, the paths close to the classical path form a *quantum bubble*, that can be interpreted as the classical (particle-like) trajectory being *dressed* with quantum (i.e., wave-like) corrections [14]. Any calculations involving path integrals automatically contain these quantum corrections, which to a certain extent, embody the action of the uncertainty principle, preventing us from accessing the classical trajectory, since this would require knowing position and momentum with infinite precision, a task that it is impossible in quantum mechanics.

For the electromagnetic field, the classical, particle-like limit is that of geometrical optics, with optical rays playing the role of classical particles. The *quantum bubble* corresponds instead to the corrections to rays generated by the wave properties of the electromagnetic field, i.e., diffraction. In light of this analogy, we can interpret our result of the propagator being expressed in terms of HG functions as being a manifestation of the (unavoidable) diffraction effects, that dress optical rays into paraxial waves.

7.1.2 Path integrals in nonlinear optics

Let us now consider a different problem, taken from the realm of nonlinear optics, to showcase how nonrelativistic path integrals can be adapted to this context and provide insightful solutions. The example that we discuss in this section is that of a parametric amplfier [15].

Before entering into the details of the calculations, it is worth noticing that we need to adopt a slight change of perspective to interpret the path integral for the electromagnetic field used in this section. In fact, rather than interpreting the propagator as a sum over all possible trajectories of an optical ray inside a material, for this example we instead look at the propagator as a sum over all possible configurations of the electromagnetic field in phase space. In a sense, this corresponds to writing the field as a bundle of optical rays in phase space, rather than in position space. As we will see, the 'optical rays' of phase space are the coherent states of the electromagnetic field.

The first task of this section is then to find a suitable representation of the propagator in terms of path integrals of coherent states. To do so, we use two ingredients: we assume the electromagnetic field to be quantised and represented in terms of creation and annihilation operators, and we represent the electromagnetic states as coherent states (see chapter 6). The propagator for the electromagnetic field evolving from the state $|\alpha_1\rangle$ at time t_1 to the state $|\alpha_2\rangle$ at time t_2 can be then written as follows:

$$G(\alpha_2, t_2; \alpha_1, t_1) = \langle \alpha_2 | \hat{U}(t_2, t_1) | \alpha_1 \rangle \equiv \langle \alpha_2, t_2 | \alpha_1, t_1 \rangle, \tag{7.27}$$

where $\hat{U}(t_2, t_1) = \hat{T} \exp\left[-i \int_{t_1}^{t_2} d\tau \ \hat{H}(\tau) \right]$ is the evolution operator generated by the Hamiltonian $\hat{H}(t)$ and $|\alpha, t\rangle = \hat{U}(t, 0)|\alpha\rangle$.

What we now need to do is to rewrite the expression above in terms of a path integral. To do so, we need to introduce the density matrix $\hat{\rho}$ of the electromagnetic field in terms of coherent states using the so-called Glauber–Sudarshan P-function as [16, 17]

$$\hat{\rho} = \int d^2\alpha \ P(\alpha)|\alpha\rangle\langle\alpha|, \tag{7.28}$$

where $d^2\alpha = d\ \mathrm{Re}\{\alpha\}d\ \mathrm{Im}\{\alpha\}$, and $P(\alpha)$ is the Glauber–Sudarshan P-function, and the expectation value of an operator in terms of this density matrix as

$$\langle \hat{A}(t) \rangle = \mathrm{Tr}\,\{\hat{\rho}\hat{A}(t)\} = \int d^2\alpha \ P(\alpha)\langle\alpha|\hat{A}(t)|\alpha\rangle. \tag{7.29}$$

Notice, that the operator $\hat{A}(t)$ introduced above evolves in time with the evolution operator, i.e., $\hat{A}(t) = \hat{U}^\dagger(t, 0)\hat{A}(0)\hat{U}(t, 0)$, as discussed in chapter 5.

To illustrate how to transform equation (7.27) into a path integral, let us take the following steps: first, recall that coherent states are an overcomplete set, i.e. [18],

$$\frac{1}{\pi} \int d^2\alpha |\alpha\rangle\langle\alpha| = 1. \tag{7.30}$$

Then, use this result to rewrite $\langle \hat{A}(t) \rangle$ in the following manner

$$
\begin{aligned}
\langle \hat{A}(t) \rangle &= \int d^2\alpha \ P(\alpha)\langle\alpha|\hat{A}(t)|\alpha\rangle = \int d^2\alpha P(\alpha)\langle\alpha|\hat{U}^\dagger(t, 0)\hat{A}(0)\hat{U}(t, 0)|\alpha\rangle \\
&= \frac{1}{\pi} \int d^2\alpha \ d^2\beta \ P(\alpha)\langle\alpha|\hat{U}^\dagger(t, 0)\hat{A}(0)|\beta\rangle\langle\beta|\hat{U}(t, 0)|\alpha\rangle \\
&= \frac{1}{\pi} \int d^2\alpha \ d^2\beta \ P(\alpha) \ \beta\langle\alpha|\hat{U}^\dagger(t, 0)|\beta\rangle\langle\beta|\hat{U}(t, 0)|\alpha\rangle \\
&= \frac{1}{\pi} \int d^2\alpha \ d^2\beta \ P(\alpha)\beta \ |G(\beta, t; \alpha, 0)|^2,
\end{aligned}
\tag{7.31}
$$

where in the second line we have inserted equation (7.30) between $\hat{A}(0)$ and $\hat{U}(t, 0)$, and to go from the second to the third line we have used the fact that coherent states are eigenstates of the annihilation operator, i.e., $\hat{A}(0)|\beta\rangle = \beta|\beta\rangle$. Notice, how the function $G(\beta, t; \alpha, 0)$ defined in equation (7.27) naturally emerges when calculating expectation values of operators in coherent state representation.

As we have seen in chapter 5, n-point correlation functions are associated to the (time-ordered) expectation value of field operators of the form $\langle a^\dagger(t_1)\hat{a}^\dagger(t_2) \cdots \hat{a}(t_{N-1})\hat{a}(t_N)\rangle$. Repeating the approach above for the general case of the expectation value of N fields then gives the following form for the propagator

$$G(\beta, t_f; \alpha, t_i) = \left(\frac{1}{\pi}\right)^N \int [d^2\alpha]_N \langle\beta, t_f|\alpha_N, t_N\rangle\langle\alpha_N, t_N|\alpha_{N-1}, t_{N-1}\rangle \cdots \langle\alpha_1, t_1|\alpha, t_i\rangle, \quad (7.32)$$

where $[d^2\alpha]_N$ is a shorthand for $d^2\alpha_1 \cdots d^2\alpha_N$, To express the propagator in terms of a path integral, we can imagine that the N terms appearing in the equation above come from a discretisation of the evolution interval $t_f - t_i$ into N steps of length $dt = (t_f - t_i)/N$. This gives us the possibility to calculate each single term $\langle\alpha_k, t_k|\alpha_{k-1}, t_{k-1}\rangle$ using the infinitesimal expression of the evolution operator, i.e.,

$$\hat{U}(t_k, t_{k-1}) \simeq 1 - i\int_{t_{k-1}}^{t_k} d\tau\, \hat{H}(\tau) + \mathcal{O}(dt^2) = 1 - i\, dt\hat{H}(t_{k-1}) + \mathcal{O}(dt^2). \quad (7.33)$$

In quantum optics terms, the equation above represents the overlap integral between two coherent states, one taken at time t_k, and the other one at time t_{k-1}. In path integral terms, instead, this overlap integral represents a single *trajectory* of the electromagnetic field in phase space [19]. Notice, morover, the formal similarity between equation (7.33) and the term $S(x_{k+1}, x_k)$ in equation (7.14).

Using equation (7.33) we can evaluate the term $\langle\alpha_k, t_k|\alpha_{k-1}, t_{k-1}\rangle$ to be

$$\langle\alpha_k, t_k|\alpha_{k-1}, t_{k-1}\rangle = \langle\alpha_k|\hat{U}(t_k, t_{k-1})|\alpha_{k-1}\rangle = |\alpha_k\rangle\left[1 - i\int_{t_{k-1}}^{t_k} d\tau\, \hat{H}(\tau)\right]|\alpha_{k-1}\rangle$$

$$= \langle\alpha_k|[1 - i\, dt\, \hat{H}(\hat{a}^\dagger, \hat{a}; t_{k-1})]|\alpha_{k-1}\rangle$$

$$= \langle\alpha_k|\left[1 - i\, dt\, H(\alpha_k^*, \alpha_{k-1}; t_{k-1})\right]|\alpha_{k-1}\rangle$$

$$= \langle\alpha_k|\alpha_{k-1}\rangle\left[1 - i\, dt\, \mathcal{H}(\alpha_k^*, \alpha_k; t_{k-1})\right] \quad (7.34)$$

$$\simeq \langle\alpha_k|\alpha_{k-1}\rangle \exp\left[-i\, dt\, \mathcal{H}(\alpha_k^*, \alpha k - 1; t_{k-1})\right]$$

$$= \exp\left[-\frac{1}{2}(|\alpha_k|^2 + |\alpha_{k-1}|^2) + \alpha_k^*\alpha_{k-1}\right]$$

$$\times \exp\left[-i\, dt\, \mathcal{H}(\alpha_k^*, \alpha_{k-1}; t_{k-1})\right].$$

Let us comment on this calculation in more detail, before proceeding further. First, we have assumed that the Hamiltonian generating the evolution can be written as a normal-ordered product of creation and annihilation operators, i.e., that

$\hat{H}(t) = \hat{H}(\hat{a}^\dagger, \hat{a}; t)$. Then, to go from the second to the third line, we have employed the so-called optical theorem [20]

$$\langle \alpha | f(\hat{a}^\dagger, \hat{a}) | \alpha \rangle = f(\alpha^*, \alpha), \tag{7.35}$$

which states that the expectation value of any function $f(\hat{a}^\dagger, \hat{a})$ of creation and annihilation operators over coherent states gives as a result the same function, with creation operators replaced by α^* and annihilation operators replaced by α. In The fourth line, we have re-summed the term in square brackets to an exponential and introduced the function

$$\mathcal{H}(\beta^*, \alpha; t) = \frac{\langle \beta | \hat{H}(\hat{a}^\dagger, \hat{a}; t) | \alpha \rangle}{\langle \beta | \alpha \rangle}, \tag{7.36}$$

for simplicity of notation. Finally, in the last line we have made use of the non-orthogonality of coherent states [18] to write

$$\langle \alpha | \beta \rangle = \exp\left[-\frac{1}{2}(|\alpha|^2 + |\beta|^2) + \alpha^* \beta \right]. \tag{7.37}$$

We can now use the result in equation (7.34) to calculate all the terms appearing in equation (7.32), then take the limit for $N \to \infty$ and arrive at the final result, i.e., the definition of the propagator for the electromagnetic field in terms of path integrals in coherent state space

$$G(\beta, t_f; \alpha, t_i) = \int \mathscr{D}\alpha \, \exp\left\{ \int_{t_i}^{t_f} d\tau \left[\frac{\alpha(\tau)\dot{\alpha}^*(\tau) - \alpha^*(\tau)\dot{\alpha}(\tau)}{2} - i\, \mathcal{H}(\alpha^*, \alpha; t) \right] \right\}, \tag{7.38}$$

with the boundary conditions chosen such that $\alpha(t_i) = \alpha$ and $\alpha(t_f) = \beta$.

7.1.2.1 Path integral for parametric amplification

Now that we have the general expression for the propagator in terms of path integral over coherent states, let us apply this formalism to the case of parametric amplification and see what insight we can get from this result. A sketch of the process is presented in figure 7.1. Essentially, an intense (pump) and a weak (signal)

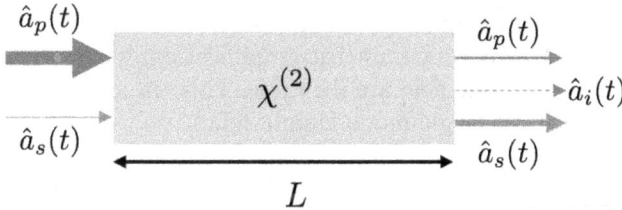

Figure 7.1. Schematic representation of the process of parametric amplification. An intense pump beam (bold, red arrow) and a weak, signal beam (thin, green arrow) impinge upon the input facet of a $\chi^{(2)}$ crystal. As a result of the nonlinear interaction, energy is exchanged between the pump and the signal beam, so that the latter can be amplified. Because of conservation of energy, however, an extra idler field (dashed, blue line) is generated in the process.

laser beam impinge upon a nonlinear crystal, which mediates the interaction between these two fields, resulting in an amplification of the signal and the generation of an ancillary laser beam (idler) because of energy conservation [15]. The Hamiltonian describing this process is given, in general, as

$$\hat{H} = \sum_{k=\{p,\,s,\,i\}} \omega_k \hat{a}(t)_k^\dagger \hat{a}(t)_k + g\left[\hat{a}_s^\dagger(t)\hat{a}_i^\dagger(t)\hat{a}_p(t) + \hat{a}_p^\dagger(t)\hat{a}_s(t)\hat{a}_i(t)\right], \tag{7.39}$$

where the subscripts $\{p,\,s,\,i\}$ refer to the pump, signal, and idler mode, respectively, ω_k is the photon energy in each mode, and g is the coupling constant encapsulating the nonlinear susceptibility. The equations of motion for the operators $\hat{a}_k(t)$ and $\hat{a}_k^\dagger(t)$ generated by this Hamiltonian can be calculated from equation (5.18) and are, in general a set of coupled, nonlinear, partial differential equations, that are not solvable analytically. Therefore, analogously to the case in the previous section, a general analytical form of the propagator is not available, and assumptions need to be made to simplify the problem down to a level, where it can be solved analytically. For the case of nonlinear optics, this means to adopt the so-called *undepleted pump approximation* [15, 21], which is frequently verified in typical nonlinear optics experiments. Essentially, this approximation assumes that the intensity (i.e., number of photons) in the pump beam does not change significantly during the nonlinear process[3]. Hence, the pump beam can be treated as a classical field, since its intensity remains, for all intent and purposes, constant during the nonlinear process. This means, that we can substitute the pump operators $\hat{a}_p(t)$ and $\hat{a}_p^\dagger(t)$ appearing in equation (7.39) with the intensity and phase of the pump beam, i.e., $\hat{a}_p \rightarrow I \exp(i\phi)$ and $\hat{a}_p^\dagger \rightarrow I \exp(-i\phi)$. This gives the following expression for the undepleted pump Hamiltonian

$$\hat{H} = \sum_{k=\{s,\,i\}} \left[\omega_i \hat{a}_k^\dagger \hat{a}_k + \tilde{g}\hat{a}_s^\dagger(t)\hat{a}_i^\dagger(t) + \tilde{g}^*\hat{a}_s(t)\hat{a}_i(t)\right], \tag{7.40}$$

where $\tilde{g} = g\,I \exp(i\phi)$ is the effective coupling constant, that contains information on the pump beam, and we have removed the free Hamiltonian of the pump $\omega_p \hat{a}_p^\dagger(t)\hat{a}_p(t)$ since with the substitution above it only amounts to an inessential constant. Notice, how equation (7.40) represents a quadratic Hamiltonian in the signal and idler operators and, as such, its dynamics can be solved analytically.

To keep the level of complexity of the calculations as low as possible, let us consider the case of *degenerate* parametric amplification, where the signal and idler modes of the electromagnetic field are the same. This reduces the undepleted pump Hamiltonian above to the single-mode Hamiltonian

$$\hat{H} = \omega \hat{a}^\dagger \hat{a} + g[\exp(i2\omega)\hat{a}^2 + \exp(-i2\omega)\hat{a}^\dagger], \tag{7.41}$$

---

[3] A typical order of magnitude for the number of photons carried by laser beams is 10^9. When a nonlinear event happens inside the nonlinear crystal, one pump photon is converted into a signal–idler pair. This reduces the number of photons in the pump beam to $10^9 - 1 \simeq 10^9$. The pump beam still retains (practically) the same number of photons, and it is therefore undepleted. Hence the origin of the name of the approximation.

where we have also made explicit the time dependence of the creation and annihilation operators, according the the Heisenberg equations of motion (5.18). Reducing the problem to a single-mode problem has also the important function to allow a direct comparison of the calculations to follow with those made in the previous section for the electromagnetic field propagating in a inhomogeneous medium. As will appear clear from the calculations, in fact, there are certain similarities between the two problems. Most of them come from the fact, that the underlying physical model is the same. In the case of section 7.1.1, the assumptions made to simplify the problem led to a quadratic Lagrangian and to a harmonic oscillator with z-dependent frequency. Here, the undepleted pump approximation (plus the degeneracy between signal and idler) led to a quadratic Hamiltonian, whose equations of motion are those of a harmonic oscillator with t-dependent frequency.

This analogy is even more evident after we make use of the Hamiltonian in equation (7.41) to calculate the expression of the propagator given by equation (7.38). This amounts, as in the case of equation (7.16), to calculating the N nested Gaussian integrals appearing in equation (7.32), and then taking the limit $N \to \infty$ to get the final expression.

Performing the N nested Gaussian integrals gives

$$G(\beta, t_f; \alpha, t_i) = \frac{1}{d_N} \exp\{i\, dt\, (a_N + b_N \alpha^2 + c_N \alpha) + X_{N+1}(\beta^*)^2 + Y_{N+1}\alpha\beta + Z_{N+1}\beta^*\}, \quad (7.42)$$

where

$$a_N = \sum_{k=1}^{N} \frac{g \exp(i2\omega t_k) Z_k^2}{D_k}, \tag{7.43a}$$

$$b_N = \sum_{k=0}^{N} \frac{g \exp(i2\omega t_k) Y_k^2}{D_k}, \tag{7.43b}$$

$$c_N = \sum_{k=0}^{N} \frac{2g \exp(i2\omega t_k) Y_k Z_k}{D_k}, \tag{7.43c}$$

$$d_N = \prod_{k=0}^{N} \sqrt{1 + 4i\, dt\, g \exp(i2\omega t_k) X_k} \equiv \prod_{k=1}^{N} D_k, \tag{7.43d}$$

and the quantities X_k, Y_k, and Z_k are defined recursively by

$$X_k \equiv X(t_k) = -i\, dt\, g \exp(-i2\omega t_k) + \frac{(1 - i\, dt\, \omega)^2 X_{k-1}}{D_{k-1}}, \tag{7.44a}$$

$$Y_k \equiv Y(t_k) = \frac{(1 - i\, dt\, \omega) Y_{k-1}}{D_{k-1}}, \tag{7.44b}$$

$$Z_k \equiv Z(t_k) = \frac{(1 - i\, dt\, \omega) Z_{k-1}}{D_{k-1}}, \tag{7.44c}$$

with the initial values $X_0 = Z_0 = 0$ and $Y_0 = 1$.

Then, as we have done in section 7.1.1, taking the limit $N \to \infty$ gives the limiting expression of the coefficients above as

$$a_\infty = \exp\left[-ig \int_{t_i}^{t_f} d\tau \, \exp(i2\omega\tau)X(\tau)\right], \tag{7.45a}$$

$$b_\infty = -ig \int_{t_i}^{t_f} d\tau \, \exp(i2\omega\tau)Z^2(\tau), \tag{7.45b}$$

$$c_\infty = -ig \int_{t_i}^{t_f} d\tau \, \exp(i2\omega\tau)Y^2(\tau), \tag{7.45c}$$

$$d_\infty = -i2g \int_{t_i}^{t_f} d\tau \, \exp(i2\omega\tau)Y(\tau)Z(\tau). \tag{7.45d}$$

Here, the auxiliary functions $X(t)$, $Y(t)$ and $Z(t)$ obey differential equations coming from the limit operation applied to equations (7.44a). For example, the recursion relation for X_k can be rewritten in the following form (we neglect terms of order dt^2)

$$\frac{X_{k-1} - X_k}{dt} = -2i\omega \, X_{k-1} - 4ig \, \exp(i2\omega t_k)X_{k-1}X_k - ig \, \exp(-i2\omega t_k), \tag{7.46}$$

which becomes, upon taking the limit $N \to \infty$,

$$\frac{dX}{dt} = -2i \, \omega \, X - 4ig \, \exp(i2\omega t)X^2 - ig \, \exp(-i2\omega t). \tag{7.47}$$

Analogously, the recursion relations for Y_k and Z_k generate the following differential equations

$$\frac{dY}{dt} = -i[\omega + 4g \, \exp(i2\omega t)X]Y, \tag{7.48a}$$

$$\frac{dZ}{dt} = -i[\omega + 4g \, \exp(i2\omega t)X]Z. \tag{7.48b}$$

The initial conditions for the differential equations above are $X(t_i) = Z(t_i) = 0$, and $Y(t_i) = 1$.

To find an explicit expression for the propagator associated with degenerate parametric amplification, we need to solve the differential equations governing the evolution of $X(t)$, $Y(t)$, and $Z(t)$, as their solutions enter in equation (7.42) through the coefficients a_∞ etc. The differential equations for $Y(t)$ and $Z(t)$ can be integrated, treating $X(t)$ as a time-dependent parameter, to give [22]

$$Y(t) = \exp\left\{-i\left[\omega \, t + 4g \int_0^t d\tau \, \exp(2i\omega\tau)X(\tau)\right]\right\}, \tag{7.49a}$$

$$Z(t) = 0 \tag{7.49b}$$

where we have set $t_i = 0$ and $t_f = t$ for convenience. Remarkably, equation (7.47) admits an analytical solution of the form

$$X(t) = -\frac{i}{2} \exp(-i2\omega\, t)\tanh(2gt). \tag{7.50}$$

Substituting this result into the first of equations (7.48a) then

$$Y(t) = \exp(-i\omega\, t)\mathrm{sech}(2gt). \tag{7.51}$$

Using these results, we can find explicit expressions for the coefficients in equations (7.45a) and substitute them into equation (7.42) to obtain the following final form of the propagator for a degenerate parametric amplifier

$$
\begin{aligned}
G(\beta,\, t_f;\, \alpha,\, t_i) &= \sqrt{\mathrm{sech}(2g\tau)}\,\exp\!\left(-\frac{|\beta|^2 + |\alpha|^2}{2}\right) \\
&\quad \times \exp\left\{\beta^*\alpha\,\exp(-i\omega\tau)\mathrm{sech}(2g\tau) - i\left(\frac{\beta^*}{\sqrt{2}}\right)^2 \exp(-i2\omega t_f)\tanh(2g\tau) \right. \\
&\qquad \left. - i\left(\frac{\alpha}{\sqrt{2}}\right)^2 \exp(-i2\omega t_i)\tanh(2g\tau)\right\}
\end{aligned}
\tag{7.52}
$$

where we have defined $\tau = t_f - t_i$. This propagator can be then used to evaluate any normal-ordered expectation value of the form $\langle \hat{a}^\dagger \cdots \hat{a}\rangle$, and therefore it can be used to fully characterise the dynamical and statistical properties of a degenerate parametric amplifier. As an example, let us calculate the time-dependent intensity of the signal mode. This can be done by taking the expectation value of the intensity operator $\hat{a}^\dagger(t)\hat{a}(t)$ using equation (7.31) and the result above for the propagator, obtaining

$$I(t) = \langle \hat{a}^\dagger(t)\hat{a}(t)\rangle = \left(\frac{1}{\pi}\right)^2 \int d^2\alpha\, d^2\beta\, (|\beta|^2 - 1)|G(\beta,\, t;\, \alpha,\, 0)|^2. \tag{7.53}$$

To evaluate this integral, let us first notice, that

$$|\exp[\beta^*\alpha\,\exp(-i\omega\, t)\mathrm{sech}(2g\, t)]|^2 = \exp[2\mathrm{sech}(2g\, t)F_1(\alpha,\, \beta)], \tag{7.54a}$$

$$\left|\exp\left[-i\left(\frac{\beta^*}{\sqrt{2}}\right)^2 \exp(-i2\omega t)\tanh(2g\, t)\right]\right|^2 = \exp\left[-\frac{\tanh(2g\, t)}{2}F_2(\alpha,\, \beta)\right], \tag{7.54b}$$

$$\left|\exp\left[-i\left(\frac{\alpha}{\sqrt{2}}\right)^2 \tanh(2g\, t)\right]\right|^2 = \exp[\tanh(2g\, t)xy], \tag{7.54c}$$

where

$$F_1(\alpha, \beta) = (x\xi + y\eta) \cos \omega\, t - (x\eta - y\xi) \sin \omega\, t, \tag{7.55a}$$

$$F_2(\alpha, \beta) = (\xi^2 - \eta^2)\sin(2\omega\, t) + 2\xi\eta \sin(2\omega\, t), \tag{7.55b}$$

with $\alpha = x + iy$ and $\beta = \xi + i\eta$. The integral in equation (7.53) is Gaussian in $\{x, y, \xi, \eta\}$ and it can be therefore evaluated explicitly. After some tedious but simple integrations, the final result is given by [5]

$$
\begin{aligned}
I(t) = \sinh^2(2g\, t) &+ \frac{1}{4}\{(2gt)^2[2\sinh^2(2g\, t) + 1] + 2gt[2\sinh(2g\, t)\cosh(2g\, t)] \\
&- 3\sinh^4(2g\, t) - 3\sinh^2(2g\, t)\}.
\end{aligned}
\tag{7.56}
$$

The time evolution of the intensity of the signal mode is shown in figure 7.2.

Another interesting feature of parametric oscillators is their ability to generate squeezing, i.e., the ability to beat the fundamental limit dictated by the uncertainty principle in one quadrature (i.e., position or momentum) of the field, allowing one to achieve better precision in measuring said quadrature, at the expenses of the other quadrature, which experiences anti-squeezing, i.e., an amplification of its uncertainty [18, 23, 24]. In our case, the squeezing parameter can be calculated as follows

$$S(t) = \langle \hat{a}^2(t) \rangle = \left(\frac{1}{\pi}\right)^2 \int d^2\alpha\, d^2\beta\, (\beta^*)^2 |G(\beta, t; \alpha, 0)|^2, \tag{7.57}$$

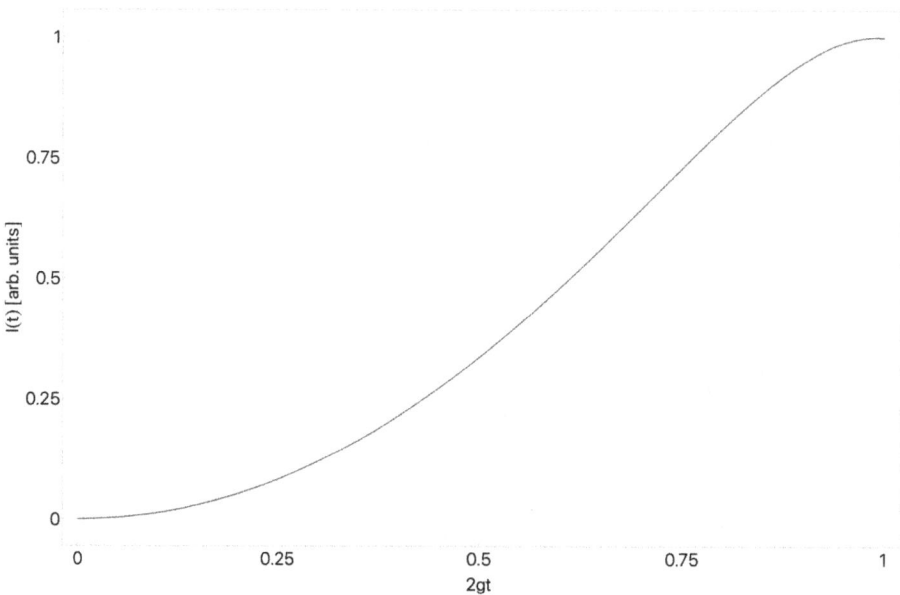

Figure 7.2. Time evolution of the intensity of the signal field as a function of the normalised time $\tau = 2gt$. This curve represents, as a function of time, how the intensity of the signal field changes, as a result of the nonlinear interaction in the crystal. As (normalised) time goes by, more pump photons are converted into signal ones, thus amplifying the signal beam.

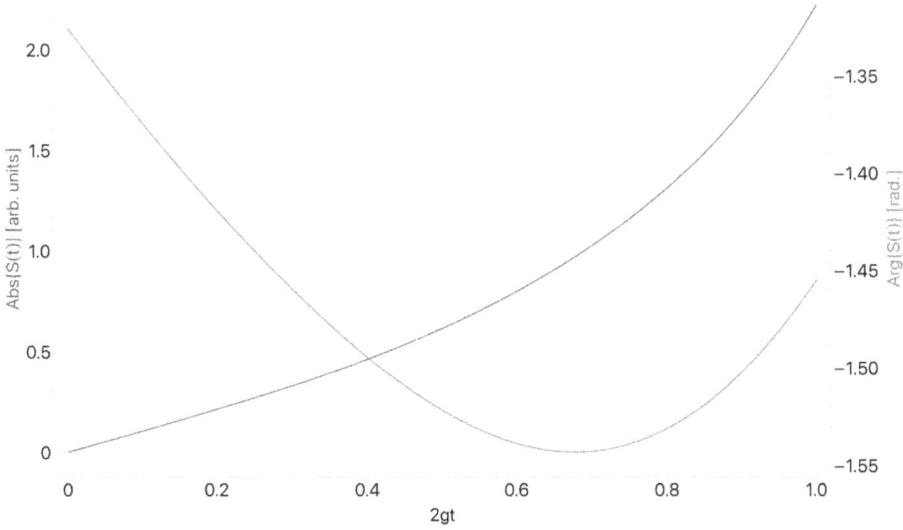

Figure 7.3. Time evolution of the modulus (blue) and phase (red) of the time-dependent squeezing parameter $S(t)$ for the degenerate parametric amplifier derived in this section. The plots are made as a function of the normalised time $\tau = 2gt$, and frequency $\omega/g = 1/2$. The modulus of the squeezing parameter represents the actual gain in precision, when measuring the correspondent quadrature (i.e., how much the variance of that quadrature is reduced, below the Heisenberg limit dictated by the uncertainty principle). The phase, on the other hand, indicates the direction, in phase space, along which the squeezing is happening. Notice, that $\mathrm{Arg}\{S(t)\} = 0$ corresponds to the position axis, while $\mathrm{Arg}\{S(t)\} = \pi/2$ to the momentum axis of the phase space.

which, once evaluated, gives

$$
\begin{aligned}
S(t) = &- i \exp(-i2\omega\ t)\sinh(2g\ t)\cosh(2g\ t) \\
&- \frac{1}{4}\exp(-i2\omega\ t)\{(2gt)^2[2\sinh(2g\ t)\cosh(2g\ t)] + gt[2\sinh^2(2g\ t) + 2] \\
&- 3\sinh^3(2g\ t)\cosh(2g\ t) - 2\sinh(2g\ t)\cosh(2g\ t)\}.
\end{aligned}
\tag{7.58}
$$

The time evolution of the modulus and phase of the time-dependent squeezing parameter are shown in figure 7.3.

7.2 Part II: quantum field theory, path integrals and photonics

The following sections use the formalism developed in chapter 6 to tackle the problem of light–matter interaction from a path integral perspective. Section 7.2.1 recasts the Huttner–Barnett model discussed in chapter 4 in terms of path integrals, to then derive, in section 7.2.2 the dielectric function, obtaining the same result as in section 4.7. The quantisation of the light–matter path integral is instead presented in section 7.2.3, where the path integral with respect to the electromagnetic field is carried out explicitly, and the expression for the dressed propagator is given. Then, section 7.2.4 derives the field operators and discusses the origin and implications of the noise currents The remaining section, instead, deal with the problem of describing the quantum interaction of the electromagnetic field in nonlinear media. Sections 7.2.5 and 7.2.6 lay out the necessary formalism to deal with nonlinearities within the path integral representation, while sections 7.2.7 and 7.2.8 present the explicit calculations for second-order nonlinear processes in the quantum and undepleted pump regime, respectively. The results of these two sections should be compared to those obtained in chapter 6 using a Hamiltonian, instead of Lagrangian, approach. Finally, third-order and cascaded nonlinearities are briefly discussed in sections 7.2.9 and 7.2.10, respectively.

7.2.1 Path integrals description of light–matter interaction

We start this section by recalling the form of the Huttner–Barnett Lagrangian, introduced in section 4.7 (equation (4.55))

$$
\begin{aligned}
\mathcal{L}_{\mathrm{HB}} = {} & \frac{1}{2}(|\mathbf{E}|^2 - |\mathbf{B}|^2) + \frac{\alpha(x)}{2\beta(x)}\left(|\partial_0 \mathbf{X}|^2 - \omega_0^2 |\mathbf{X}|^2\right) \\
& + \frac{\alpha(x)\rho(x)}{2}\int_0^\infty d\Omega\,(|\partial_0 \mathbf{Y}(\Omega)|^2 - \Omega^2|\mathbf{Y}(\Omega)|^2) \\
& - \alpha(x)(\mathbf{E}\cdot\partial_0\mathbf{X}) - \int_0^\infty d\Omega\, f(\Omega)\mathbf{X}\cdot\partial_0\mathbf{Y}(\Omega) \\
\equiv {} & \mathcal{L}_{\mathrm{em}} + \mathcal{L}_{\mathrm{mat}} + \mathcal{L}_{\mathrm{res}} + \mathcal{L}_{\mathrm{int}} + \mathcal{L}_{\mathrm{mr}}.
\end{aligned}
\tag{7.59}
$$

The various terms in the expression above have been discussed in section 4.7 already, but we briefly recall them here, for the sake of simplicity. The first three terms are the free Lagrangians for the electromagnetic, matter, and reservoir field, respectively. The fourth term represents the light–matter interaction term in the usual electric dipole approximation. The last term, instead, represents the interaction between the matter and reservoir field. A scheme of the system described by the equation above, and the various interactions between the fields is shown in figure 7.4(a).

Contrary to the Huttner–Barnett Lagrangian defined in equation (4.55), here we assume that the density $\rho(x)$ and the polarisability $\beta(x)$ can be position-dependent functions, to be more general. Moreover, we introduce the material function $\alpha(x)$, so that $\alpha(x) = 1$ inside the material, and $\alpha(x) = 0$ elsewhere.

Since we have three fields, i.e., the electromagnetic, matter and reservoir field, interacting with each other, to fully quantise the Huttner–Barnett in the path integral formalism as described in chapter 6, we need to introduce three source terms, namely $\mathbf{J}(x)$ for the electromagnetic field, $\mathbf{J}_X(x)$ for the matter field, and $\mathbf{J}_Y(\omega)$ for the reservoir field. This allows us to define the partition function of the quantised Huttner–Barnett model as

$$
Z(J, J_X, J_Y) = \int \mathcal{D}\mathbf{A}\mathcal{D}\mathbf{X}\mathcal{D}\mathbf{Y}\,\exp\left\{iS_{\mathrm{HB}} + i\int d^4x \left[J^\mu A_\mu + J_X^\mu X_\mu + J_Y^\mu Y_\mu\right]\right\},
\tag{7.60}
$$

where $S_{\mathrm{HB}} = \int d^4x\,\mathcal{L}_{\mathrm{HB}}$ is the Huttner–Barnett action. The three source terms introduced will describe, in the fully quantised theory, the dynamics of the correspondent quantum fields, i.e., photons for the electromagnetic current $\mathbf{J}$, polaritons

(a)

(b)

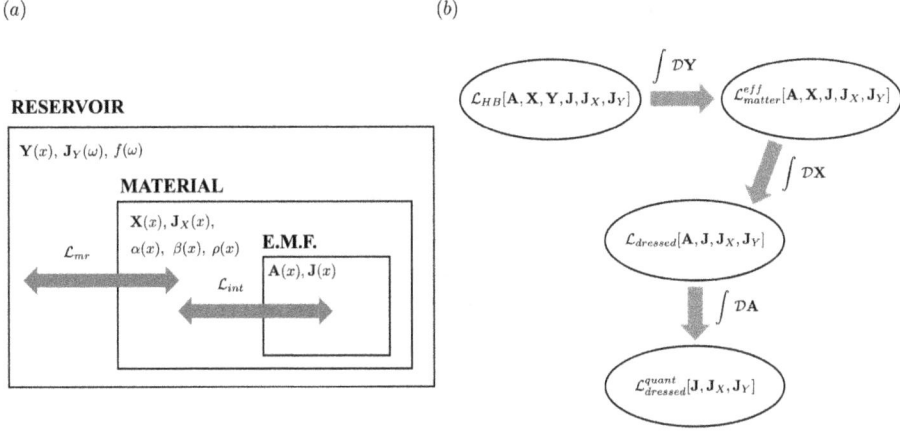

Figure 7.4. (a) Schematic representation of the hierarchy between the fields involved in the Huttner–Barnett model. The electromagnetic field is described by its vector potential $\mathbf{A}(x)$ and the source current $\mathbf{J}(x)$. It can interact with matter through the dipole interaction term represented by $\mathscr{L}_{\text{int}}$ (fourth term in equation (7.59)). The material is instead described by the matter field $\mathbf{X}(x)$ and the matter current $\mathbf{J}_X(x)$, with $\alpha(x)$, $\beta(x)$, and $\rho(x)$ being its constitutive parameters (respectively, material function, polarisability, and density). The matter field interacts with the electromagnetic field through the dipole interaction and with the resevoir field through the matter–reservoir interaction term $\mathscr{L}_{\text{mr}}$ (fifth term in equation (7.59)). Finally, the reservoir is described by the field $\mathbf{Y}(\omega)$ and its correspondent current $\mathbf{J}_Y(\omega)$, and characterised by the spectral coupling $f(\omega)$. The reservoir field interacts with the matter field through the interaction term $\mathscr{L}_{\text{mr}}$, which, effectively, models losses for the material. (b) Schematic representation of the steps made to quantise the Huttner–Barnett model in path integral formalism. Taking into account the hierarchy described in panel (a), the Huttner–Barnett Lagrangian $\mathscr{L}_{\text{HB}}$ is firstly integrated with respect to the reservoir field $\mathbf{Y}(\omega)$, giving an effective matter Lagrangian $\mathscr{L}_{\text{matter}}^{\text{eff}}$. This first step allows one to have an effective description of losses in the material. Then, an integration over the matter degrees of field $\mathbf{X}(x)$ results in the dressed Lagrangian $\mathscr{L}_{\text{dressed}}$, corresponding to the effective Lagrangian for a classical free electromagnetic field, dressed by the presence of a dispersive and lossy medium. Finally, by integrating over the electromagnetic field $\mathbf{A}(x)$, one obtains the quantum dressed Lagrangian $\mathscr{L}_{\text{dressed}}^{\text{quant}}$ and the correspondent quantum partition function $Z(\mathbf{J}, \mathbf{J}_X, \mathbf{J}_Y)$, which, in general, contains information about photon dynamics (generated by the electromagnetic current $\mathbf{J}$), polariton dynamics (through the matter current $\mathbf{J}_X$), and phonon dynamics (through the reservoir current $\mathbf{J}_Y$).

for the matter current $\mathbf{J}_X$, and phonons for the reservoir current $\mathbf{J}_Y$. A schematic representation of the quantisation steps involving path integration with respect to the different fields appearing in the Huttner–Barnett model is given in figure 7.4(b).

In this section, we focus on the classical problem, i.e., that of finding a suitable effective (or dressed) electromagnetic partition function describing the electromagnetic field as freely propagating in a 'free space' dressed by the properties of the medium contained in equation (7.59). This can be done by taking the first two steps shown in figure 7.4(b), i.e., integrating the term $\exp(iS_{\text{HB}})$, i.e., the integrand of equation (7.60), with respect to the reservoir and matter degrees of freedom, to obtain

$$\int \mathscr{D}\mathbf{X}\mathscr{D}\mathbf{Y} \exp\left[iS_{\text{HB}}\right] \equiv \exp\left[i\int d^4x \; \mathscr{L}_{\text{dressed}}\right], \qquad (7.61)$$

where also, for the sake of simplicity, we have assumed $J_X^\mu = 0 = J_Y^\mu$, as we are only interested in photon dynamics[4].

The goal of this section is then to calculate the explicit expression of the dressed Lagrangian above. To do that, we need to perform path integration with respect to both the reservoir and matter fields. This will result in having an effective Lagrangian for the electromagnetic field resembling that of a free field (i.e., of the form $J_\mu G^{\mu\nu} J_\nu$), but with the propagator being more complicated than its free-space counterpart, as it now contains information not only on the propagation of the electromagnetic field, but also on the (linear) optical properties of the material it is propagating through.

This will give us access to the dielectric properties of the material (see next section), as well as to the dressed partition function

$$Z_{\text{dressed}}(J) = \int \mathcal{D}\mathbf{A} \, \exp\left\{\int d^4x [\mathcal{L}_{\text{dressed}} + J^\mu A_\mu]\right\}, \tag{7.62}$$

which will be used in section (7.2.3) as a starting point to quantise the electromagnetic field. We can already anticipate, that since the electromagnetic field freely propagates in the material (i.e., there are no nonlinear interactions), we expect the quantum dressed partition function to have a form similar to that of equation (6.99), i.e.,

$$Z(J) = \exp\left[\frac{i}{2}\int d^3r d^3r' d\omega \, J_\mu(\mathbf{r}, \omega) G_{\text{eff}}^{\mu\nu}(\mathbf{r} - \mathbf{r}', \omega) J_\nu(\mathbf{r}', \omega)\right], \tag{7.63}$$

where ω is the angular frequency, and $G_{\text{eff}}^{\mu\nu}(\mathbf{r} - \mathbf{r}', \omega)$ is the dressed Feynman propagator for the electromagnetic field, which accounts for the (linear) optical properties of the material the electromagnetic field is propagating into. The explicit expression for the dressed propagator is calculated explicitly in section 7.2.3.

As shown in figure 7.4(b), the necessary steps towards obtaining the dressed Lagrangian are path integration first with respect to the reservoir and then the matter field. Doing calculations in this order allows us to make losses naturally appear as a renormalisaiton of matter polarisaiton and also gives us the ossibility to derive an integro-differential equation for the dressed electromagnetic propagator (see equation (7.111)).

7.2.1.1 Step 1: path integral for the reservoir field
From the Huttner–Barnett model (7.59) we can extract the reservoir Lagrangian as

$$\mathcal{L}_{\text{reservoir}} = \mathcal{L}_{\text{res}} + \mathcal{L}_{\text{mr}} = \frac{\alpha(x)\rho(x)}{2}\int_0^\infty d\Omega \, (|\partial_0 \mathbf{Y}(\Omega)|^2 - \Omega^2|\mathbf{Y}(\Omega)|^2) \\ - \int_0^\infty d\Omega \, f(\Omega)\mathbf{X} \cdot \partial_0 \mathbf{Y}(\Omega). \tag{7.64}$$

---

[4] Generalising the results below to the case of polaritons or phonon dynamics only amounts to add back the correspondent current terms in the relevant steps and keep them in the path integration. The steps for the matter current are shown in section 7.2.3. The same approach could also be used to keep count of the phonon currents $\mathbf{J}_Y$.

A closer inspection of the expression above reveals that the reservoir Lagrangian is, essentially, a quadratic form in the reservoir field $\mathbf{Y}(\mathbf{x}, t; \Omega)$, with a linear interaction term that contains the matter field $\mathbf{X}$. This suggests, that the path integration with respect to the reservoir field can be carried out using the technique of Gaussian integration described in chapter 6. However, we need some preparatory work to rewrite the reservoir Lagrangian explicitly as a quadratic form in $\mathbf{Y}(\mathbf{x}, t; \Omega)$. To do so, let us manipulate the term appearing in the first line above as follows

$$
\begin{aligned}
&\int dt \int_0^\infty d\Omega \, (|\partial_0 \mathbf{Y}(\Omega)|^2 - \Omega^2 |\mathbf{Y}(\Omega)|^2) \\
&= \int dt \int_0^\infty d\Omega \left[\partial_0(|\mathbf{Y}(\Omega)\partial_0\mathbf{Y}(\Omega)|) - |\mathbf{Y}(\Omega)\partial_0^2\mathbf{Y}(\Omega)| - \Omega^2|\mathbf{Y}(\Omega)|^2 \right] \\
&= \int dt \int_0^\infty d\Omega[|\partial_0\mathbf{Y}(\Omega)|^2 - \Omega^2|\mathbf{Y}(\Omega)|^2] \\
&= \int dt\, dt' \int_0^\infty d\Omega \, |\mathbf{Y}(\mathbf{r}, t'; \Omega)|\left[\left(\partial_0^2 - \Omega^2\right)\delta(t - t')\right]|\mathbf{Y}(\mathbf{r}, t; \Omega)|.
\end{aligned}
$$
(7.65)

To go from the first to the second line we have used the identity

$$
\frac{\partial}{\partial t}\left[f \frac{\partial f}{\partial t} \right] = \left(\frac{\partial f}{\partial t} \right)^2 + f \frac{\partial^2 f}{\partial t^2},
$$
(7.66)

to rewrite the term $|\partial_0 \mathbf{Y}(\Omega)|^2$. To go from the second to the third line we have instead integrated the first term by parts, obtaining

$$
|\partial_0(\mathbf{Y}\partial_0\mathbf{Y})| - |\mathbf{Y}|\,|\partial_0^2\mathbf{Y}| = |\partial_0\mathbf{Y}||\partial_0\mathbf{Y}| + |\mathbf{Y}|\,|\partial_0^2\mathbf{Y}| - |\mathbf{Y}|\,|\partial_0^2\mathbf{Y}| \rightarrow |\mathbf{Y}\partial_0^2\mathbf{Y}|.
$$
(7.67)

Finally, in the last line we have used the definition of the Dirac delta function $\int dx\, \delta(x - x_0) = 1$ to rewrite the expression above explicitly in terms of a quadratic form of the type $\mathbf{F}(t') \cdot [\hat{A}(t - t')\mathbf{F}(t)]$.

Using these result and introducing the quantities

$$
\hat{A}(t - t') = \frac{i\rho(x)\alpha(x)}{2}\left[\left(\partial_0^2 - \Omega^2\right)\delta(t - t')\right],
$$
(7.68a)

$$
\mathbf{B}(\mathbf{r}, t') = -i\alpha(x)f(\Omega)\delta(t - t')\partial_0\mathbf{X}(\mathbf{r}, t'),
$$
(7.68b)

allows us to rewrite equation (7.64) as a quadratic form in $\mathbf{Y}$ and to calculate the correspondent path integral using Gaussian integration as follows:

$$
\begin{aligned}
I_Y &= \int \mathscr{D}\mathbf{Y} \exp\left\{ -\frac{1}{2}\int d^3r\,dt\,dt' \int_0^\infty d\Omega[|\mathbf{Y}(\mathbf{r}, t'; \Omega)|\hat{A}(t - t') + \mathbf{B}(t') \cdot |\mathbf{Y}(\mathbf{r}, t; \Omega)|] \right\} \\
&= \mathscr{N}_Y \exp\left\{ -\frac{i}{2}\int d^3r\,dt \frac{\alpha(x)}{\rho(x)}\left[\int_0^\infty d\,\Omega|f(\Omega)|^2\,|\mathbf{X}(\mathbf{r}, t)|^2 \right.\right. \\
&\quad \left.\left. + \int dt'\mathbf{X}(\mathbf{r}, t')G_{Fy}(\mathbf{r}, t - t')\mathbf{X}(\mathbf{r}, t) \right] \right\},
\end{aligned}
$$
(7.69)

where $\mathcal{N}_\gamma$ is a normalisation constant proportional to $\det \hat{A}$, and $G_{F\gamma}(\mathbf{r}, t - t')$ is the time-domain Green's function of the reservoir field (associated with the inverse of $\hat{A}(t - t')$), defined as

$$G_{F\gamma}(\mathbf{r}, t - t') = , \int_0^\infty d\,\Omega\,\Omega^2\,|f(\Omega)|^2 D_F(t - t', \Omega), \qquad (7.70)$$

where $D_F(t - t', \Omega)$ is the Feynman propagator for the reservoir field, whose explicit expression is given by equation (5.40). The presence of the matter polarisation term $\mathbf{X}(\mathbf{r}, t)$ in equation (7.69), however, means that the effect of the reservoir, as we will see in the next paragraph, is that of renormalising the matter degrees of freedom to account for losses. Hence, to complete the picture and obtain a dressed propagator for the electromagnetic field, we need to also perform the integration with respect to the matter degrees of freedom. This, as we will see, gives rise to photon-polariton propagators.

7.2.1.2 Step 2: path integral for the matter field

The next step of our calculation towards the dressed electromagnetic Lagrangian is to calculate, according to figure 7.4, the path integral with respect to the matter degrees of freedom. The integration with respect to the reservoir generates the following effective matter Lagrangian[5]

$$\begin{aligned}
\mathscr{L}^{\text{eff}}_{\text{matter}} = \frac{\alpha(x)}{2\beta(x)} & \left[|\partial_0 \mathbf{X}|^2 - \left(\omega_0^2 + \frac{\beta(x)}{\rho(x)} \int_0^\infty d\omega\,|f(\Omega)|^2 \right) |\mathbf{X}|^2 \right] \\
& - \frac{\alpha(x)}{2\rho(x)} \int dt'\,\mathbf{X}^*(t')G_{F\gamma}(\mathbf{r}, t - t')\mathbf{X}(t) - \alpha(x)\mathbf{E}\cdot\mathbf{X}.
\end{aligned} \qquad (7.71)$$

Notice, how comparing the Lagrangian above to the matter part from the Huttner–Barnett model in equation (7.59) reveals how the integration over the reservoir field generated two extra terms in the matter Lagrangian. One acts as a correction term to the characteristic frequency of the matter field (the ω-integral proportional to $|\mathbf{X}|^2$, while the other one is a correction to the kinetic term, expressed as a convolution between the matter field and the reservoir propagator (first term in the second line of the expression above).

It is then useful to redefine the characteristic frequency of the matter field as follows, before proceeding further with calculations

$$\tilde{\omega}_0^2 = \omega_0^2 \left[1 + \frac{\beta(x)}{\rho(x)} \int_0^\infty d\Omega\,|f(\Omega)|^2 \right]. \qquad (7.72)$$

---

[5] In writing equation (7.71), we have introduced the complex conjugate of the matter field in the convolution term, so that it can be written in a quadratic form similar to $|\partial_0 \mathbf{X}|^2 = (\partial_0 \mathbf{X}^*)(\partial_0 \mathbf{X})$. We will use this convention also for the electromagnetic field, so that everything remains consistent with the initial assumption of considering the electromagnetic, matter, and reservoir fields as complex vector fields.

The physical meaning of this renormalisation operation is to assign an imaginary part to the frequency, making it complex, which is reminiscent of the fact that the material contains loss channels, modelled by the reservoir field. The particular type of losses that one could model is tied with the explicit expression of the spectral coupling $f(\Omega)$. We will discuss this more in detail later, when deriving the dielectric function from the effective Lagrangian model, and we will see that different choices of $f(\Omega)$ can lead to different materials being described by this model.

Next, we can transform the effective Lagrangian (7.71) into a quadratic form $\mathbf{X} \cdot [\hat{B}(\mathbf{r}, t - t')\mathbf{X}(t')] - i\mathbf{C}(\mathbf{r}, t) \cdot \mathbf{X}(t)$ following the same line of reasoning of the previous calculation step, where now

$$\hat{B}(\mathbf{r}, t - t') = \frac{i\alpha(x)}{\omega_0^2 \beta(x)}\left(\frac{\partial^2}{\partial t^2} + \tilde{\omega}_0^2\right)\delta(t - t') - \frac{i\alpha(x)}{\rho(x)}G_{F_y}(\mathbf{r}, t - t'), \qquad (7.73a)$$

$$\mathbf{C}(\mathbf{r}, t) = -i\alpha(x)\mathbf{E}(\mathbf{r}, t). \qquad (7.73b)$$

This allows us to use once again Gaussian integration[6] to calculate the path integral, obtaining the following result

$$\int \mathscr{D}\mathbf{X} \exp\left[\int d^4x \mathscr{L}_{\text{matter}}^{\text{eff}}\right] = \mathscr{N} \exp\left[\frac{i}{2}\int d^3x dt dt' \, \alpha(x)\mathbf{E}(t')\Gamma(\mathbf{r}, t - t')\mathbf{E}(t)\right], \quad (7.74)$$

where $\Gamma(\mathbf{r}, t - t')$ is a rank-2 tensor, representing the inverse of the operator $\hat{B}(\mathbf{r}, t - t')$, and satisfies the integro-differential equation

$$\frac{1}{\omega_0^2 \beta(x)}\left(\frac{\partial^2}{\partial t^2} + \tilde{\omega}_0^2\right)\Gamma(\mathbf{r}, t - t') - \frac{1}{\rho(x)}\int d\tau \, G_{F_y}(\mathbf{r}, t - \tau)\Gamma(\mathbf{r}, \tau - t') = \delta(t - t') \qquad (7.75)$$

The propagator $\Gamma(\mathbf{r}, t - t')$ represents the correction to the free-space propagator of the electromagnetic field due to the presence of matter. Notice, moreover, how the result in equation (7.74) only involves a quadratic form in the electric field, while the magnetic field is not present. This is due to the fact, that the Huttner–Barnett model describes electromagnetic interactions in non-magnetic materials, so the magnetic field is not affected by the material properties. In the more general case of a material with magnetic properties, the correspondent Huttner–Barnett Lagrangian will contain a term proportional to the magnetic field (a Lorentz-force-like term, for example), which will result in the appearance of a term of the kind $\mathbf{B} \cdot \Gamma_{\text{M}} \cdot \mathbf{B}$ in equation (7.74), which will contain the correction to the free-space propagation of the magnetic field due to the presence of matter. Terms of the form $\mathbf{E} \cdot \Gamma_{\text{EM}} \cdot \mathbf{B}$ might also arise, if the material allows for electric–magnetic interactions, which will still correspond to (more complicated) corrections to free-space propagation, once represented in the vector potential, rather than electromagnetic, basis.

---

[6] The integration steps are fully analogous to those in equation (7.69) and are left to the reader as an exercise.

We can make this correction to free-space propagation appear explicitly if we write the complete expression of the dressed Lagrangian, by adding the free electromagnetic Lagrangian to the result above. This gives the following result

$$\mathscr{L}_{\text{dressed}} = \frac{1}{2} \int dt' \left\{ \mathbf{E}(t')[\varepsilon_0 \delta(t - t') + \alpha(x)\Gamma(\mathbf{r}, t - t')]\mathbf{E}(t) - \frac{1}{\mu_0}|\mathbf{B}|^2 \right\}. \quad (7.76)$$

To obtain this result, we have rewritten the electric part of the electromagnetic Lagrangian as follows

$$\frac{\varepsilon_0}{2}|\mathbf{E}(t)|^2 = \frac{\varepsilon_0}{2}\mathbf{E}(t) \cdot \mathbf{E}(t) = \frac{\varepsilon_0}{2} \int dt' \, \mathbf{E}(t')\delta(t - t') \, \mathbf{E}(t) \quad (7.77)$$

to make appear explicitly, through the Dirac delta function, the dependence on the integration variable t', so that we can sum this contribution to that coming from equation (7.74).

The dressed Lagrangian above is the final result for the classical theory of light–matter interaction and contains the same amount of information as the Hamiltonian in equation (4.64). As an example of this, in the next section we derive the dielectric function $\varepsilon(\omega)$ of the material from the dressed Lagrangian above.

7.2.2 Dielectric function from path integrals

The dressed Lagrangian (7.76) contains all information on both the dynamics of the electromagnetic field inside a material, and the optical properties of the material itself. In particular, the quantity inside square brackets in equation (7.76) is closely related to the dielectric permittivity of the material. To see this in an easy manner, we recall that the dielectric properties of a material are encoded in the constitutive relation between the electric field and the displacement vector field [10], i.e.,

$$\mathbf{D}(\mathbf{r}, t) = \int d\tau \, \varepsilon(\mathbf{r}, t - \tau)\mathbf{E}(\mathbf{r}, \tau), \quad (7.78)$$

or, in the more familiar form in frequency domain

$$\mathbf{D}(\mathbf{r}, \omega) = \varepsilon(\mathbf{r}, \omega)\mathbf{E}(\mathbf{r}, \omega). \quad (7.79)$$

Standard electromagnetic field theory allows us to calculate the displacement vector field from the dressed Lagrangian as [25]

$$\mathbf{D}(\mathbf{r}, t) = \frac{\partial \mathscr{L}_{\text{dressed}}}{\partial \mathbf{E}(t)}. \quad (7.80)$$

In our case, however, it is more insightful (and it also allows us to work on more familiar ground) to use a similar relation for the frequency domain counterpart of the dressed Lagrangian (7.76). This can be easily obtained by introducing the Fourier transform of the electric (and magnetic) field as

$$\mathbf{E}(\mathbf{r}, t) = \int d\omega \, \mathbf{E}(\mathbf{r}, \omega)\exp(-i\omega t), \quad (7.81)$$

to transform, using the relation $\mathbf{E}(\mathbf{r}, -\omega) = \mathbf{E}^*(\mathbf{r}, \omega)$ (and a similar one for the magnetic field), equation (7.76) into the following Fourier-space dressed Lagrangian

$$\mathscr{L}_{\text{dressed}} = \frac{1}{2}\mathbf{E}^*(\omega)[\varepsilon_0 + \tilde{\Gamma}(\mathbf{r}, \omega)]\mathbf{E}(\omega) - \frac{1}{\mu_0}|\mathbf{B}(\omega)|^2, \tag{7.82}$$

where $\tilde{\Gamma}(\mathbf{r}, \omega)$ is the Fourier transform of the matter propagator, which can be derived from equation (7.75) and whose explicit form is given below. In Fourier space, equation (7.80) now becomes

$$\mathbf{D}(\mathbf{r}, \omega) = \frac{\partial \mathscr{L}_{\text{dressed}}}{\partial \mathbf{E}^*(\omega)}, \tag{7.83}$$

and, when applied to equation (7.82), gives the following result

$$\mathbf{D}(\mathbf{r}, \omega) = \varepsilon_0[1 + \alpha(x)\tilde{\Gamma}(\mathbf{r}, \omega)]\mathbf{E}(\omega) \equiv \varepsilon_0\varepsilon(\omega)\mathbf{E}(\omega), \tag{7.84}$$

and allows one to define the dielectric function $\varepsilon(\omega)$ as

$$\varepsilon(\omega) = 1 + \frac{\alpha(x)}{\varepsilon_0}\tilde{\Gamma}(\mathbf{r}, \omega). \tag{7.85}$$

The dielectric permittivity is then proportional to the Green's function of the matter field $\tilde{\Gamma}(\mathbf{r}, \omega)$. To find its explicit expression, we can take the Fourier transform of equation (7.75) to get

$$\tilde{\Gamma}(\mathbf{r}, \omega) = \frac{\omega_0^2\beta(x)}{\tilde{\omega}_0^2 - \omega^2 - \dfrac{\omega_0^2\beta(x)}{\rho(x)}\tilde{G}_{F_y}(\mathbf{r}, \omega)}, \tag{7.86}$$

where $\tilde{G}_{F_y}(\mathbf{r}, \omega)$ is the Fourier transform of the reservoir Green's function, whose explicit expression can be readily found by manipulating equation (7.70) as follows

$$G_{F_y}(\mathbf{r}, t) = \int_0^\infty d\Omega\, \Omega^2|f(\Omega)|^2 D_F(t, \Omega) = \int_0^\infty d\Omega\, \Omega^2|f(\Omega)|^2 \int \frac{d\omega}{2\pi}\frac{\exp(i\omega t)}{\Omega^2 - \omega^2 + i\varepsilon}$$
$$= \int d\omega \left[\int_0^\infty \frac{d\Omega}{2\pi}\frac{\Omega^2|f(\Omega)|^2}{\Omega^2 - \omega^2 + i\varepsilon}\right]\exp(i\omega t), \tag{7.87}$$

from which we can directly read the expression of $\tilde{G}_{F_y}(\mathbf{r}, \omega)$ as the expression in square brackets, i.e.,

$$\tilde{G}_{F_y}(\mathbf{r}, \omega) = \int_0^\infty \frac{d\Omega}{2\pi}\frac{\Omega^2|f(\Omega)|^2}{\Omega^2 - \omega^2 + i\varepsilon}$$
$$= \int_0^\infty \frac{d\Omega}{2\pi}|f(\Omega)|^2 + \frac{\omega^2}{2\pi}\int_0^\infty d\Omega\frac{|f(\Omega)|^2}{\Omega^2 - \omega^2 + i\varepsilon}, \tag{7.88}$$

where to pass from the first to the second line we have used the relation $\Omega^2 = (\Omega^2 - \omega^2) + \omega^2$ to rewrite the Ω^2 factor at the numerator, and then set $\varepsilon = 0$ in the term containing $(\Omega^2 - \omega^2)$ to allow for simplification.

We can use this result to rewrite the denominator of equation (7.86) is a simpler form as follows

$$\tilde{\omega}_0^2 - \omega^2 - \frac{\omega_0^2 \beta(x)}{\rho(x)} \tilde{G}_{F_y}(\mathbf{r}, \omega) = \tilde{\omega}_0^2 - \omega^2 - \frac{\omega_0^2 \beta(x)}{\rho(x)} \left[\int_0^\infty d\Omega \, |f(\Omega)|^2 \right.$$
$$\left. + \frac{\omega^2}{2\pi} \int_0^\infty d\Omega \, \frac{|f(\Omega)|^2}{\Omega^2 - \omega^2 + i\varepsilon} \right] \tag{7.89}$$
$$= -\omega^2 + \omega_0^2 - \omega^2 \lambda(\omega).$$

To get from the second to the third line we have used equation (7.72) to eliminate $\tilde{\omega}_0^2$, and we have introduced the spectral function

$$\lambda(\omega) = \frac{1}{\rho^2(x)} \int_0^\infty d\Omega \, \frac{|g(\Omega)|}{\Omega^2 - \omega^2 + i\varepsilon}, \tag{7.90}$$

with

$$g(\Omega) = \sqrt{\frac{\omega_0^2 \beta(x) \rho(x)}{2\pi}} f(\Omega). \tag{7.91}$$

The physical meaning of the spectral function $\lambda(\omega)$ is better grasped if we take a look at the final form of the dielectric function, obtained combining the result above with equations (7.86) and (7.85)

$$\varepsilon(\mathbf{r}, \omega) = 1 + \alpha(x) \frac{\omega_0^2 \beta(x)}{\omega_0^2 - \omega^2 [1 + \lambda(\omega)]}. \tag{7.92}$$

Here, we see how $\lambda(\omega)$ acts as a (possibly frequency-dependent) correction factor to the ω^2-term in the denominator of $\tilde{\Gamma}(\mathbf{r}, \omega)$. This means, that its main function is to modify the resonance condition for the dielectric function from $\omega = \pm\omega_0$ to a more complicated one, depending on the explicit form of $\lambda(\omega)$. The simplest form of modification, in line with the fact that its expression contains information about the spectral coupling of the reservoir field, is to introduce losses in the material. This can be simply seen by defining $\lambda(\omega) = i\gamma/\omega$, so that the equation above becomes

$$\varepsilon(\mathbf{r}, \omega) = 1 + \frac{\omega_0^2 \beta(x)}{\omega_0^2 - \omega^2 - i\gamma\omega}, \tag{7.93}$$

which is the frequency representation of a damped harmonic oscillator [22]. However, as already discussed in section 4.7, the model presented here cannot describe Lorentz-like materials, since the function $\lambda(\omega)$ cannot take the value γ/ω (see the discussion after equation (4.69) or reference [26]).

We can, however, have a closer look at the form and properties of the (generally complex) function $\lambda(\omega)$ and see under which circumstances it can assume explicit forms, which, as we will see, correspond to metals, or epsilon-near-zero materials.

The guiding principle to investigate the forms $\lambda(\omega)$ can take is that the dielectric function $\varepsilon(\omega)$ (7.85) must always remain a causal function, as this is a fundamental requirement for electrodynamics [10]. Mathematically, this translates in requiring that the analytic continuation of $\varepsilon(\omega)$ to the complex plane is holomorphic in the upper-half complex plane, i.e., it doesn't have any poles there.

To check the causality of $\varepsilon(\omega)$ we then need to have a closer look at the denominator of equation (7.92). First, however, let us rewrite $\lambda(\omega)$ in an easier manner by noticing that

$$\frac{1}{\Omega^2 - \omega^2 + i\varepsilon} = \frac{1}{2\omega}\left(\frac{1}{\Omega - \omega + i\varepsilon} + \frac{1}{\Omega + \omega - i\varepsilon}\right). \tag{7.94}$$

Substituting this result in the definition of $\lambda(\omega)$ (equation (7.90)) and recalling that the function $g(\Omega)$ introduced in equation (7.91) is an even function of frequency [i.e., $g(-\Omega) = g(\Omega)$] then gives

$$\begin{aligned}
\lambda(\omega) &= \frac{1}{\rho^2(x)}\int_0^\infty d\Omega\,\frac{|g(\Omega)|^2}{\Omega^2 - \omega^2 + i\varepsilon} = \frac{1}{2\omega\rho^2(x)}\left[\int_0^\infty d\Omega\frac{|g(\Omega)|^2}{\Omega - \omega + i\varepsilon}\right. \\
&\left. + \int_{-\infty}^0 d(-\Omega)\,\frac{|g(-\Omega)|^2}{-\Omega + \omega - i\varepsilon}\right] = \frac{1}{2\omega\rho^2(x)}\int_{-\infty}^{+\infty} d\Omega\frac{|g(\Omega)|^2}{\Omega - \omega - i\varepsilon}.
\end{aligned} \tag{7.95}$$

Cast in this form, it is easy to see that $\lambda(\omega)$ is analytic in the upper-half complex plane.

The denominator of the dielectric function can be then written as

$$\omega_0^2 - \omega^2[1 + \lambda(\omega)] \equiv D(\omega), \tag{7.96}$$

and the condition $D(\omega) \neq 0$ for all $\omega > 0$ must hold to enforce causality on $\varepsilon(\omega)$. To check this condition, let us assume $\omega = \omega_r + i\omega_i$ and consider first the case, where *omega* lies entirely on the imaginary axis, i.e., $\omega_r = 0$. In this case, the expression above becomes

$$\begin{aligned}
D(i\omega_i) &= \omega_0^2 + \omega_i^2 + \frac{\omega_i^2}{\rho^2(x)}\int_0^\infty d\Omega\,\frac{|g(\Omega)|^2}{\Omega^2 - (i\omega_i)^2 + i\varepsilon} \\
&= \omega_0^2 + \omega_i^2 + \frac{\omega_i}{\rho^2(x)}|g(\omega_i^2)|^2.
\end{aligned} \tag{7.97}$$

Notice that $D(i\omega_i) \in \mathbb{R}$, which means that, on the imaginary axis, $\mathrm{Im}\{D(\omega)\} = 0$. For the general case with $\omega_r \neq 0$, the equation above modifies to

$$D(\omega) = \omega_0^2 - (\omega_r + i\omega_i)^2 - (\omega_r + i\omega_i)\frac{|g(\omega)|^2}{\rho^2(x)}, \tag{7.98}$$

showing that if $D(\omega)$ has some zeros, they necessarily need to lie on the imaginary axis, where, however, we just proved that $D(i\omega_i) > 0$. This argument is then enough to guarantee, that the permittivity $\varepsilon(\omega)$ is a causal function of ω. Notice, moreover, that the same observations done in section 4.7 are still valid for the permittivity derived from path integrals, since it is based on the same microscopic model.

7.2.3 Dressed electromagnetic propagator

Now that we have derived the dressed Lagrangian and looked at the classical properties of the electromagnetic field in a dispersive and lossy medium, the next step is to look at its quantum properties. To do so, according to figure 7.4 (b), we need to integrate the dressed model with respect to the vector potential $\mathbf{A}(x)$. In doing so, however, only taking the expression of the dressed Lagrangian in equation (7.76) is not enough, as we need to include the matter current $\mathbf{J}_X$ in our calculation, as they carry information on the so-called *noise currents*, fundamental for quantising the electromagnetic field in a lossy medium in the correct manner. The interested reader can find extensive information on this in the book by Vogel and Welsch [27]. Here, we will defer discussion of this issue to a later point, when we will have access to the explicit form of the dressed propagator.

The first step towards the full quantum dressed partition function $Z_0[J]$ is then fairly easy: we need to include the matter sources in the dressed Lagrangian in equation (7.76). To do so, we retrace our calculation leading to equation (7.76) and notice that the introduction of a source term for the matter field of the form $\int dt\, d^3r\, \alpha(x)\mathbf{J}_X(\mathbf{r},t) \cdot \mathbf{X}(\mathbf{r},t)$ only amounts to a change in equation (7.73b), i.e., $-i\alpha(x)\mathbf{E}(\mathbf{r},t) \rightarrow -i\alpha(x)[\mathbf{E}(\mathbf{r},t) - \mathbf{J}_X(\mathbf{r},t)]$. This, in turn, changes the expression of the dressed Lagrangian to

$$\mathscr{L}_{\text{dressed}} = \mathscr{L}_{\text{em}} + \frac{1}{2}\int dt\, dt'\, d^3r\, \alpha(x)[\mathbf{E}(\mathbf{r},t) - \mathbf{J}_X(\mathbf{r},t)]\Gamma(t-t',\mathbf{r})[\mathbf{E}(\mathbf{r},t') - \mathbf{J}_X(\mathbf{r},t')], \quad (7.99)$$

where $\mathscr{L}_{\text{em}}$ is the free electromagnetic Lagrangian.

The second task consists in rewriting the expression above and, in particular, the free electromagnetic Lagrangian, in terms of the vector potential solely. There are several ways to do this. For example, one could use the relativistic form of the free electromagnetic Lagrangian, i.e., $\mathscr{L}_{\text{em}} = -F_{\mu\nu}F^{\mu\nu}/4$ and then attach the appropriate gauge fixing term through the Faddeev–Popov trick (see section 6.3.1). This will give us a partition function that can describe the quantum properties of the dressed electromagnetic field in any gauge of choice. Another, simpler, approach is that to fix a gauge (thus removing the necessity of the Faddeev–Popov quantisation scheme) and use the definition of electric and magnetic fields in terms of the vector potential to rewrite the dressed electromagnetic Lagrangian as a quadratic form in the vector potential, and then employ gaussian integration to calculate the path integral.

Here, we choose the latter approach. Our choice of gauge is the so-called Weyl (or temporal) gauge, $A_0 = \phi = 0$. Notice that in this gauge the vector potential is in general not a transverse vector, since the condition $\nabla \cdot \mathbf{A} = 0$ is not enforced by the

Weyl gauge[7]. In the Weyl gauge, the electric and magnetic field are connected to the vector potential through the usual relations

$$\mathbf{E} = -\partial_t \mathbf{A}, \tag{7.100a}$$

$$\mathbf{B} = \nabla \times \mathbf{A}, \tag{7.100b}$$

where now $\mathbf{A}(\mathbf{r}, t)$ is the actual vector potential (i.e., the spatial part of the 4-vector potential $A_\mu(x)$, since $A_0 = 0$. Moreover, since now there is no ambiguity anymore between the spacetime coordinate x and the spatial coordinate $\mathbf{x}$, we use $(\mathbf{x}, \mathbf{y})$ instead of $(\mathbf{r}, \mathbf{r}')$ as integration variables.

Substituting these into the electromagnetic Lagrangian, equation (4.1a), gives the following form for the electromagnetic action (see equation (4.4))

$$S_{\text{em}} = \frac{1}{2} \int dt \, d^3x \left[\varepsilon_0 \, |\partial_t \mathbf{A}|^2 - \frac{1}{\mu_0} |\nabla \times \mathbf{A}|^2 \right]. \tag{7.101}$$

The first term can be readily integrated by part with respect to time to give, in terms of the spatial components $A_i(\mathbf{r}, t)$ of the vector potential[8]

$$\varepsilon_0 \int dt \, d^3x \, |\partial_t \mathbf{A}|^2 = -\varepsilon_0 \int dt \, d^3r A_i \left(\partial_t^2 \delta_{ij} \right) A_j. \tag{7.102}$$

The second term can be rewritten using a similar argument but needs a bit more work. First, we write the curl in terms of components as [22]

$$\nabla \times \mathbf{A} = \varepsilon_{ijk} \hat{\mathbf{x}}_i \partial_j A_k, \tag{7.103}$$

where $\hat{\mathbf{x}}_i = \{\hat{\mathbf{x}}, \hat{\mathbf{y}}, \hat{\mathbf{z}}\}$, and ε_{ijk} is the totally antisymmetric Levi-Civita symbol [22]. Using this expression, we can calculate the square of the curl as follows

$$\begin{aligned} (\nabla \times \mathbf{A}) \cdot (\nabla \times \mathbf{A}) &= \varepsilon_{ijk} \varepsilon_{\ell mn} \hat{\mathbf{x}}_i \cdot \hat{\mathbf{x}}_\ell \partial_j A_k \partial_m A_n \\ &= (\delta_{jm} \delta_{kn} - \delta_{jn} \delta_{km}) \partial_j A_k \partial_m A_n \\ &= \partial_m A_n \partial_m A_n - \partial_n A_m \partial_m A_n. \end{aligned} \tag{7.104}$$

To go from the first to the second line we have used the fact that $\hat{\mathbf{x}}_i \cdot \hat{\mathbf{x}}_\ell = \delta_{i\ell}$ and the property of the Levi-Civita tensor $\varepsilon_{jki} \varepsilon_{mni} = \delta_{jm} \delta_{kn} - \delta_{jn} \delta_{km}$. Substituting the result above into the expression of the action, and integrating by parts with respect to space gives the final result

---

[7] Some observations here are in order. First, the Weyl gauge is incomplete, meaning that we have enough freedom to impose, if necessary, further constraints to the form of the vector potential. This freedom can be used, for example, to restore transversality of $\mathbf{A}$ after quantisation. Another important consequence of using the Weyl gauge is that it naturally suggests using Gupta–Bleuer quantisation for the electromagnetic field, i.e., representing the quantum field not with two orthogonal polarisation states, but with four polarisation states, two corresponding to the usual, transverse, polarisation, one to the longitudinal polarisation, and the fourth one to the so-called scalar polarisation [28–30].

[8] Remember that we are working, in time domain, with reals fields, so $|\mathbf{A}|^2 = \mathbf{A} \cdot \mathbf{A}$.

$$S_{em} = -\frac{1}{2} \int dt\, d^3x\, A_i \left[\left(\varepsilon_0 \partial_t^2 - \frac{1}{\mu_0} \nabla^2 \right) \delta_{ij} + \frac{1}{\mu_0} \partial_i \partial_j \right] A_j$$

$$\equiv -\frac{1}{2} \int dt\, dt'\, d^3x\, d^3y\, A_i(\mathbf{x}, t) R_{ij}(t - t', \mathbf{x} - \mathbf{y}) A_j(\mathbf{y}, t'),$$

(7.105)

where the electromagnetic Kernel $R_{ij}(t - t', \mathbf{x} - \mathbf{y})$ is defined as

$$R_{ij}(t - t', \mathbf{x} - \mathbf{y}) = \left[\left(\varepsilon_0 \partial_t^2 - \frac{1}{\mu_0} \nabla^2 \right) \delta_{ij} + \frac{1}{\mu_0} \partial_i \partial_j \right] \delta(t - t') \delta(\mathbf{x} - \mathbf{y}). \quad (7.106)$$

To transform the second term in equation (7.99), we write $\mathbf{E} = -\partial_t \mathbf{A}$ as defined above and then integrate by parts twice, one for $\mathbf{E}(\mathbf{x}, t)$ and one for $\mathbf{E}(\mathbf{x}, t')$. This will make terms proportional to both $\partial_t^2 \Gamma$ and $\partial_t \Gamma$ appear. Finally, we add the electromagnetic source term $J_i(\mathbf{x}, t) A_i(\mathbf{x}, t)$ so that the dressed action becomes

$$S_{dressed} = -\frac{1}{2} \int dt\, dt'\, d^3x\, d^3y\, A_i(\mathbf{x}, t) \left[R_{ij}(t - t', \mathbf{x} - \mathbf{y}) + \alpha(\mathbf{x}) \partial_t^2 \Gamma(t - t', \mathbf{x} - \mathbf{y}) \delta_{ij} \right] A_j(\mathbf{y}, t')$$

$$+ \frac{1}{2} \int dt\, dt'\, d^3x\, d^3y\, \alpha(\mathbf{x}) J_{X,i}(\mathbf{x}, t) \partial_t \Gamma_{ij}(t - t', \mathbf{x} - \mathbf{y}) J_{X,j}(\mathbf{y}, t')$$

$$- \frac{1}{2} \int dt\, dt'\, d^3x\, d^3y\, \alpha(\mathbf{x}) A_i(\mathbf{x}, t) \partial_t \Gamma_{ij}(t - t', \mathbf{x} - \mathbf{y}) J_{X,j}(\mathbf{y}, t') \qquad (7.107)$$

$$- \frac{1}{2} \int dt\, dt'\, d^3x\, d^3y\, \alpha(\mathbf{x}) J_{X,i}(\mathbf{x}, t) \partial_t \Gamma_{ij}(t - t', \mathbf{x} - \mathbf{y}) A_j(\mathbf{y}, t')$$

$$+ \int dt\, d^3x\, J_i(\mathbf{x}, t) A_i(\mathbf{x}, t),$$

where

$$\Gamma(t - t', \mathbf{x} - \mathbf{y}) = \Gamma(t - t', \mathbf{x}) \delta(\mathbf{x} - \mathbf{y}). \tag{7.108}$$

and the terms proportional to $\partial_t \Gamma$ and $\partial_t^2 \Gamma$ are defined accordingly. The term appearing in the second line of the equation above does not depend on the vector potential (this term, in fact, corresponds to the propagator for polaritons) and it can be therefore treated as a multiplicative constant and absorbed in the overall normalisation constant for the path integral.

We are now ready to perform the path integral with respect to the (gauged) vector potential $A_i(\mathbf{x}, t)$ using Gaussian integration to obtain

$$Z(J, J_X) = \int \mathscr{D}\mathbf{A}\, \exp\left[\frac{i}{\hbar} S_{dressed} \right] = \mathscr{N} \exp\left[\frac{i}{\hbar} \Sigma(J, J_X) \right]. \tag{7.109}$$

The functional $\Sigma(J, f)$ can be written symbolically as

$$\Sigma(J, J_X) = \frac{1}{2}(\partial_t \Gamma \cdot \mathbf{J}_X - \mathbf{J})_i\, G_{ij}\, (\partial_t \Gamma \cdot \mathbf{J}_X - \mathbf{J})_j, \tag{7.110}$$

where $\partial_t \Gamma \cdot \mathbf{J}_X$ stands for the term appearing in the third line of equation (7.105) and $(\cdots)_i$ means taking the ith component of the vector inside the brackets. The quantity G_{ij} is the inverse of the Kernel $(R_{ij} + \alpha \partial_t^2 \Gamma \delta_{ij})$ appearing in the first line of equation

(7.105) and represents the Green's function of the (free) dressed electromagnetic field, which is the solution of the following integro-differential equation

$$\left[\left(\varepsilon_0 \partial_t^2 - \frac{1}{\mu_0}\nabla^2\right)\delta_{ij} + \frac{1}{\mu_0}\partial_i\partial_j\right]G_{jk}(t - t', \mathbf{x}, \mathbf{y})$$

$$+ \alpha(x)\int d\tau\, \partial_t^2\Gamma(t - \tau, \mathbf{x})G_{ik}(t - \tau, \mathbf{x}, \mathbf{y}) = \delta_{ik}\delta(t - t')\delta(\mathbf{x} - \mathbf{y}). \tag{7.111}$$

If we take the Fourier transform with respect to time, we get, using the results of the previous section to make the dielectric function appear (see equation (7.85))

$$(-\nabla^2 \delta_{ij} + \partial_i\partial_j)\tilde{G}_{jk}(\mathbf{x}, \mathbf{y}, \omega) - \frac{\omega^2}{c^2}\varepsilon(\mathbf{x}, \omega)\tilde{G}_{ik}(\mathbf{x}, \mathbf{y}, \omega) = \mu_0\delta_{ik}\delta(\mathbf{x} - \mathbf{y}), \tag{7.112}$$

which is the defining equation of the dressed photon propagator in frequency domain, often also called the dyadic Green's function of the electromagnetic field in a medium characterised by the (generally complex) permittivity $\varepsilon(\mathbf{x}, \omega)$ [27].

Notice that in general we can't apply a Fourier transform in space, since the system we are considering is not translation invariant, as it comprises a region of space (corresponding to $\alpha(x) = 1$) where the material is located, plus a region of space (corresponding to $\alpha(x) = 0$) where there is no material. We can, however, find a solution to the equation above using a very common method used in QFT, known as the Schwinger–Dyson equation [30]. This method will allow us to write the solution to equation (7.112) in terms or a recursive integral equation, from which we can then deduce some physical properties. We do the explicit derivation in appendix B and report here only the end result, which reads[9]

$$\tilde{G}_{ik}(\mathbf{x}, \mathbf{y}, \omega) = \tilde{N}_{ik}(\mathbf{x}, \mathbf{y}, \omega) + \mu_0\tilde{G}_{ik}^{(0)}(\mathbf{x}, \mathbf{y}, \omega)$$

$$+ \frac{\omega^2}{c^2}\int d^3z\, \tilde{G}_{ij}^{(0)}(\mathbf{x}, \mathbf{z}, \omega)[\varepsilon(\mathbf{z}, \omega) - 1]\tilde{G}_{jk}(\mathbf{y}, \mathbf{z}, \omega), \tag{7.113}$$

where $\tilde{G}_{ik}^{(0)}(\mathbf{x}, \mathbf{y}, \omega)$ is the free propagator of the electromagnetic field (i.e., the inverse of the Kernel R_{ij} defined in equation (7.106)) and $\tilde{N}_{ik}(\mathbf{x}, \mathbf{y}, \omega)$ is a solution of the homogeneous equation associated to equation (7.112) and it accounts for the noise currents generated in the material by the vacuum field fluctuations, as discussed below.

Before proceeding further, notice that, as discussed in reference [6], the explicit form of both the dressed propagator $\tilde{G}_{ij}(\mathbf{x}, \mathbf{y}, \omega)$ and the noise function $\tilde{N}_{ij}(\mathbf{x}, \mathbf{y}, \omega)$ depend on the particular boundary conditions imposed by the problem at hand. This level of generality derives, essentially, from the fact that we have chosen the Weyl

---

[9] The reader familiar with many particle systems or QFT will recognise equation (7.113) as a self-energy-like equation. Indeed, the conceptual steps behind it are the same as deriving the self-energy correction for electrons in condensed matter and here have the physical meaning of interpreting the dielectric function as a 'self-energy' correction to the free-photon propagator. We will find a similar situation in chapter 7, where this claculation will be used explicitly to derive the form of the conductivity of a 2D material from its effective action.

gauge to compute the partition function. If we had chosen a different gauge, such as the Coulomb gauge, for example, there would have been more stringent conditions on the propagator (in this example, transversality), and equation (7.112) would have reflected this choice, giving an easier form for the solution.

To keep the discussion simple, let us make the following set of assumptions, that will lead to a simpler form of the dressed propagator, that we could then use in the next section to calculate the field operators and give an explicit expression to the noise currents.

(1) The electromagnetic field is propagating in an infinitely extended material. This endows the system with translation invariance and allows us to take a Fourier transform with respect to spatial variables too.

(2) Because the electromagnetic field is propagating freely in the vacuum dressed by the material (i.e., the dressed action in equation (7.107) is quadratic in the vector potential), we can assume a form of the dressed propagator similar to that of equation (6.99) and, by virtue of that, only consider the transverse part of the propagator as the only one carrying the relevant information about the quantised field (see equation (6.111) and reference [31])[10].

(3) The Weyl gauge is an incomplete gauge, meaning that once the gauge condition has been enforced, there is still more freedom of choosing more constraints to limit the possible form of the vector potential in this gauge [32]. We can then use this extra freedom to enforce transversality to the propagator.

(4) We set, for the moment, $\tilde{N}_{ij}(\mathbf{x}, \mathbf{y}, \omega) = 0$, since the noise term doesn't enter the definition of the dressed propagator. We will restore this term in the next section, where we will give an explicit expression to it, and discuss its physical meaning.

(5) Without loss of generality, we also set $\mathbf{J}_X = 0$ in equation (7.109). This corresponds to only considering the photon dynamics, disregarding therefore the dynamics of polaritons in the material.

The first assumption allows us to take the spatial Fourier transform of equation (7.112), while the second and third ones allow for the following ansatz for the dressed propagator

$$G_{ij}^T(\mathbf{k}, \omega) = G(\mathbf{k}, \omega)\left(\delta_{ij} - \frac{k_i k_j}{k^2}\right), \qquad (7.114)$$

where $G(\mathbf{k}, \omega)$ is a function to be determined, and the term in brackets reflects the transverse nature of the dressed propagator, coming from the assumptions above. Substituting this ansatz into equation (7.113) and using assumption (4) then gives

---

[10] In general, the electromagnetic field can always be decomposed into its transverse and longitudinal part, with the former being associated to the actually propagating field, and the latter to the presence of field sources, such as charges and currents [10].

$$G(\omega, \mathbf{k}) = -\frac{1}{\varepsilon_0} \frac{1}{\omega^2 \varepsilon(\omega) - c^2 k^2}, \qquad (7.115)$$

where $k = |\mathbf{k}|$. Equation (7.115) is not surprising, since it generalises the result for the free-space propagator to the case of a material characterised by a dielectric constant $\varepsilon(\omega)$.

Finally, assumption (5) allows us to rewrite the dressed electromagnetic partition function as

$$Z_0[J] = \exp\left[\frac{i}{2\hbar} \int dt\, dt'\, d^3x\, d^3y\, J_i(\mathbf{x}, t) G_{ij}^T(t - t', \mathbf{x} - \mathbf{y}) J_j(\mathbf{y}, t')\right], \quad (7.116)$$

from which according to equation (6.111), we can define the dressed propagator through the relation

$$G_{ij}^T(t - t', \mathbf{x} - \mathbf{y}) = i\hbar \frac{\partial^2 Z_0[J]}{\partial J_i(\mathbf{x}, t) \partial J_j(\mathbf{y}, t')} \bigg|_{\mathbf{J}=0}. \qquad (7.117)$$

This relation, together with the definition of the transverse dressed propagator, will be useful in the next section to derive the field operators.

7.2.4 Field operators and noise currents

We have now all the necessary ingredients to derive the expression of the field operator from the results above. Before doing that, it is instructive to notice that the partition function in equation (7.116) is fully quantum, since quantisation has been taken care of by path integration. In fact, according to equations (5.30) and (5.56), derivatives of the partition function $Z_0[J]$ in equation (7.116) generate all possible expectation values.

To see this from a more familiar perspective, it is useful to show how to construct the usual field operators starting from equation (7.117). To do so, let us first write the general expression for the (transverse) vector potential in terms of the usual creation and annihilation operators[11] as (see equation (6.11)(a))

$$\hat{\mathbf{A}}(\mathbf{x}, t) = \sum_\lambda \int_0^\infty d\omega \int \frac{d^3k}{(2\pi)^3} \hat{\mathbf{e}}_\lambda(\mathbf{k})[\phi(\omega, \mathbf{k})\, \hat{a}_\lambda(\omega, \mathbf{k}) \exp\left[i(\mathbf{k} \cdot \mathbf{x} - \omega t)\right] + \text{h.c.}], \quad (7.118)$$

where $\hat{\mathbf{e}}_\lambda(\mathbf{k})$ are the local unit vectors accounting for the field's polarisation ($\lambda = 1$ for TE polarisation, and $\lambda = 2$ for TM polarisation), defined as discussed in chapter 6, $\phi(\omega, \mathbf{k})$ is a complex scalar field to be determined and can be interpreted as *mode function* for the field, and the creation and annihilation operators obey the usual bosonic commutation rules

$$[\hat{a}_\lambda(\omega, \mathbf{k}), \hat{a}_\mu^\dagger(\Omega, \mathbf{K})] = \delta_{\lambda\mu} \delta(\omega - \Omega) \delta(\mathbf{k} - \mathbf{K}). \qquad (7.119)$$

---

[11] Notice however, that in this context $\hat{a}^\dagger$ and $\hat{a}$ are dressed creation and annihilation operators, i.e., they create a photon inside the material, rather than in free space.

The connection between field operators and propagator is given by equation (6.111), i.e.,

$$G_{ij}^{T}(t - t', \mathbf{x} - \mathbf{y}) = \langle 0|\hat{T}\hat{A}_i(\mathbf{x}, t)\hat{A}_j(\mathbf{y}, t')|0\rangle = i\hbar \frac{\partial^2 Z[J]}{\partial J_i(\mathbf{x}, t)\partial J_j(\mathbf{y}, t')} \Big|_{\mathbf{J}=0}. \quad (7.120)$$

Substituting the expression of the transverse propagator given by equations (7.114) and (7.115), and taking the Fourier transform of the vector potential operator in equation (7.118) leads, after a little algebra[12], to the following integral equation for the complex scalar field $\phi(\omega, \mathbf{k})$

$$\frac{1}{\hbar(2\pi)^3} \int_0^\infty d\Omega \frac{\Omega |\phi(\Omega, \mathbf{k})|^2}{\Omega - \omega - i\varepsilon} = -\frac{1}{\varepsilon_0} \frac{\omega}{\omega^2 \varepsilon(\omega) - c^2 k^2}. \quad (7.121)$$

This equation can be solved using the Cauchy integral representation formula for holomorphic functions [22]

$$f(z) = \frac{1}{\pi} \int_{i\infty}^\infty dw \frac{\text{Im}\{f(w)\}}{w - z - i\varepsilon}, \quad (7.122)$$

which is valid for a complex function $f(z)$ analytic in the upper-half plane and such that $\lim_{z\to\infty} f(z) = 0$ with $\text{Im}\{z\} > 0$. To apply this formula we then need to make sure that the denominator on the right-hand side of equation (7.121) has no zeros in the upper-half plane. This can be easily done following a similar argument to the one used for proving the causality of the dielectric function in section 7.2.3[13]. Then, we can finally write the explicit expression of the complex scalar field $\phi(\omega, \mathbf{k})$ as

$$\phi(\omega, \mathbf{k}) = \sqrt{\frac{(2\pi)^3 \hbar}{\varepsilon_0 \pi}} \frac{\omega\sqrt{\text{Im}\{\varepsilon(\omega)\}}}{\omega^2 \varepsilon(\omega) - c^2 k^2}. \quad (7.123)$$

Substituting the above expression into equation (7.118), and restoring the tensor dependence $(\delta_{ij} - k_i k_j/|\mathbf{k}|^2)$ of the propagator, as in equation (7.114) we arrive, after taking the inverse Fourier transform to revert the result back to space and time, at the following result

$$\hat{\mathbf{A}}(t, \mathbf{x}) = \sqrt{\frac{\hbar}{\pi\varepsilon_0}} \sum_\lambda \int_0^\infty d\omega \, \omega\sqrt{\text{Im}\{\varepsilon(\omega)\}} \int d^3y \, G_{ij}^{(\lambda)}(\mathbf{x}, \mathbf{y}, \omega)\hat{a}_\lambda(\mathbf{y}, \omega), \quad (7.124)$$

---

[12] The steps to take to derive this result are as follows: (1) notice that $\hat{a}_\lambda(\omega, \mathbf{k})$ annihilates the ground state $|0\rangle$ and use equation (5.36) to write the time-ordered product of two vector potential operators in equation (7.120). (2) assume that the complex scalar field $\phi(\omega, \mathbf{k})$ is an even function of the wave vector $\mathbf{k}$ so that $\phi(\omega, -\mathbf{k}) = \phi(\omega, \mathbf{k})$ holds. (3) Assume that the analytical continuation of $|\phi(\mathbf{k}, \omega)|^2$ is an odd function of ω to ensure causality. (4) Factor out the tensorial dependence of the propagator using equation (7.114) and concentrate only on the scalar part of the propagator, for simplicity. (5) Use equation (7.94) to get the integral form appearing in equation (7.121).

[13] The interested reader can consult appendix C of [6] or [33] for further details.

where

$$G_{ij}^{(\lambda)}(\mathbf{x}, \mathbf{y}, \omega) = \int \frac{d^3k}{(2\pi)^3} [\hat{\mathbf{e}}_\lambda(\mathbf{k})]_i \frac{\left(\delta_{ij} - \dfrac{k_i k_j}{k^2}\right)}{\omega^2 \varepsilon(\omega) - c^2 k^2}, \qquad (7.125)$$

is the propagator associated with the polarisation λ. The above expression of the vector potential is, essentially, proportional to $\sqrt{\mathrm{Im}\{\varepsilon(\omega)\}}\, \overset{\leftrightarrow}{G}(\mathbf{x}, \mathbf{y}, \omega)$, which is consistent with the usual definition of the vector potential operator in dispersive and lossy materials [27], where $\overset{\leftrightarrow}{G}(\mathbf{x}, \mathbf{y}, \omega)$ is usually understood as the dyadic Green's function of the electromagnetic field, which is the solution of the Helmholtz equation inside the material with permittivity $\varepsilon(\omega)$, i.e., equation (7.111) [27].

Let us then relax assumptions (4) and (5) above and derive the explicit expression of the noise current. First, if we restore $\mathbf{J}_X$ in equation (7.109) we can see that assumptions (1)–(3) made for the dressed photon propagator are valid for the polariton field as well. Therefore, we can write the propagator for the free polariton field using the ansatz (7.114) as

$$L_{ij}^T(\mathbf{k}, \omega) = P(\mathbf{k}, \omega)\left(\delta_{ij} - \frac{k_i k_j}{k^2}\right). \qquad (7.126)$$

The scalar function $L(\mathbf{k}, \omega)$ can be determined from the definition of the polariton propagator, similar to equation (7.117) but with the derivatives taken with respect to the components of the polariton current $\mathbf{J}_X$ and the partition function taken as in equation (7.109). This generates a Schwinger–Dyson equation for $L_{ij}(\omega, \mathbf{x}, \mathbf{y})$ of the form

$$\tilde{L}_{ij}(\omega, \mathbf{x}, \mathbf{y}) = \delta_{ij}\tilde{\Gamma}(\omega, \mathbf{x}, \mathbf{y}) + \omega^2 \int d^3z\, d^3v\; \tilde{\Gamma}(\omega, \mathbf{x}, \mathbf{z})\tilde{G}_{ij}(\omega, \mathbf{z}, \mathbf{v})\tilde{\Gamma}(\omega, \mathbf{v}, \mathbf{y}), \qquad (7.127)$$

from which it follows, substituting equation (7.126) into the expression above,

$$L(\mathbf{k}, \omega) = \varepsilon_0 \chi(\omega)\frac{\omega^2 - c^2 k^2}{\omega^2 \varepsilon(\omega) - c^2 k^2}, \qquad (7.128)$$

where $\chi(\omega) = \varepsilon(\omega) - 1$ is the linear susceptibility of the material [10]. Similarly to what we did for the electromagnetic field above, we now introduce the matter polarisation operator as

$$\hat{\mathbf{P}}(\mathbf{x}, t) = \sum_\lambda \int_0^\infty d\omega \int \frac{d^3k}{(2\pi)^3} \hat{\mathbf{e}}_\lambda(\mathbf{k})[p(\omega, \mathbf{k})\, \hat{a}_\lambda(\omega, \mathbf{k})\exp\left[i(\mathbf{k}\cdot\mathbf{x} - \omega t)\right] + \text{h.c.}]. \qquad (7.129)$$

We can then proceed in the same way as did for $\mathbf{A}(\mathbf{x}, t)$, i.e., use equation (7.117) (with $\mathbf{J}_X$ as the polariton current and $L_{ij}(\mathbf{k}, \omega)$ as the polariton propagator) to establish a connection between the polariton propagator and the matter polarisation field. This will give, as before, an equation for $|p(\mathbf{k}, \omega)|$. To completely define $p(\mathbf{k}, \omega)$ we need an auxiliary condition, to be able to determine its phase. This can be constructed from the constitutive relation $\mathbf{C} = \varepsilon_0 \mathbf{E} + \mathbf{P}$, using the relations

$\mathbf{E} = -\partial_t \mathbf{A}$ and $\mathbf{B} = \nabla \times \mathbf{A}$ to make the vector potential appear explicitly, and $\nabla \times \mathbf{B} = \mu_0 \partial_t \mathbf{D}$ to relate $\mathbf{P}$ and $\mathbf{A}$. Combining these relations we get the following equation for $\mathbf{P}(\mathbf{x}, t)$

$$\frac{\partial \mathbf{P}}{\partial t} = \varepsilon_0 \frac{\partial^2 \mathbf{A}}{\partial t^2} + \frac{1}{\mu_0} \nabla \times \nabla \times \mathbf{A}, \tag{7.130}$$

which gives, after a little algebra,

$$p(\mathbf{k}, \omega) = -i\varepsilon_0 \omega \left(1 - \frac{c^2 k^2}{\omega^2}\right)\phi(\mathbf{k}, \omega). \tag{7.131}$$

Substituting this result into equation (7.129) gives the explicit expression for the matter polarisation operator. We can then use this expression together with equation (7.118) to write down the explicit expression of the displacement vector operator, following its definition given by the constitutive relation $\mathbf{D} = \varepsilon_0 \mathbf{E} + \mathbf{P} = -\varepsilon_0 \partial_t \mathbf{A} + \mathbf{P}$. First, notice that calculating the quantity $-\varepsilon_0 \partial_t \mathbf{A} + \mathbf{P}$ leads to an integrand of the form

$$-\varepsilon_0 \partial_t \mathbf{A} + \mathbf{P} \to i\omega\varepsilon_0 \phi(\mathbf{k}, \omega) = i\omega\varepsilon_0 \left(1 - \frac{c^2 k^2}{\omega^2}\right)\phi(\mathbf{k}, \omega)$$

$$= i\omega\varepsilon_0 \phi(\mathbf{k}, \omega)\left(\frac{c^2 k^2}{\omega^2}\right) \tag{7.132}$$

$$= i\omega\varepsilon_0 \phi(\mathbf{k}, \omega)\left(\varepsilon(\omega) - \frac{\omega^2 \varepsilon(\omega) - c^2 k^2}{\omega^2}\right).$$

To go from the second to the third line, we have used the (trivial) equality

$$\frac{1}{\omega^2 \varepsilon(\omega) - c^2 k^2} = \frac{1}{\omega^2 \left[\varepsilon(\omega) - \dfrac{c^2 k^2}{\omega^2}\right]}, \tag{7.133}$$

to find an explicit expression of $c^2 k^2 / \omega^2$ in terms of the two terms appearing in the third line. This allows us to write the explicit form for the displacement operator as[14]

$$\hat{\mathbf{D}}(\mathbf{x}, t) = i\sqrt{\frac{(2\pi)^3 \hbar \varepsilon_0}{\pi}} \sum_\lambda \int_0^\infty d\omega \int \frac{d^3 k}{(2\pi)^3} \hat{\mathbf{e}}_\lambda(\mathbf{k})\left\{\left[\frac{\omega^2 \varepsilon(\omega)\sqrt{\text{Im}\{\varepsilon(\omega)\}}}{\omega^2 \varepsilon(\omega) - c^2 k^2}\right.\right. \tag{7.134}$$

$$\left.\left. - \sqrt{\text{Im}\{\varepsilon(\omega)\}}\right]\hat{a}_\lambda(\mathbf{k}, \omega)\exp[i(\mathbf{k} \cdot \mathbf{x} - \omega t)] - \text{h.c.}\right\}.$$

In classical electrodynamics, the displacement vector is proportional to the electric field through the dielectric permittivity, i.e., $\mathbf{D} = \varepsilon_0 \mathbf{E} + \mathbf{P} = \varepsilon_0 \varepsilon(\omega)\mathbf{E}$. The first term

---

[14] Notice, that equation (7.134) is quivalent to equation (89) of reference [6]. However, equation (89) of reference [6] contains some misprints, i.e., the prefactor is missing an exponent of $1/2$, making it a square root, and the unit vector $\hat{\mathbf{e}}_\lambda \mathbf{k}$ is also missing.

in the square brackets in the expression above states, in fact, exactly this and allows us to write the electric field operator insider the material as

$$\hat{\mathbf{E}}(\mathbf{x},\,t) = i\sqrt{\frac{(2\pi)^3\hbar\varepsilon_0}{\pi}}\sum_\lambda \int_0^\infty d\omega \int \frac{d^3k}{(2\pi)^3}\hat{\mathbf{e}}_\lambda(\mathbf{k})\left[\frac{\omega^2\sqrt{\mathrm{Im}\{\varepsilon(\omega)\}}}{\omega^2\varepsilon(\omega) - c^2k^2}\right] \qquad (7.135)$$
$$\times \hat{a}_\lambda(\mathbf{k},\,\omega)\exp[i(\mathbf{k}\cdot\mathbf{x} - \omega t)] - \text{h.c.}].$$

The second term in equation (7.134), instead, has no classical counterpart, since the constitutive relation for **D** in classical electrodynamics does not admit a zero-field term. It is, in fact, a noise term (frequently called *noise polarisation*) accounting for the response of the material to the vacuum fluctuations of the electric field inside it. Its explicit expression is then given by

$$\hat{\mathbf{N}}(\mathbf{x},\,t) = -i\sqrt{\frac{(2\pi)^3\hbar\varepsilon_0}{\pi}}\sum_\lambda \int_0^\infty d\omega \int \frac{d^3k}{(2\pi)^3}\hat{\mathbf{e}}_\lambda(\mathbf{k})\left[\sqrt{\mathrm{Im}\{\varepsilon(\omega)\}}\right. \qquad (7.136)$$
$$\times \hat{a}_\lambda(\mathbf{k},\,\omega)\exp[i(\mathbf{k}\cdot\mathbf{x} - \omega t)] - \text{h.c.}].$$

The presence of this term then generalises the constitutive relation for the displacement vector to the quantum case as $\mathbf{D} = \varepsilon_0\mathbf{E} + \mathbf{N}$. This is in accordance with the standard results of the quantum electromagnetic field in lossy media [27, 34–36].

To better understand the physical origin and implications of the noise polarisation, let's go back to the wave equation (2.4) for the electric field inside a medium and rewrite it, after taking the Fourier transform in both space and time, for the case of a classical field, where the normal constitutive relation $\mathbf{D} = \varepsilon_0\mathbf{E} + \mathbf{P}$ holds and for the quantum field, when instead $\mathbf{D} = \varepsilon_0\mathbf{E} + \mathbf{P} + \mathbf{N}$ holds instead. In the first case, after introducing the usual relation $\mathbf{P} = \varepsilon_0\chi\mathbf{E}$, we get

$$[\omega^2\varepsilon(\omega) - c^2k^2]\mathbf{E} = 0, \qquad (7.137)$$

while in the second case we get

$$[\omega^2\varepsilon(\omega) - c^2k^2]\hat{\mathbf{E}}(\mathbf{k},\,\omega) = -\frac{\omega^2}{\varepsilon_0}\hat{\mathbf{N}}(\mathbf{k},\,\omega). \qquad (7.138)$$

Since $\varepsilon(\omega)$ is a complex function, equation (7.137) cannot be satisfied with both ω and k real numbers. Hence, a dispersion relation $\omega(\mathbf{k})$ for lossy media does not exist. One could argue that by promoting either the frequency of the wave vector to a complex quantity, a solution to equation (7.137) could be found, thus restoring the concept of dispersion relation. This, however, is not possible, since here ω and k are real quantities, deriving from a Fourier transform!

The correct way to solve this issue is then to make equation (7.137) inhomogeneous, by introducing a noise term as in equation (7.138). In this case, a solution to equation (7.138) with real ω and k exists, and no relation between ω and k is necessary to fullfill this equation. This also means, that the concept of dispersion relation is lost, when looking at the optical properties of lossy media.

Lastly, as it can be seen from equation (7.136), the noise polarisation depends on the material properties only, through $\varepsilon(\omega)$. As discussed in section 7.2.2, the permittivity contains information about both the matter and the reservoir field, through the propagator $\Gamma(\mathbf{x}, \mathbf{y}, t - t')$. The origin of $\hat{\mathbf{N}}(\mathbf{x}, t)$ can be then traced back to the origin of $\Gamma(\mathbf{x}, \mathbf{y}, t - t')$. As we have seen in section 7.2.1, the propagator $\Gamma(\mathbf{x}, \mathbf{y}, t - t')$ dresses the matter field with the reservoir. In particular, since the reservoir is modelled as a continuum of harmonic oscillators, with this dressing (i.e., thanks to the path integration with respect to the reservoir) the matter field inherits losses, which model the noise originating from the coupling with the reservoir. Then, when the electromagnetic field is dressed with the matter field (i.e., after path integration with respect to the matter field), it also inherits this noise. However, as long as we remain at the classical level, this noise term only manifests in the material properties, i.e., the permittivity $\varepsilon(\omega)$ through losses. When we quantise the field, however, this noise term manifests explicitly through the noise currents $\tilde{N}_j(\mathbf{x}, \mathbf{y}, \omega)$ in the Schwinger–Dyson equation (7.113) or, equivalently, with the noise polarisation defined in equation (7.136). Thus, the physical origin of the noise polarisation can be traced back to the existence of losses in the medium, which justifies why they are not present in a conventional, lossless material, even when the field is quantised.

This discussion concludes this section and the investigation of the linear properties of the dressed electromagnetic field. To conclude and summarise, notice that all the results presented here can also be obtained using the standard Hamiltonian formalism discussed in chapter 6. Indeed, the book of Vogel and Welsch [27] takes this latter perspective to quantise the electromagnetic field in lossy media. Here, on the other hand, it is interesting to see how everything comes out naturally from the path integral description, without the need to make specific assumptions on the form of the operators etc. The dressed propagator derived in section 7.2.3 already contains all the relevant information on the material, including its losses. The connection to the field operators presented in this section makes this evident, and serves as a bridge between the two approaches.

7.2.5 Quantum nonlinear optics using path integrals

From this section onwards, we turn our attention to the nonlinear interaction of the dressed electromagnetic field, and how we can describe them within the path integral formalism. From this perspective, then, this and the following sections mirror part II of chapter 6, i.e., section 6.2. They, in fact, present the same calculations for second-order nonlinearities in the undepleted pump approximation, and extend them to the fully quantum case as well.

To describe nonlinearities within the path integral framework we have essentially two possibilities. Either we introduce a suitable interaction term, describing the nonlinear interaction, in the classical dressed Lagrangian (7.82), or we introduce the same interaction Lagrangian in the fully quantised dressed Lagrangian (7.105). In the former case, since the electromagnetic field is not quantised, we would describe the classical theory of nonlinear optics[15]. In the latter case, on the other hand, we can describe quantum nonlinear optical effects, since the electromagnetic field is fully quantised. In this section, we choose to operate in the latter situation, and we therefore deal with nonlinear interactions of the dressed electromagnetic field in the quantum regime.

We have already seen in section 6.2.1, that in order to describe usual optical nonlinearities (i.e., second- and third-order nolinearities), we need to define an interaction Hamiltonian proportional to an appropriate power of the electric field[16] (see, for exmaple, equation (6.38)).

For the path integral case we adopt a similar approach, adding an interaction Lagrangian that contains an appropriate power of the electric field, consistent with the order of nonlinearity we want to describe. Following the standard calculations in nonlinear optics, we do this addition to the dressed Lagrangian in frequency domain, rather than in time domain, so that we can express the nonlinear matter polarisation operator in terms of powers of the frequency-dependent electric field, rather than nested field convolutions [15, 34]. Following reference [7], and the standard procedure of QFT [31], the interaction Lagrangian including all orders of (perturbative) electromagnetic nonlinear interactions inside a medium can be written, in frequency domain, as

$$\mathscr{L}_{\text{int}}(\mathbf{x}, \omega) = \sum_{n=2}^{\infty} \int [d\omega]_n \frac{i^{n+1}}{(n+1)!} \chi^{(n)}([\omega]_{n+1}, \mathbf{x}) \cdot \mathbf{A}(\omega, \mathbf{x}) \cdots \mathbf{A}(\omega_{n+1}, \mathbf{x}) \quad (7.139)$$

where $[d\omega]_n = d\omega_1 \cdots d\omega_n$, $[\omega]_{n+1} = \{\omega, \omega_1, \ldots, \omega_n\}$, $(n + 1)!$ counts the number of different ways to arrange $n + 1$ fields into a product [31], and $\chi^{(n)}(\omega, \mathbf{x})$ is the nth order nonlinear susceptibility [15]. Written in this form, the interaction Lagrangian reproduces the usual nonlinear interaction of the electromagnetic field, typically represented with a susceptibility being a function of $(n + 1)$ frequencies, and enabling the interaction of $(n + 1)$ fields at frequencies $\{\omega, \omega_1, \ldots, \omega_n\}$.

It is important to point out at this level of analysis, that the interaction Lagrangian above is only valid in the perturbative regime, where $\chi^{(n+1)}(\omega, \mathbf{x})/\chi^{(n)}(\omega, \mathbf{x}) \ll 1$, i.e., when the nonlinearity only acts as a perturbation over the linear permittivity. Although this is the case for many different materials, and the implicit assumption of standard nonlinear optics [15], there are cases, such as 2D materials and epsilon-near-zero materials, for which this approach fails to give physically meaningful results, since the perturbative approximation is not valid anymore [37].

With the addition of the interaction Lagrangian above to the dressed Lagrangian of the electromagnetic field, the partition function in equation (7.109) gets modified as

$$Z[J] = \int \mathscr{D}\mathbf{A} \exp\left[\frac{i}{\hbar} \int d\omega \, d^3x (\mathscr{L}_{\textit{eff}}(\mathbf{x}, \omega) + \mathscr{L}_{\text{int}}(\mathbf{x}, \omega) + \mathbf{J}(\mathbf{x}, \omega) \cdot \mathbf{A}(\mathbf{x}, \omega))\right] \quad (7.140)$$

---

[15] see reference [15] for a detailed discussion of standard nonlinear optics.
[16] We neglect magnetic nonlinearities here, since in normal materials used for traditonal nonlinear optics they are negligible.

In the presence of the interaction term $\mathscr{L}_{\text{int}}$ the path integral cannot be calculated analytically anymore, since the integrand in the expression above in non-Gaussian, i.e., it contains terms containing the exponential of more than two fields. if we want to find an analytical expression for the nonlinear partition function, we need to find a way to relate the expression above to the free partition function $Z_0[J]$, defined in time domain in equation (7.116). First, we take the Fourier transform of equation (7.116) to get

$$Z_0[J] = \exp\left[\frac{i}{2\hbar} \int d\omega \, d^3x \, d^3y \, J^\mu(\mathbf{x}, \omega) G_{\mu\nu}(\mathbf{x} - \mathbf{y}, \omega) J^\nu(\mathbf{y}, \omega)\right]. \quad (7.141)$$

With this expression at hand, we can then rewrite equation (7.140) in terms of $Z_0[J]$ using a well-known method from QFT [38], which consists in finding a differential equation satisfied by the partition function $Z[J]$, and then solving it in terms of $Z_0[J]$. The idea of it is essentially given in equation (5.58), and the general method applied for a scalar field is treated in standard QFT books, such as references [38, 39], for example. We then apply this method to the electromagnetic field (and give the detailed calculations in appendix 7.2.10) to get

$$Z[J] = \mathscr{N} \exp\left[\frac{1}{\hbar} \int d\omega \, d^3x \, \mathscr{L}_{\text{int}}\left(\frac{\hbar}{i} \frac{\partial}{\partial J}\right)\right] Z_0[J], \quad (7.142)$$

where we have used equation (5.55) to write the interaction Lagrangian in terms of current-derivatives, instead of powers of the vector potential. The advantage of writing the partition function in this form, is that under the assumption that the nonlinear interaction described by $\mathscr{L}_{\text{int}}$ is perturbative, the exponential term in the right-hand-side of the above equation can be expanded in a power series, and the effect of the nonlinearity can be then written simply as a perturbative expansion of derivatives of the free partition function. This leads, in fact, to the following expression for $Z[J]$

$$Z[J] = Z_0[J] + \sum_{\ell=2}^{\infty} Z_\ell[J] + Z_{\text{cross}}[J], \quad (7.143)$$

where

$$Z_\ell[J] = \sum_{n=1}^{\infty} \frac{i^{\ell+1}}{n!(\ell+1)!} \left(\frac{i}{\hbar}\right)^n \left[\int [d\omega]_{n+1} \, d^3x \, \chi^{(\ell)}([\omega]_{n+1}, \mathbf{x}) \cdot \left(\frac{\hbar}{i} \frac{\partial}{\partial J}\right)^{\ell+1}\right]^n Z_0[J] \quad (7.144)$$

represents the correction up to order $\chi^{(\ell)}$ to the free propagation of the electromagnetic field inside the material, due to the nonlinear interaction, and $Z_{\text{cross}}[J]$ is a term accounting for the interplay between the different orders of nonlinearity (e.g. containing terms proportional to $\Pi_\ell \chi^{(\ell)} \chi^{(\ell+1)} \chi^{(\ell+2)} \ldots$). In practical nonlinear optics calculations, $Z_{\text{cross}}[J]$ can be neglected, since it is of higher order than the term $Z_\ell[J]$ corresponding to the specific nonlinear process considered. For example, for the case of third-order nonlinearities we have that $Z_3[J] \simeq |\chi^{(3)}|$, while $Z_{\text{cross}}[J] \simeq |\chi^{(2)}||\chi^{(3)}| \ll |\chi^{(3)}|$, which represents the possibility to have third-order processes generated by the mutual interaction of second- and third-order nonlinearities. These processes, although in principle present in the material, are either rigorously absent, if the material symmetry does not allow both second- and third-order processes to coexist, or, if they are allowed, they have an extremely low probability of occurrence, and cannot be excited with the present level of electric field intensity that can be reached by state-of-the-art laser systems. To have a feeling about this, let us consider the following thought experiment: let us imagine a material whose values for the second- and third-order nonlinear susceptibilities have the somehow typical average values of $|\chi^{(2)}| \simeq 10^{-12}$ m V^{-1}, and $|\chi^{(3)}| \simeq 10^{-20}$ m^{-2} V^{-2} [15]. Let us also imagine that we want to compare the purely third-order processes, ruled by $|Z_3[J]| \propto |\chi^{(3)}E^2|$, with the cross-term $|Z_{\text{cross}}[J]| \propto |\chi^{(2)}\chi^{(3)}E^3|$. To have a non-negligible contribution of the cross-term we need to require that $|Z_{\text{cross}}[J]/Z_3[J]| \simeq 1$, i.e., that $E \simeq 10^{12}$ V/m. To put things in perspective, the highest peak electric field amplitude reported, corresponding to the intensity of a tightly focussed state-of-the-art petawatt laser [40], is $E \simeq 10^{14}$ V m^{-1}. This corresponds to a peak intensity of the order of $I \simeq 10^{14}$ GW cm^{-2}.

Standard experiments in nonlinear optics, on the other hand, run with peak intensities typically in the range 1–100 GW cm^{-2}, meaning that the peak field amplitude is around $E \simeq 10^{8\vee}10^9$ V m^{-1}. This corresponds to $|Z_{\text{cross}}[J]/Z_3[J]| \simeq 10^{-3}$, and therefore the impact of the cross-terms is de facto negligible.

7.2.6 Feynman diagrams from path integrals for interacting theories

In section 5.4 we introduced Feynman rules and Feynman diagrams for an interacting theory, based on its Hamiltonian representation, and in section 5.6 we extended their definition to interacting theories in the path integral formalism, with equation (5.60) being the central equation that defines Feynman diagrams from path integrals. Equation (7.144) has the same form as of equation (5.60), with n in equation (7.144) playing the role of the number of vertices V in equation (5.60), and the number of propagators P appears explicitly in equation (7.144) after writing the free partition function $Z_0[J]$ in a power series form. If we do so, we can rewrite equation (7.144) in the following, diagram-ready, form

$$
Z[J] = Z_0[J] + \sum_{\ell=2}^{\infty} \sum_{n=1}^{\infty} \sum_{p=0}^{\infty} \frac{i^{\ell+1}(i/\hbar)^n}{p!n!(\ell+1)!} \left[\int [d\omega]_{n+1} \, d^3x \, \chi^{(\ell)}([\omega]_{n+1}, \mathbf{x}) \cdot \left(\frac{\hbar}{i} \frac{\partial}{\partial \mathbf{J}} \right)^{\ell+1} \right]^n
$$
$$
\times \left[\frac{i}{2\hbar} \int d\omega \, d^3y \, d^3z \, J^\mu(\mathbf{y}, \omega) G_{\mu\nu}(\mathbf{y} - \mathbf{z}, \omega) J^\nu(\mathbf{z}, \omega) \right]^p,
$$
(7.145)

where now ℓ counts the order of the nonlinear interaction, n the number of vertices in the Feynman diagram (i.e., the order of the perturbative expansion that describes the interaction), and p the number of propagators involved in the interaction, i.e., the number of photons involved in the nonlinear interaction (at the Lagrangian level). Notice that with this convention we need to impose the constraint $p = \ell + 1$, since the number of photons participating in a nonlinear interaction of order ℓ is dictated by the interaction itself to be $\ell + 1$. For example, for second-order nonlinear processes, $\ell = 2$, and the first nonlinear contribution (at the single vertex level, i.e., $n = 1$) to the interaction comes from the term with $p = \ell + 1 = 3$ propagators, i.e., a three-photon interaction term.

We now specify our analysis to the case of second-order nonlinearities, deriving the explicit analytical expressions for the interaction terms and their Feynman diagram representation. First, we calculate the transition probability, i.e., the probability of a second-order nonlinear event to occur, in the fully quantum regime. Then, to mirror the results obtained in part II of chapter 6, we also show how the undepleted pump approximation is retrieved, and discuss the differences between the two regimes.

7.2.7 Second-order nonlinearities—quantum pump

Second-order nonlinearities are characterised by the interaction of three photons, conventionally called, in nonlinear optics jargon [15], pump (P), signal (S), and idler (I). The names derive from the setup for the classical nonlinear optics experiments involving second-order nonlinearities, where a weak electromagnetic field (the signal) with carrier frequency ω_s is amplified via energy transfer from a high-intensity field (the pump) with carrier frequency ω_p. Because of energy conservation imposed by the three-field interaction, a third, auxiliary field (the idler), with a carrier frequency ω_i, is generated during the process, such that $\omega_p = \omega_s + \omega_i$ (i.e., energy conservation) would hold at any time. Additionally, also momentum conservation, i.e., $k(\omega_p) = k(\omega_s) + k(\omega_i)$, where $k(\omega) = k_0 n(\omega)$ must hold at any time[17]. For a detailed description of classical nonlinear optics, the reader is referred to any standard book on the subject, such as the book by Boyd [15] or Butcher and Cotter [21].

The interaction Lagrangian associated with second-order nonlinearities is given by equation (7.139) with $n = 2$ as the only term contributing to the sum, i.e.,

$$\mathscr{L}_2(\mathbf{x}, \omega) = -\frac{i}{3!} \int d\omega_1 \, d\omega_2 \, d^3x \, \chi^{(2)}_{\mu\nu\sigma}(\omega, \omega_1, \omega_2, \mathbf{x}) \, A_\mu(\omega_1, \mathbf{x}) A_\nu(\omega_2, \mathbf{x}) A_\sigma(\omega, \mathbf{x}). \quad (7.146)$$

Notice, that the three photons participating in the interaction are characterised by the three frequencies $\{\omega, \omega_1, \omega_2\}$. Identifying each of these with signal, idler, and pump, gives all the second-order nonlinear processes described in figure 6.2.

To describe these second-order nonlinearities within the framework of path integrals, we make use of Feynman diagrams, i.e., equation (7.145). In particular, we look only at the first order in the perturbation expansion of the nonlinear partition function, i.e., we set $n = 1$ in equation (7.145). Moreover, we also set $\ell = 2$, as we are dealing with second-order nonlinearities, and $p = \ell + 1 = 3$, since the interaction involves three photons (pump, signal, and idler). Neglecting the higher orders is a fair assumption for most of the experimental situations dealing with nonlinear optics, as the peak intensities of the involved electromagnetic field are, as discussed above, too low to trigger higher-order processes. For the sake of clarity, in what follows we leave out the linear terms, proportional to $Z_0[J]$, i.e., the first term

---

[17] $k(\omega) = k_0 n(\omega)$ is the wave vector inside a material with refractive index $n(\omega)$.

appearing in equation (7.145), and only concentrate on the terms arising from the nonlinear interaction.

Substituting $\ell = 2$, $p = 3$, and $n = 1$ into equation (7.145), and neglecting the linear term, gives us the following expression

$$Z_2[J] = \frac{1}{8(3!)^2\hbar} \int [d\omega]_3 \, d^3x \, \chi^{(2)}_{\mu\nu\sigma}([\omega]_3, \mathbf{x}) \frac{\partial^3}{\partial J_\mu(\mathbf{x}, \omega_1)\partial J_\nu(\mathbf{x}, \omega_2)\partial J_\sigma(\mathbf{x}, \omega_3)}$$
$$\times \left[\int d\omega \, d^3y \, d^3z \, J^\mu(\mathbf{y}, \omega)G_{\mu\nu}(\mathbf{y} - \mathbf{z}, \omega)J^\nu(\mathbf{z}, \Omega) \right]^3. \tag{7.147}$$

To find the explicit expression of $Z_2[J]$ one needs to calculate the derivatives above. First, let us recall that functional derivatives follow the same product rule that normal derivatives, namely

$$\frac{\partial}{\partial J}F^k[J] = k\frac{\partial F}{\partial J}F^{k-1}[J]. \tag{7.148}$$

Using this, we can calculate the functional derivative of the integral inside the square brackets in equation (7.147) as

$$\frac{\partial}{\partial J^\mu(\mathbf{x}, \omega)} \int d\Omega \, d^3y \, d^3z \, J^\alpha(\mathbf{y}, \Omega)G_{\alpha\beta}(\mathbf{y} - \mathbf{z}, \Omega)J^\beta(\mathbf{z}, \Omega)$$
$$= 2 \int d^3y \, G_{\mu\beta}(\mathbf{x} - \mathbf{y}, \omega)J^\beta(\mathbf{y}, \omega). \tag{7.149}$$

We can use this result to calculate all three derivatives in equation (7.147). After a little algebra, and the renaming of dummy indices and integration variables, the final expression for $Z_2[J]$ is

$$Z_2[J] = \frac{1}{3!\hbar} \int [d\omega]_3 \, [d^3x]_4 \, \chi^{(2)}_{\mu\nu\sigma}([\omega]_3, \mathbf{x})G_{\mu\alpha}(x - y, \omega_1)G_{\nu\beta}(x - z, \omega_2)G_{\sigma\lambda}(x - s, \omega_3)$$
$$\times J^\alpha(y, \omega_1)J^\beta(z, \omega_2)J^\lambda(s, \omega_3) + \frac{1}{3!\,\hbar} \int [d\omega]_3 \, d^3x \, d^3z \, \chi^{(2)}_{\mu\nu\sigma}([\omega]_3, \mathbf{x}) \, G_{\sigma\nu}(0, \omega_1) \tag{7.150}$$
$$\times G_{\mu\rho}(x - z, \omega_3)J^\rho(z, \omega_3) \int d\omega \, d^3y \, d^3s \, J^\alpha(y, \omega)G_{\alpha\beta}(y - s, \omega) \, J^\beta(s, \omega),$$

where $[d^3x]_4 = d^3x \, d^3y \, d^3z \, d^3s$.

To better understand the physical processes described by the partition function above, we can rewrite the expression above in terms of Feynman diagrams, following the prescription of section 5.4.2. To this aim, let us introduce the following diagram symbols for the photon propagator $G_{\mu\nu}(\mathbf{x} - \mathbf{y}, \omega)$, the source current $J^\mu(x, \omega)$, and the interaction vertex $(i/3!\hbar) \int d\omega \, d^3x \, \chi^{(2)}_{\mu\nu\sigma}(\mathbf{x}, \omega)$

$$\sim\!\!\sim\!\!\sim\!\!\sim \quad \rightarrow \quad G_{\mu\nu}(\mathbf{x} - \mathbf{y}, \omega), \tag{7.151}$$

$$\otimes \quad \rightarrow \quad i\int d^3x J^\mu(\mathbf{x}, \omega), \tag{7.152}$$

$$\bullet \quad \rightarrow \quad \frac{i}{3!\hbar} \int [d\omega]_3 \, d^3x \, \chi^{(2)}_{\mu\nu\sigma}([\omega]_3, \mathbf{x}), \tag{7.153}$$

so that the second-order partition function $Z_2[J]$ above becomes

$$Z_2[J] \quad = \tag{7.154}$$

where the term 'permutations' refers to the 3! way to arrange the three currents (represented by the coloured markers red, blue, and teal, for visual help) in each of the diagrams. The first diagram corresponds to the usual second-order process of one photon (the pump photon, in red) being annihilated at the vertex, to generate two other photons (signal, in blue, and idler, in teal), compatibly with the energy conservation constraint $\omega_p = \omega_s + \omega_i$. Because of permutation symmetry, there are six equivalent diagrams, each representing a different second-order process (see figure 6.2).

The second term in the expression above corresponds to a so-called *tadpole* diagram, which essentially consists of a loop correction to the photon propagator, as can be seen in the diagram representing it (the loop corresponds to the term proportional to $G_{\sigma\nu}(0, \omega_1)$ appearing in equation (7.150)), and it is a typical diagram appearing from interacting theories with an odd number of fields involved, such as second-order nonlinearities. We can, however, neglect this term, as it can be shown that it does not contribute actively to the probability amplitude for any second-order process, as it only amounts to a renormalisation of the free propagator[18]. Neglecting the tadpole leads then to the following form for the (1-vertex) second-order partition function

---

[18] The interested reader can find the proper way to deal with tadpole terms in any QFT book dealing with ϕ^3-theories, such as the books by Srednicki [39], Ryder [38], or Maggiore [41].

$$Z_2[J] = \qquad + \text{permutations}$$

$$= \frac{1}{3!\hbar} \int [d\omega]_3 \, [d^3x]_4 \, \chi^{(2)}_{\mu\nu\sigma}([\omega]_3, x) G_{\mu\alpha}(x-y, \omega_1) G_{\nu\beta}(x-z, \omega_2) G_{\sigma\lambda}(x-s, \omega_3)$$

$$\times \; J^\alpha(y, \omega_1) J^\beta(z, \omega_2) J^\lambda(s, \omega_3)$$

$$(7.155)$$

As discussed in section 5.3, to calculate probability amplitudes for interacting fields it is enough to calculate the correspondent n-point correlation function (see equation (5.30)). For the case of second-order nonlinearities, this amounts to calculating the three-point correlation function $\langle A_\mu(\mathbf{x}_p, \omega_p) A_\nu(\mathbf{x}_s, \omega_s) A_\sigma(\mathbf{x}_i, \omega_i)\rangle$, as there are three photons involved in the interaction[19]. Using equation (5.56) we can write the three-point correlation function in terms of $Z_2[J]$ as

$$\langle A_\mu(\mathbf{x}_p, \omega_p) A_\nu(\mathbf{x}_s, \omega_s) A_\sigma(\mathbf{x}_i, \omega_i)\rangle = \frac{\partial^3 Z_2[J]}{\partial J^\mu(\mathbf{x}_p, \omega_p) \partial J^\nu(\mathbf{x}_s, \omega_s) \partial J^\sigma(\mathbf{x}_i, \omega_i)} \Big|_{J=0}. \quad (7.156)$$

Substituting equation (7.155) into the expression above we get, after a little algebra and, as usual, renaming of dummy indices and integration variables, the following result

$$\langle A_\mu(\mathbf{x}_p, \omega_p) A_\nu(\mathbf{x}_s, \omega_s) A_\sigma(\mathbf{x}_i, \omega_i)\rangle = $$

$$(7.157)$$

$$= \frac{1}{\hbar} \int d^3x \, \chi^{(2)}_{\alpha\beta\lambda}(\mathbf{x}, \omega_p, \omega_s, \omega_i) G_{\alpha\mu}(\mathbf{x} - \mathbf{x}_p, \omega_p) G_{\beta\nu}(\mathbf{x} - \mathbf{x}_s, \omega_s)$$

$$\times \; G_{\lambda\sigma}(\mathbf{x} - \mathbf{x}_i, \omega_i),$$

where in writing the second line we have made use of the standard notation of writing the frequency dependence of the nonlinear susceptibility in terms of the various involved frequencies, i.e., pump, signal and idler in this case.

---

[19] Notice how we have used the subscripts $\{p, s, i\}$ to match the standard notation of calling the three fields involved in the interaction pump, signal, and idler.

From the expression above we see one of the advantages of doing calculations with Feynman diagrams. In fact, each functional derivative $\partial/\partial J$ appearing in equation (7.156) amounts to removing a current label, i.e., $\otimes$, and replacing it with a label containing spatial and spectral coordinates of the current appearing in the derivative. For example, the source term $\otimes$ in equation (7.155) is replaced in equation (7.157) with $(\mathbf{x}_i, \omega_i)$ (and the index σ), and this operation corresponds to taking the functional derivative $\partial/\partial J^\sigma(\mathbf{x}_i, \omega_i)$. This process leaves only the three propagators, corresponding to the three interacting photons, i.e., the pump (red), signal (blue), and idler (teal) photons, which all meet at the vertex $\bullet$, where their nonlinear interaction takes place. For the case of the diagram used in equation (7.157), the pump photon gets annihilated to create a signal and idler photon, thus corresponding to the process of sum frequency generation (SFG). The result in equation (7.157) represents the probability amplitude for an SFG process to occur.

The actual probability density (or scattering cross-section, in the language of QFT) is then found by simply taking the modulus square of such quantity, i.e.,

$$\sigma^{SFG}_{\mu\nu\sigma} = |\langle A_\mu(\mathbf{x}_p, \omega_p) A_\nu(\mathbf{x}_s, \omega_s) A_\sigma(\mathbf{x}_i, \omega_i)\rangle|^2. \qquad (7.158)$$

Notice, moreover, that since the diagrams shown in figure 6.2 are equivalent, under permutation symmetry, to that appearing in equation (7.157), their scattering cross-section is also equivalent. What actually distinguishes between the various processes, as discussed in chapter 6, are the boundary conditions. For SFG, for example, it is necessary to have a single pump photon at the beginning of the interaction. For a difference frequency generation process (DFG), instead, both the pump and signal (or idler) photons are needed, in order to make this process happen.

The result above only allows us to say something about the occurrence probability, and not about the actual energy transfer between the pump, signal and idler fields. To account for this, one would need to derive the equations of motion of the interacting system, i.e., compute the Euler–Lagrange equation using $\mathscr{L}_2(\mathbf{x}, \omega)$ as the interaction Lagrangian and obtain therefore a set of three coupled, nonlinear equations for the amplitudes of the pump, signal and idler fields. These equations are fully equivalent with the ordinary, classical nonlinear coupled mode equations that are described in nonlinear optics textbooks [15, 21][20].

Instead of doing that, we present a simple example, to familiarise the reader with this formalism. Let us consider a very simple system, namely a one-dimensional, homogeneous, isotropic optical waveguide of length L, characterised by a refractive index $n(\omega)$ and a second-order susceptibility $\chi^{(2)}_{\mu\nu\sigma}([\omega]_3, \mathbf{x}) \equiv \chi$. To calculate equation (7.157) we need the expression for the propagator of the dressed electromagnetic field in such a waveguide. Since we are dealing with a one-dimensional material, its explicit expression can be readily calculated as [22]

---

[20] This is done by substituting equations (7.146) and (7.105) into equation (3.6), which is left to the reader as an exercise.

$$G(x - y, \omega) = \frac{1}{2ik(\omega)}\{\Theta(x - y)\exp[ik(\omega)(x - y) - i\omega t]$$
$$+ \Theta(y - x)\exp[-ik(\omega) + i\omega t]\}, \tag{7.159}$$

where $\Theta(x)$ is the Heaviside step function [12], and $k(\omega) = k_0 n(\omega)$. Knowing the explicit for of the photon propagator allows us to calculate equation (7.157) explicitly as

$$\langle A_\mu(\mathbf{x}_p, \omega_p)A_\nu(\mathbf{x}_s, \omega_s)A_\sigma(\mathbf{x}_i, \omega_i)\rangle = \frac{i\chi \exp(i\Delta\omega t)}{4\hbar k_p k_s k_i}L\mathrm{sinc}(\Delta kL) \tag{7.160}$$
$$+ \text{ non-resonant terms},$$

where $k_a = k(\omega_a)$, $\mathrm{sinc}(x) = \sin x/x$, $\Delta\omega = \omega_p - \omega_s - \omega_i$ is the energy conservation constraint (if seen as $\Delta\omega = 0$), $\Delta k = k_p - k_s - k_i$ is the phase mismatch relation (with the usual phase matching condition [15] obtained as $\Delta k = 0$), and the non-resonant terms are all those contributions, where the sum of the frequencies or wave vectors do not match $\Delta\omega$ and Δk, respectively.

From the result above, the probability for a second-order nonlinear process to happen is then given by equation (7.158), i.e.,

$$\sigma_2^{\text{quantum pump}}(L) = \left(\frac{\chi L}{4\hbar k_p k_i k_s}\right)^2 \mathrm{sinc}^2(\Delta kL), \tag{7.161}$$

which is in agreement with standard results in nonlinear optics, that second-order processes grow (at least ideally) with the square of the length of the nonlinear crystal [15, 21]. Notice also, that contrary to the usual results in nonlinear optics, the expression above does not contain a dependence on the intensity of the pump. This is due to the fact that all three fields, namely pump, signal and idler, are treated as quantum fields. This also justifies the superscript *quantum*. The opposite, and more common in usual nonlinear optics experiments, situation is that where the pump field is considered a classical field. This, as seen in chapter 6, is known as the *undepleted pump approximation*, and it is the subject of the next section.

7.2.8 Second-order nonlinearities—undepleted pump

Usual experimental situations consist in pumping the nonlinear material with a high-intensity laser beam (the pump beam), which contains many photons and it is normally treated, at the quantum level, as a coherent, rather than a Fock, state. The presence of many photons in the pump beam makes it so that when the nonlinear interaction takes place and one pump photon is annihilated to create a signal–idler pair, this only marginally affects the intensity (i.e., number of photons) remaining on the pump field, as in the large photon limit $N \simeq N - 1$, and the quantum evolution of the pump field becomes irrelevant, since the nonlinear interaction practically does not affect its quantum state. This is what is normally referred as the *undepleted pump approximation*. Physically, this means that the three-photon problem described by the interaction Lagrangian in equation (7.146) becomes effectively a two-photon process (i.e., the signal and the idler) in an effective vacuum dressed by the presence of the pump, which is treated as a classical field.

To account for this in our calculations, we can introduce an effective second-order susceptibility, which incorporates the effects of the classical pump (whose only relevant degrees of freedom now are amplitude and phase) as follows

$$\lambda_{\mu\nu}(\omega; \omega_1, \omega_2, \mathbf{x}) = \chi^{(2)}_{\mu\nu\sigma}(\omega, \omega_1, \omega_2)A_\sigma(\omega, \mathbf{x}), \qquad (7.162)$$

so that the interaction Lagrangian for second-order processes in the undepleted pump approximation becomes

$$\mathscr{L}_2^{(\text{und})}(\mathbf{x}, \omega) = -\frac{i}{3!} \int d\omega_1 \, d\omega_2 \, d^3x \, \lambda_{\mu\nu}(\omega; \omega_1, \omega_2, \mathbf{x})A_\mu(\omega_1, \mathbf{x})A_\nu(\omega_2, \mathbf{x}). \quad (7.163)$$

Notice, how now this is quadratic in the signal and idler fields. As discussed in chapter 3, quadratic Lagrangians correspond to linear equations of motion for the fields. Within the undepleted pump approximation, the signal and idler fields can be treated as free fields, and the nonlinear Lagrangian above can be simply seen as a correction to their free propagation. This can be seen by combining the result above with equation (7.105) to obtain the 'free' action for the unepleted pump approximation as

$$S^{(\text{und})} = -\frac{1}{2} \int d\omega \, d\omega_1 \, d\omega_2 \, d^3x \, d^3y \, \mathscr{L}_{\text{free}}^{(\text{und})}(\omega, \omega_1, \omega_2, \mathbf{x}, \mathbf{y}), \qquad (7.164)$$

where

$$\mathscr{L}_{\text{free}}^{(\text{und})}(\omega, \omega_1, \omega_2, \mathbf{x}, \mathbf{y}) = A_\mu(\omega_1, \mathbf{x})\Big\{ R_{\mu\nu}(\mathbf{x} - \mathbf{y}, \omega)\delta(\omega - \omega_1)\delta(\omega - \omega_2)$$
$$- \frac{i}{3}\lambda_{\mu\nu}(\omega; \omega_1, \omega_2, \mathbf{y})\delta(\mathbf{x} - \mathbf{y}) \Big\}A_\nu(\omega_2, \mathbf{y}) \qquad (7.165)$$
$$\equiv A_\mu(\omega_1, \mathbf{x})\mathcal{R}_{\mu\nu}(\mathbf{x} - \mathbf{y}, \omega_1, \omega_2)A_\nu(\omega_2, \mathbf{y}),$$

where $R_{\mu\nu}(\mathbf{x} - \mathbf{y}, \omega)$ is the Fourier transform of the electromagnetic Kernel defined in equation (7.106). The expression above has a very simple physical interpretation: $\mathcal{R}_{\mu\nu}(\mathbf{x} - \mathbf{y}, \omega_1, \omega_2)$ can in fact be seen as the 'free' propagation of both the signal and idler fields, in a vacuum that is dressed by the presence of the pump field, which essentially regulates the nonlinear interaction between signal and idler. Because in this case both signal and idler are evolving in a dressed vacuum, every nonlinear interaction can be understood as originating from the dressed vacuum (i.e., from the classical pump). This is even more evident in terms of Feynman diagrams. Introducing the undepleted pump approximation into equation (7.155), in fact, leads to the following diagrammatic representation for $Z_2[J]$ (neglecting tadpole contributions as before)

$$J^\alpha(y, \omega)$$

$$Z_2[J] \quad = \quad \text{------} \qquad + \text{permutations}$$

$$J^\beta(z, \omega)$$

$$= \frac{1}{3!\hbar} \int d\omega_1 \, d\omega_2 \, d^3x \, d^3y \, d^3z \, \lambda_{\mu\nu}(\omega; \omega_1, \omega_2, \mathbf{x}) G_{\mu\alpha}(x - y, \omega_1)$$

$$\times \quad G_{\nu\beta}(x - z, \omega_2) J^\alpha(y, \omega_1) J^\beta(z, \omega_2).$$

$$(7.166)$$

The dashed line indicates the dressed vacuum $|0\rangle \equiv |0_s, 0_i; \omega_p\rangle$, i.e., the state with no signal and idler photons, but with a classical pump in it. The advantage of having an effective free theory, instead of a fully interacting one, like the one discussed in the previous section for the quantum pump, is that now it is enough to calculate the two-point correlation function, i.e., the dressed propagator, to actually calculate the nonlinear interaction and its cross-section. We have in fact, that

$$(x_s, \omega_s)$$

$$\langle A_\mu(\mathbf{x}_s, \omega_s) A_\nu(\mathbf{x}_i, \omega_i)\rangle = \quad \text{------}$$

$$(7.167)$$

$$(x_i, \omega_i)$$

$$= \frac{1}{\hbar} \int d^3x \, \lambda_{\alpha\beta}(\omega; \omega_s, \omega_i, \mathbf{x}) G_{\alpha\mu}(\mathbf{x} - \mathbf{x}_s, \omega_s) G_{\beta\nu}(\mathbf{x} - \mathbf{x}_i, \omega_i)$$

$$= \frac{1}{\hbar} \mathscr{B}^{(2)}(\mathbf{x}_s - \mathbf{x}_i, \omega_s, \omega_i),$$

where the last line is the definition of the biphoton (or two-mode) propagator. In QFT terms, the diagram above describes the spontaneous generation of a signal–idler pair from the vacuum. In quantum optics terms, instead, it describes the spontaneous generation (frequently called spontaneous parametric down-conversion (SPDC) in nonlinear optics [15]) of a signal–idler photon pair from the classical pump inducing the nonlinear interaction in the material. The correspondent equations of motion for this kind of process are then linear in the signal and idler fields, and can be solved analytically (see, e.g., references [15, 18, 21, 24] for an extensive discussion about them).

We close this section by applying this result to the example of a one-dimensional waveguide, as done in the previous section. We use the same expression and conventions of equation (7.159) for calculating the propagator for the signal and idler modes, and go through the same calculations as before. Additonally, we need to replace the nonlinear susceptibility χ with the effective one $\lambda = \chi A_p \exp(i\varphi_p)$, where A_p and φ_p are the amplitude and phase of the pump beam. Putting everything together gives the following explicit form for the biphoton propagator

$$\mathcal{B}(x_s - x_i, \omega_s, \omega_i) = \Theta(x_s - x_i)\frac{\chi A_p L}{4\hbar k_s k_i}\exp\left[i(k_s x_s + k_i x_i + \varphi_p + \Delta\omega t)\right]\mathrm{sinc}(\Delta kL), \quad (7.168)$$

where $\Delta\omega$ and Δk are defined as before. The probability of a second-order effect to occur in the undepleted pump approximation then becomes

$$\sigma_2^{\text{undepleted pump}}(L, A_p) = \left(\frac{\chi A_p L}{4\hbar k_s k_i}\right)^2 \mathrm{sinc}^2(\Delta kL). \quad (7.169)$$

Notice that now σ also depends on the intensity of the pump field (i.e., A_p^2). This simply comes from the substitution $\chi \to \lambda \propto \chi A_p$ in equation (7.161), due to the undepleted pump approximation. Although minimal, this change is important, since it means that one can use the intensity of the pump (i.e., the number of photons contained in the pump field) to amplify the nonlinear response. Comparing the result above with equation (7.161), in fact, leads to the following result

$$\sigma_2^{\text{undepleted pump}}(L, A_p) = k_p^2 A_p^2 \sigma_2^{\text{quantum pump}}(L), \quad (7.170)$$

which essentially means that using a classical pump amplifies the probability of a second-order nonlinear interaction to occur of a factor of, essentially, A_p^2. This is essentially due to the fact, that instead of looking for single interaction events, using a classical pump can be seen as inducing A_p^2 simultaneous events, therefore enhancing the probability to observe such events accordingly.

7.2.9 Third-order nonlinearities

Another class of nonlinear interactions that are frequently encountered in photonics are the so-called $\chi^{(3)}$-nonlinearities, i.e., the third-order ones, which involve four fields, conventionally called simply $\{\omega_1, \omega_2, \omega_3, \omega_4\}$. For this class of nonlinear effects, the interaction Lagrangian (7.139) takes the form

$$\mathscr{L}_3(\mathbf{x}, \omega) = \frac{1}{4!}\int [d\omega]_3 \, \chi^{(3)}_{\mu\nu\sigma\tau}(\omega; [\omega]_3, \mathbf{x})A_\mu(\omega_1, \mathbf{x})A_\nu(\omega_2, \mathbf{x})A_\sigma(\omega_3, \mathbf{x})A_\tau(\omega, \mathbf{x}), \quad (7.171)$$

and the third-order partition function can be written in terms of Feynman diagrams as follows

$$Z_3[J] \quad = \quad \begin{array}{c} J^\beta(x_2, \omega_2) \qquad J^\rho(x_4, \omega_4) \\[2pt] \\ J^\alpha(x_1, \omega_1) \qquad J^\lambda(x_3, \omega_3) \end{array} \quad + \text{permutations}$$

$$= \frac{1}{4!\hbar} \int [d\omega]_4 [d^3 x]_4 \chi^{(3)}_{\mu\nu\sigma\tau}(\omega; [\omega]_3, x) G_{\mu\alpha}(x - x_1, \omega_1) G_{\nu\beta}(x - x_2, \omega_2)$$

$$\times \quad G_{\sigma\lambda}(x - x_3, \omega_3) G_{\tau\rho}(x - x_4, \omega_4) J^\alpha(x_1, \omega_1) J^\beta(x_2, \omega_2) J^\lambda(x_3, \omega_3) J^\rho(x_4, \omega_4)$$

$$(7.172)$$

where the equivalent diagrams through permutations are now 24, corresponding to the 4! ways to arrange the red, blue, teal, and orange currents $\otimes$ appearing above. Notice, also, that for third-order nonlinear interactions there are no tadpoles, since an even number of propagators (i.e., photons) meet at a single vertex [39], and the diagram above is the only relevant one at the 1-vertex approximation.

The cross-section for third-order processes can be calculated as the four-point correlation function or by simply removing the current sources $\otimes$ in the diagram above and replacing them with the correspondent label, i.e.,

$$\langle A_\mu(x_1, \omega_1) A_\nu(x_2, \omega_2) A_\sigma(x_3, \omega_3) A_\tau(x_4, \omega_4) \rangle = \begin{array}{c} (x_2, \omega_2) \qquad (x_4, \omega_4) \\[2pt] \\ (x_1, \omega_1) \qquad (x_3, \omega_3) \end{array} \quad + \text{permutations}$$

$$= \frac{1}{\hbar} \int d^3 x \chi^{(3)}_{\alpha\beta\lambda\rho}(\omega_1, \omega_2, \omega_3, \omega_4, x) G_{\alpha\mu}(x - x_1, \omega_1) G_{\beta\nu}(x - x_2, \omega_2) G_{\lambda\sigma}(x - x_3, \omega_3)$$

$$\times \quad G_{\rho\tau}(x - x_4, \omega_4).$$

$$(7.173)$$

Analogously to the case of second-order nonlinearities, we can demote one of the fields to be a classical field, thus being able to describe the third-order processes in the undepleted pump approximation as effective second-order nonlinear processes. Moreover, by considering two of the fields to be degenerate (e.g. $\omega_1 = \omega_2 \equiv \omega_p$), one could mimic second-order processes using third-order nonlinearities. Choosing the right diagram, i.e., the right ordering of the four fields, gives access to all the four-wave mixing processes usually described by classical nonlinear optics. Again, as discussed before, calculating the n-point correlation function only gives access to the cross-section of the process (i.e., the probability of that particular process to occur). Using instead equations (7.171) and (7.105) together with (3.6) allows us to derive the equations of motion for the four fields involved in the interactions (or fewer, if the undepleted pump approximation is used), which gives access to the energy transfer dynamics.

7.2.10 High-order and cascaded nonlinearities

The formalism of path integrals and Feynman diagrams can also be useful in evaluating quickly the cross-section for higher-order nonlinearities, as well as cascaded nonlinear processes. For the former, an approach similar to the previous sections can be used, with an interaction Lagrangian defined by equation (7.139) and n chosen accordingly with the order of the nonlinearity. For the case of cascaded nonlinearities, instead, one needs to distinguish two cases, before writing down their diagrammatic representation. The first case represents cascaded nonlinearities of the same kind. In this case, it is enough to extend the power series expansion in equation (7.145) to higher orders in n, in order to consider n-vertex diagrams, which automatically take into account cascaded processes of the same kind. For example, for the case of two cascaded $\chi^{(2)}$-processes, a possible Feynman diagram describing the interaction is obtained from equation (7.145) with $n = 2$ and reads

$$Z_2^{cascaded}[J] \quad = \quad \text{(diagram)} \quad + \text{permutations}, \qquad (7.174)$$

which represents the interaction of a signal (blue) and idler (teal) photon to create a pump (red) photon, which then decays again in another signal–idler pair.

The second category of cascaded processes, instead, involve different nonlinearities, such as, for example, a $chi^{(2)}$-process followed by a $\chi^{(3)}$-one. In this case, these terms are contained in the $Z_{cross}[J]$ term discussed in section 7.2.5 and can also be represented in terms of Feynman diagrams (if their occurrence is to be accounted for, and not neglected as we assumed in section 7.2.5) by simply juxtaposing the Feynman diagrams for the correspondent nonlinear processes. For example, a $\chi^{(2)} - \chi^{(3)}$ cascaded diagram could look something like this

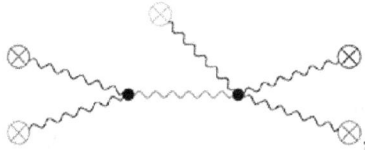

where now the first dot represents the second-order interaction, where a signal (blue) and an idler (teal) photon interact to generate a pump (red) photon, which then interacts with another (orange) photon through a third-order nonlinear process to generate a (black–violet) photon pair. Following the diagram rules explained in section 7.2.5 one can then easily write down the expression for the partition function of interest for the nonlinear process at hand, and then derive the expression for the cross-section of that particular process by simply eliminating the current sources $\otimes$ and relabelling them accordingly, as shown in the previous sections.

Appendix A: Derivation of equations (7.20a)

In this appendix we give an explicit derivation of equations (7.20a) in terms of the complex function $\xi(z)$, and discuss its origin. To do so, we first need to rewrite the expression for a_N, b_N, and c_N in an easier form, from which we can immediately derive a discretised version of equation (7.9). To do so, let us introduce the quantity

$$\Lambda_k = \prod_{j=1}^{k} \frac{\beta}{2\sigma_j} \equiv \prod_{j=1}^{k} \frac{1}{\lambda_j}, \qquad (7.175)$$

so that equations (7.17a) become

$$a_N = \frac{\beta}{2}\left(1 - \sum_{k=1}^{N-1} \Lambda_k \Lambda_{k-1}\right), \qquad (7.176a)$$

$$b_N = \frac{\beta}{2}\left(1 - \frac{\Lambda_{N-1}}{\Lambda_{N-2}}\right), \qquad (7.176b)$$

$$c_N = \beta \Lambda_{N-1}, \qquad (7.176c)$$

and the quantities λ_k obey the following recursion relations, that can be derived from the definition of β and α_k

$$\lambda_k = 2[1 - f^2(z_k)(d\zeta)^2] - \frac{1}{\lambda_{k-1}}. \qquad (7.177)$$

This recursion relation suggests us to rewrite λ in terms of a new function ξ_k as

$$\lambda_k = \frac{\xi_{k+1}}{\xi_k}. \qquad (7.178)$$

By doing so, and substituting the expression above into equation (7.177) we get the following recursion relation for ξ_k

$$\xi_{k+1} = 2\left[1 - \frac{1}{2}f^2(z_k)(d\zeta)^2\right]\xi_k - \xi_{k-1}. \qquad (7.179)$$

This recursion relation is what we need to relate the coefficients a_N, b_N, and c_N with the complex function $\xi(z)$. A closer inspection of it, in fact, reveals that the recursion relation above can be recast in the following form

$$\frac{\xi_{k+1} - 2\xi_k + \xi_{k-1}}{(d\zeta)^2} = -f^2(z_k)\xi_k, \qquad (7.180)$$

together with the boundary condition $\xi_0 = 0$. This is none other than the discretised version of equation (7.9), so if we can relate the coefficients a_N, b_N, and c_N to ξ_k and then take the limit $N \to \infty$ (which corresponds to set $d\zeta \to 0$), we can obtain the expressions of the limiting factors a_∞, b_∞, and c_∞ as a function of $\xi(z) = R(z)\sin[\varphi(z) - \varphi(z_a)]$, which is the continuous limit of the solution of equation (7.180).

Substituting equation (7.178) into equations (7.176) gives us the expression of the coefficients a_N, b_N, and c_N as a function of the discretised complex function ξ_k, i.e.,

$$a_N = \frac{\beta}{2}\left(1 - \sum_{k=1}^{N-1} \frac{\xi_1}{\xi_{k+1}} \frac{\xi_1}{\xi_k}\right),$$

(7.181a)

$$b_N = \frac{\beta}{2} \frac{\xi_N - \xi_{N-1}}{\xi_N},$$

(7.181b)

$$c_N = \beta \frac{\xi_1}{\xi_N}.$$

(7.181c)

To illustrate how to take the limit, let us first consider the coefficients b_N and c_N, since their limit is immediate. In clalculating the limit, it is worth keeping in mind that $\lim_{N\to\infty} \xi_N = \lim_{N\to\infty} \xi(z_N) = \xi(z_b)$, i.e., the end point of the propagation, $\lim_{N\to\infty} \xi_1 = \lim_{N\to\infty} \xi(z_1) = \xi(z_a)$, i.e., the starting point of the propagation, and that $\beta = n_0/\lambdabar(d\zeta)$. Thus, we get

$$\lim_{N\to\infty} b_N = \frac{n_0}{2\lambdabar} \lim_{d\zeta\to 0} \frac{\xi_N - \xi_{N-1}}{d\zeta} \frac{1}{\xi_N} = \frac{n_0}{2\lambdabar} \frac{\dot{\xi}(z_b)}{\xi(z_b)},$$

(7.182a)

$$\lim_{N\to\infty} c_N = \frac{n_0}{\lambdabar} \lim_{d\zeta\to 0} \frac{\xi_1 - \xi_0}{d\zeta} \frac{1}{\xi_N} = \frac{n_0}{\lambdabar} \frac{\dot{\xi}(z_a)}{\xi(z_b)},$$

(7.182b)

where the dot indicates derivation with respect to z. Notice, that to make the derivative appear we have added ξ_0 at the numerator of c_N, but this is not a problem since $\xi_0 = 0$ by definition. Substituting the expression of $\xi(z)$ into equation (7.182a) gives

$$\frac{\dot{\xi}(z_b)}{\xi(z_b)} = \frac{\dot{\xi}(z_b)}{\xi(z_b)} + \dot{\varphi}(z_b)\cot \Phi(\zeta)$$

$$= \frac{d}{dz} \log R(z)|_{z=z_b} + \dot{\varphi}(z)\cot \Phi(\zeta).$$

(7.183)

Substituting instead the expression of $\xi(z)$ into equation (7.182b) leads to the following result

$$\frac{\dot{\xi}(z_a)}{\xi(z_b)} = \frac{R(z_a)\dot{\varphi}(z_a)}{R(z_b)\sin \Phi(\zeta)}.$$

(7.184)

This result is equivalent to equation (7.21), but to show that we need to manipulate the expression above a little further, and express $R(z_a)/R(z_b)$ in terms of phase factors only. To do so, let us first substitute the expression of $\xi(z)$ into equation (7.9) and separate the terms proportional to sine and cosine to obtain the following set of coupled differential equations

$$\ddot{R} - R(\dot{\varphi})^2 + f^2(z)R = 0,$$

(7.185a)

$$2\dot{R}\dot{\varphi} + R\ddot{\varphi} = 0.$$

(7.185b)

We can solve the second equation right away by first defining $x(z) = \dot\varphi(z)$, so that it becomes a first-order ordinary differential equation with z-dependent coefficients, namely

$$\dot{x} + \frac{2\dot{R}}{R}x = 0, \tag{7.186}$$

whose solution can be readily written as [22]

$$x(z) = \exp\left(-\int dz\frac{2\dot{R}}{R}\right) = \frac{c}{R^2}, \tag{7.187}$$

where c is an arbitrary constant. Reverting back to φ we get the sought-after relation $\dot\varphi(z) = c/R(z)^2$, which we can now invert (and choose $c = 1$ for convenience) to get $R(z) = \sqrt{\dot\varphi(z)}$. Substituting this into equation (7.185) gives then the final reuslt

$$\frac{\dot\xi(z_a)}{\xi(z_b)} = \frac{R(z_a)\dot\varphi(z_a)}{R(z_b)\sin \Phi(\zeta)}$$

$$= \frac{\sqrt{\dot\varphi(z_a)\dot\varphi(z_b)}}{\sin \Phi(\zeta)}, \tag{7.188}$$

which coincides with equation (7.20c).

Calculating the limit for a_N requires instead a bit more care. First, let us rewrite a_N by taking out the term corresponding to $k = 1$ as follows

$$a_N = \frac{\beta}{2}\left(1 - \frac{\xi_1}{\xi_2} - \sum_{k=2}^{N-1}\frac{\xi_1^2}{\xi_k\xi_{k+1}}\right). \tag{7.189}$$

Then, using the fact that $\xi_{k+1} = \xi(z_a + (k+1)d\zeta) \simeq \xi(z_a + kd\zeta) + \dot\xi(z_a + kd\zeta)d\zeta + \cdots = \xi_k + \dot\xi_k d\zeta + \cdots$, we can rewrite the sum as follows

$$\sum_{k=2}^{N-1}\frac{\xi_1^2}{\xi_k\xi_{k+1}} = \sum_{k=2}^{N-1}\frac{\xi_1^2}{\xi_k[\xi_k + d\zeta(\cdots)]} \simeq \sum_{k=2}^{N-1}\frac{\xi_1^2}{\xi_k^2}, \tag{7.190}$$

where in the last passage we have neglected terms proportional to $d\zeta$ in the denominator, since they will go to zero once we take the limit. If we now put everything together, we get the following result for a_N

$$a_N = \frac{n_0}{2\lambda}\frac{\xi_2 - \xi_1}{d\zeta}\frac{1}{\xi_2} - \frac{n_0}{2\lambda}\frac{1}{d\zeta}\sum_{k=2}^{N-1}\frac{\dot\xi^2(d\zeta)^2}{\xi_k^2}$$

$$\rightarrow -\frac{n_0}{2\lambda}\frac{\dot\xi(z_a)}{\xi(z_a)} - \frac{n_0}{2\lambda}\lim_{d\zeta\to 0}d\zeta\int_{z_a+d\zeta}^{z_b}dz\frac{\dot\xi^2(z_a)}{\xi^2(z)} \tag{7.191}$$

$$= -\frac{n_0}{2\lambda}\frac{\dot\xi(z_a)}{\xi(z_a)} = \frac{d}{dz}\log R(z)|_{z=z_a} + \dot\varphi(z_a)\cot \Phi(\zeta).$$

Appendix B: Schwinger–Dyson equation for the dressed propagator

In this appendix we show how to derive the Schwinger–Dyson equation (7.113) from the dressed partition function (7.109). This equation is the solution to the integro-differential equation (7.111), which is generated, in frequency domain, by the Kernel $\tilde{R}_{ij}(\mathbf{x}, \mathbf{y}, \omega) - \omega^2\tilde{\Gamma}_{ij}(\mathbf{x}, \mathbf{y}, \omega)$ (see equation (7.107)). This means, that for the purpose of the calculations in this appendix, we can focus on the quadratic term in the vector potential in the dressed action (7.107) and neglect the other terms. We can therefore set $J_X = 0$ in equation (7.109) and rewrite the relevant part of the partition function as follows

$$
\begin{aligned}
Z(J) = \exp[iS(J)] &= \int \mathscr{D}A \exp\left[\frac{i}{\hbar}S_{\text{free}} + i\int d^3x\, J_i(\mathbf{x})A_i(\mathbf{x})\right] = \\
&= \int \mathscr{D}A \exp\left\{-\frac{i}{2\hbar}\int [dx]A_i(\mathbf{x}, \omega)\tilde{R}_{ij}(\mathbf{x}, \mathbf{y}, \omega)A_j(\mathbf{y}, \omega)\right. \\
&\quad \left. + i\int d^3x\, J_i(\mathbf{x}, \omega)A_i(\mathbf{x}, \omega)\right\},
\end{aligned}
\tag{7.192}
$$

where $[dx] = d^3x\,d^3y\,d\omega$, and $\tilde{R}_{ij}(\mathbf{x}, \mathbf{y}, \omega) = \tilde{R}_{ij}^{(0)}(\mathbf{x}, \mathbf{y}, \omega) - \omega^2\tilde{\Gamma}(\mathbf{x}, \mathbf{y}, \omega)$ is defined in equation (7.107). $R_{ij}^{(0)}(\mathbf{x}, \mathbf{y}, \omega)$ is the free electromagnetic Kernel defined in equation (7.106), whose operator inverse is the free propagator $G_{ij}^{(0)}(\mathbf{x}, \mathbf{y}, \omega)$, in the sense that

$$
R_{ij}^{(0)}(\mathbf{x}, \mathbf{y}, \omega)G_{ij}^{(0)}(\mathbf{x}, \mathbf{y}, \omega) = \delta_{ij}\delta(\mathbf{x} - \mathbf{y}).
\tag{7.193}
$$

The equations of motion for the free action S_{free} are simply given by[21]

$$
\frac{\partial S_{\text{free}}}{\partial A_i(\mathbf{x}, \omega)} + J_i(\mathbf{x}, \omega) = 0,
\tag{7.194}
$$

and can be recast in terms of an equation involving the partition function by integrating them with respect to A and substituting any occurrence of A in $\partial S_{\text{free}}/\partial A_i(\mathbf{x}, \omega)$ with the correspondent functional derivative $-i\partial/\partial J_i(\mathbf{x}, \omega)$. This gives

$$
\begin{aligned}
\int \mathscr{D}A &\left[\frac{\partial S_{\text{free}}}{\partial A_i(\mathbf{x}, \omega)}\left(A_i(\mathbf{x}, \omega) \to -i\frac{\partial}{\partial J_i(\mathbf{x}, \omega)}\right) + J_i(\mathbf{x}, \omega)\right]\exp[iS(J)] \\
&= \left[\frac{\partial S_{\text{free}}}{\partial A_i(\mathbf{x}, \omega)}\left(A_i(\mathbf{x}, \omega) \to -i\frac{\partial}{\partial J_i(\mathbf{x}, \omega)}\right) + J_i(\mathbf{x}, \omega)\right]Z[J] = 0,
\end{aligned}
\tag{7.195}
$$

---

[21] This form of the equations of motion is fully equivalent to the Euler–Lagrange equations. The reader can check this as an exercise.

which, upon using equation (7.192) for $Z[J]$, leads to[22]

$$-i \int d^3y \, \tilde{R}_{ij}(\mathbf{x}, \mathbf{y}, \omega) \frac{\partial S}{\partial J_j(\mathbf{y}, \omega)} + J_i(\mathbf{x}, \omega) = 0. \tag{7.196}$$

To solve this equation, it is useful to introduce the Legendre transform of the functional $S(J)$ appearing in equation (7.192) as

$$S(J) = i \, F(A) + i \int d^3x \, J_i(\mathbf{x}, \omega) \, A_i(\mathbf{x}, \omega). \tag{7.197}$$

The functional $F(A)$ is frequently referred to in QFT literature as the *effective action* [23], and its physical meaning will be clear in a moment.

From the definition above, we can derive the following relations

$$A_i(\mathbf{x}, \omega) = -i \frac{\partial S}{\partial J_i(\mathbf{x}, \omega)}, \tag{7.198a}$$

$$J_i(\mathbf{x}, \omega) = -\frac{\partial F}{\partial A_i(\mathbf{x}, \omega)}, \tag{7.198b}$$

and use them to transform equation (7.196) into

$$\frac{\partial F}{\partial A_i(\mathbf{x}, \omega)} = \int d^3y \, \tilde{R}_{ij}(\mathbf{x}, \mathbf{y}, \omega) \, A_j(\mathbf{y}, \omega). \tag{7.199}$$

If we now take the functional derivative of the expression above with respect to the vector potential $A_j(\mathbf{x}, \omega)$, we arrive at the following result

$$\frac{\partial^2 F}{\partial A_j(\mathbf{y}, \omega) \partial A_i(\mathbf{x}, \omega)} = \tilde{R}_{ij}(\mathbf{x}, \mathbf{y}, \omega) = \tilde{R}_{ij}^{(0)}(\mathbf{x}, \mathbf{y}, \omega) - \omega^2 \tilde{\Gamma}(\mathbf{x}, \omega) \delta_{jk}, \tag{7.200}$$

where $\tilde{\Gamma}(\mathbf{x}, \omega)$ is given by equation (7.86).

All that is left to do to derive the Schwinger–Dyson equation is to show that the left-hand side of the equation above is the inverse of the dressed propagator. To do so, let us first consider the following relation

$$\frac{\partial}{\partial J_j(\mathbf{y}, \omega)} \left(\frac{\partial F}{\partial A_i(\mathbf{x}, \omega)} \right) = -\delta_{ij} \delta(\mathbf{x} - \mathbf{y}). \tag{7.201}$$

To obtain this result we have used equation (7.198b) to write $\partial F/\partial A_i(\mathbf{x}, \omega)$ in terms of the current. Let us now use equation (7.198a) to say that the vector potential $A_i(\mathbf{x}, \omega)$ is a function of $J_i(\mathbf{x}, \omega)$, and, therefore, also the effective action is implicitly a function of $J_i(\mathbf{x}, \omega)$. This allows us to rewrite the functional derivative above as

$$\frac{\partial}{\partial J_j(\mathbf{y}, \omega)} \left(\frac{\partial F}{\partial A_i(\mathbf{x}, \omega)} \right) = \int d^3z \, \frac{\partial A_k(\mathbf{z}, \omega)}{\partial J_j(\mathbf{y}, \omega)} \frac{\partial F}{\partial A_i(\mathbf{x}, \omega)}. \tag{7.202}$$

---

[22] Some renaming of indices might be necessary
[23] See, for example, the book by Peskin and Schroeder [42] for a nice introduction on the topic.

We can then use this result to rewrite equation (7.201) as

$$
\begin{aligned}
\delta_{ij}\delta(\mathbf{x} - \mathbf{y}) &= - \int d^3z \, \frac{\partial A_k(\mathbf{z}, \omega)}{\partial J_j(\mathbf{y}, \omega)} \frac{\partial^2 F}{\partial A_k(\mathbf{z}, \omega) A_i(\mathbf{x}, \omega)} \\
&= i \int d^3z \, \frac{\partial}{\partial J_j(\mathbf{y}, \omega)} \left(\frac{\partial S}{\partial J_k(\mathbf{z}, \omega)} \right) \left(\frac{\partial^2 F}{\partial A_k(\mathbf{z}, \omega) A_i(\mathbf{x}, \omega)} \right) \\
&= i \int d^3z \, \left(\frac{\partial^2 S}{\partial J_j(\mathbf{y}, \omega)\partial J_k(\mathbf{z}, \omega)} \right) \left(\frac{\partial^2 F}{\partial A_k(\mathbf{z}, \omega) A_i(\mathbf{x}, \omega)} \right) \\
&= \int d^3z \, G_{jk}(\mathbf{y}, \mathbf{z}, \omega) \left(\frac{\partial^2 F}{\partial A_k(\mathbf{z}, \omega)\partial A_i(\mathbf{x}, \omega)} \right),
\end{aligned}
\tag{7.203}
$$

where in the second line we used equation (7.198a) to write the field A in terms of the derivative $\partial S/\partial A$, and in the last line we used the definition of the propagator in terms of derivatives of the partition function $Z[J] = \exp(iS)$.

A closer inspection of equation (7.203) reveals that in order for it to hold, the term $\partial^2 F/\partial A_k \partial A_j$ needs to be the inverse of the propagator G_{jk}, i.e., it needs to be the differential operator $L_{ki}(\mathbf{z}, \mathbf{x})$ generating the propagator, such that

$$
\int d^3z \, L_{ik}(\mathbf{x}, \mathbf{z})G_{kj}(\mathbf{z}, \mathbf{y}) = \delta_{ij}\delta(\mathbf{x} - \mathbf{y}).
\tag{7.204}
$$

Another way to understand this is to rewrite equation (7.203) in terms of its matrix representation. In doing so, both $\partial^2 S/\partial J^2$ and $\partial^2 F/\partial A^2$ become matrices, the l.h.s of equation (7.203) then corresponds to the identity matrix, and integration plus summation over indices corresponds to matrix multiplication. Doing so allows us to rewrite equation (7.203) as

$$
\left(\frac{\partial^2 S}{\partial J \partial J} \right)_{jk} \left(\frac{\partial^2 F}{\partial A \partial A} \right)_{ki} = \mathbb{I},
\tag{7.205}
$$

from which it is easier to interpret $\partial^2 F/\partial A^2$ as the inverse of the propagator. With this in mind, we then write

$$
\left(\frac{\partial^2 F}{\partial A_k(\mathbf{z}, \omega)\partial A_i(\mathbf{x}, \omega)} \right)^{-1} = G_{jk}(\mathbf{y}, \mathbf{z}, \omega),
\tag{7.206}
$$

with the understanding that expressions containing products of propagators will also have to contain summation over repeated indices and integration [42].

Analogously, we can also interpret the term $R_{ij}^{(0)}(\mathbf{x}, \mathbf{y}, \omega)$ appearing in equation (7.200) as the inverse of the free-photon propagator $G_{jk}(\mathbf{y}, \mathbf{z}, \omega)$, in the sense of equation (7.204). Substituting this and equation (7.206) into equation (7.200) gives the formal equation

$$
\frac{1}{G_{jk}(\mathbf{y}, \mathbf{z}, \omega)} = \frac{1}{\mu_0 G_{jk}^{(0)}(\mathbf{y}, \mathbf{z}, \omega)} - \omega^2 \tilde{\Gamma}(\mathbf{x}, \mathbf{y}, \omega)\delta_{jk},
\tag{7.207}
$$

which can be formally solved for the dressed propagator to give

$$
G_{jk}(\mathbf{y}, \mathbf{z}, \omega) = \mu_0 G_{jk}^{(0)}(\mathbf{y}, \mathbf{z}, \omega) + \mu_0 \omega^2 \int d^3x \, G_{jb}^{(0)}(\mathbf{y}, \mathbf{x}, \omega)\tilde{\Gamma}(\mathbf{x}, \omega)G_{bk}(\mathbf{x}, \mathbf{z}, \omega),
\tag{7.208}
$$

where we have introduced the integration over d^3x as part of the 'matrix multi-plicaiton' rule between the two propagators appearing in the second term. This is the Schwinger–Dyson equation. By adding a solution to the homogeneous version of equation (7.112) we then obtain equation (7.113).

Appendix C: Derivation of equation (7.142)

In this appendix we sketch how to derive equation (7.142) for an interacting field. The basic idea of this method, as will appear clear, is to construct a differential equation for the partition function $Z[J]$ for the interacting theory, find the corresponding Green's function, and show that the solution is of the form (7.142). For this derivation, we follow the steps presented in reference [38].

Let us start by the expression of the partition function for an interacting theory (equation (7.140))

$$Z[J] = \mathcal{N} \int \mathcal{D}A_\mu \exp\left[iS + i\int d^4x\, J^\mu A_\mu\right], \tag{7.209}$$

where $S = \int d^3x\,(\mathcal{L} + \mathcal{L}_{int})$ and $\mathcal{N} = \left[\int \mathcal{D}A_\mu \exp(iS)\right]^{-1}$ is a normalisation constant. For later convenience, let us call $Z_0[J]$ the partition function of the free theory, i.e., the above equation with $\mathcal{L}_{int} = 0$, which, according to equation (7.116), reads[24]

$$Z_0[J] = \exp\left[\frac{i}{2}\int d^4x\, d^4y\, J^\mu(x)G_{\mu\nu}(x-y)J^\nu(y)\right], \tag{7.210}$$

where $G_{\mu\nu}(x-y)$ is the propagator (Green's functions) of the free theory. For the partition function above, the following relation is valid

$$\frac{1}{i}\frac{\partial Z_0}{\partial J^\mu(x)} = \int d^4y\, G_{\mu\nu}(x-y)J^\nu(y)Z_0[J] = \frac{\delta_{\alpha\nu}}{\hat{H}_{\nu\mu}}J^\nu(x)Z_0[J], \tag{7.211}$$

where $\hat{H}_{\alpha\mu} = (\nabla\times\nabla - \omega^2\varepsilon(\omega)/c^2)_{\alpha\mu}$ is the Helmholtz operator, and to go from the second to the third equality we have used equation (7.111), ie., the definition of the Green's function. Rearranging the result above, we arrive at the differential equation satisfied by $Z_0[J]$, i.e.,

$$\hat{H}^{\nu\mu}\left(\frac{1}{i}\frac{\partial Z_0[J]}{\partial J^\mu(x)}\right) = J^\nu(x)Z_0[J]. \tag{7.212}$$

we now try to do the same thing, but for the interacting partition function in equation (7.209). We then have

$$\frac{1}{i}\frac{\partial Z[J]}{\partial J^\mu(x)} = \mathcal{N}\int \mathcal{D}A_\mu\, A_\mu \exp\left[i\,S + i\int d^4x\, J^\mu A_\mu\right]. \tag{7.213}$$

To derive a differential equation for $Z[J]$, it is useful to rewrite equation (7.209) as a functional Fourier transform by introducing the auxiliary functional

---

[24] We use natural units for convenience

$$\bar{Z}[A] = \mathcal{N} \exp(i\,S), \qquad (7.214)$$

so that equation (7.209) becomes

$$Z[J] = \int \mathcal{D}A_\mu\, \bar{Z}[A] \exp\left(i \int d^4x\, J^\mu A_\mu\right). \qquad (7.215)$$

We can then proceed by taking the derivative of $\bar{Z}[A]$ with respect to the field $A_\mu(x)$, to obtain

$$i\frac{\partial \bar{Z}[A]}{\partial A_\mu(x)} = \left[\hat{H}^{\mu\nu} A_\nu(x) - \frac{\partial \mathcal{L}_{int}}{\partial A_\mu(x)}\right]\bar{Z}[A], \qquad (7.216)$$

where to calculate the derivative above we have written the action of the electromagnetic field as $S = -(1/2)\int d^4x\,d^4y [\, A_\mu(x)\hat{H}^{\mu\nu}(x-y)A_\nu(y) + \mathcal{L}_{int}]$, which, for $\mathcal{L}_{int} = 0$, is equivalent to the effective action (7.105).

The next step consists in taking the functional Fourier transform of the result above, to obtain

$$\mathcal{F}\left\{i\frac{\partial \bar{Z}[A]}{\partial A_\mu(x)}\right\} = \int \mathcal{D}A_\mu\left[i\frac{\partial \bar{Z}[A]}{\partial A_\mu(x)}\right]\exp\left(i\int d^4x\, J^\mu A_\mu\right)$$

$$= \int \mathcal{D}A_\mu\left[\hat{H}^{\mu\nu}A_\nu - \frac{\partial \mathcal{L}_{int}}{\partial A_\mu}\right]\bar{Z}[A]\exp\left(i\int d^4x\, J^\mu A_\mu\right)$$

$$= \int \mathcal{D}A_\mu \hat{H}^{\mu\nu}A_\nu Z[J] - \int \mathcal{D}A_\mu \frac{\partial \mathcal{L}_{int}}{\partial A_\mu}\bar{Z}[A]\exp\left(i\int d^4x\, J^\mu A_\mu\right)$$

$$= \hat{H}^{\mu\nu}\left(\frac{1}{i}\frac{\partial Z[J]}{\partial J_\nu}\right) - \mathcal{L}_{int}'\left(\frac{1}{i}\frac{\partial}{\partial J_\mu}\right)Z[J], \qquad (7.217)$$

where the prime in $\mathcal{L}_{int}$ indicates the derivative with respect to its argument. Notice that we have used equation (7.215) to go from the second to the third line, and equation (7.213) to rewrite the first term in the third line in terms of the derivative of $Z[J]$ in the fourth line. More importantly, to transform the last term of the third line into the one in the fourth one, we have first used equation (5.55) to write the interaction Lagrangian in terms of derivatives of the current, and we have then interpreted the term $\mathcal{L}_{int}'$ as an operator acting on the functional $\bar{Z}[A]\exp\left(i\int d^4x\, J^\mu A_\mu\right)$. At this point, since $\mathcal{L}_{int}'$ does not depend on the fields A_μ anymore, we have brought it outside of the integral and used again equation (7.215) to make $Z[J]$ explicitly appear. The result above is nearly our final result. To complete it, we also need to take care of the left-hand side of the equation above, and compute the integral, which can be easily calculated using integration by parts, giving the following result

$$\mathcal{F}\left\{i\frac{\partial \bar{Z}[A]}{\partial A_\mu(x)}\right\} = \int \mathcal{D}A_\mu\left[i\frac{\partial \bar{Z}[A]}{\partial A_\mu(x)}\right]\exp\left(i\int d^4x\, J^\mu A_\mu\right)$$

$$= \left\{i\bar{Z}[A]\exp\left(i\int d^4x\, J^\mu A_\mu\right)\right\}\bigg|_{A_\mu \to \infty} + J^\mu(x)Z[J] \qquad (7.218)$$

$$= J^\mu(x)Z[J],$$

where we have made use again of equation (7.215) to rewrite the second term in the second line in terms of $Z[J]$ and used the fact that the fields at infinity go to zero to eliminate the surface term at infinity.

Collecting the results from equations (7.217) and (7.218) we can finally write the differential equation satisfied by $Z[J]$ as

$$\hat{H}^{\mu\nu}\left(\frac{1}{i}\frac{\partial Z[J]}{\partial J_\nu}\right) - \mathscr{L}_{\text{int}}'\left(\frac{1}{i}\frac{\partial}{\partial J_\mu}\right)Z[J] = J^\nu(x)Z[J]. \tag{7.219}$$

We now show, that the solution to the above differential equation can be written in the form

$$Z[J] = \mathscr{N} \exp\left[i\int d^4x\ \mathscr{L}_{\text{int}}\left(\frac{1}{i}\frac{\partial}{\partial J_\mu}\right)\right]Z_0[J], \tag{7.220}$$

which is equivalent to equation (7.142). To do so, let us first derive a useful identity. Le us start by calculating the commutator between the current $J^\nu(x)$ and the interaction Lagrangian, i.e.,

$$\begin{aligned}
[\hat{C},\hat{B}] &= \left[\int d^4y\,\mathscr{L}_{\text{int}}\left(\frac{1}{i}\frac{\partial}{\partial J_\mu(y)}\right),\ J^\nu(x)\right] \\
&= \sum_{n=0}^{\infty}\frac{1}{n!}\mathscr{L}_{\text{int}}^{(n)}(0)\int d^4y\left[\left(\frac{1}{i}\frac{\partial}{\partial J_\mu(y)}\right)^n,\ J^\nu(x)\right] \\
&= \sum_{n=0}^{\infty}\frac{1}{n!}\mathscr{L}_{\text{int}}^{(n)}(0)\int d^4y\left\{\left(\frac{1}{i}\frac{\partial}{\partial J_\mu(y)}\right)^{n-1}\left[\frac{1}{i}\frac{\partial}{\partial J_\mu(y)},\ J^\nu(x)\right]\right. \\
&\quad\left. + \left[\left(\frac{1}{i}\frac{\partial}{\partial J_\mu(y)}\right)^{n-1},\ J^\nu(x)\right]\left(\frac{1}{i}\frac{\partial}{\partial J_\mu(y)}\right)\right\} \\
&= \sum_{n=0}^{\infty}\frac{1}{n!}\mathscr{L}_{\text{int}}^{(n)}(0)\int d^4y\left\{-i\delta_{\mu\nu}\delta(x-y)\left(\frac{1}{i}\frac{\partial}{\partial J_\mu(y)}\right)^{n-1}\right. \\
&\quad\left. + \left[\left(\frac{1}{i}\frac{\partial}{\partial J_\mu(y)}\right)^{n-1},\ J^\nu(x)\right]\left(\frac{1}{i}\frac{\partial}{\partial J_\mu(y)}\right)\right\} \\
&= \cdots \\
&= \sum_{n=0}^{\infty}\frac{1}{n!}\mathscr{L}_{\text{int}}^{(n)}(0)\int d^4y\left\{-i\delta_{\mu\nu}\delta(x-y)\,n\left(\frac{1}{i}\frac{\partial}{\partial J_\mu(y)}\right)^{n-1}\right\} \\
&= -i\mathscr{L}_{\text{int}}'\left(\frac{1}{i}\frac{\partial}{\partial J_\mu(y)}\right),
\end{aligned} \tag{7.221}$$

where first we have written the interaction Lagrangian in a power series expansion, then to go from the second to the third line we have used the property of the commutator $[\hat{A}\hat{B}, \hat{C}] = \hat{A}[\hat{B}, \hat{C}] + [\hat{A}, \hat{C}]\hat{B}$ and repeated it n times to calculate the various commutators appearing above. Moreover, we have also made use of the 'functional position–momentum' commutation relation $[(1/i)(\partial/\partial J_\mu(x)), J_\nu(y)] = -i\delta_{\mu\nu}\delta(x - y)$. Finally, we have re-summed the second to last line of the above equation, resulting in the derivative of the interaction Lagrangian

Next, we use the Baker–Campbell–Hausdorff formula [43]

$$\exp(\hat{C})\hat{B}\exp(-\hat{C}) = \hat{B} + [\hat{C}, \hat{B}] + \frac{1}{2!}[\hat{C}, [\hat{C}, \hat{B}]] + \cdots, \tag{7.222}$$

with $\hat{C} = -i\int d^4y\,\mathscr{L}_{\text{int}}$ and $\hat{B} = J^\nu(x)$. Notice, that $[\hat{C}, \hat{B}]$ only contains terms proportional to $\partial/\partial J$ (i.e., to $\hat{C}$), and therefore all the higher-order nested commutators are zero, since they are all proportional to terms like $[(\partial/\partial J)^n, (\partial/\partial J)^m] = 0$. We then have

$$\exp\left[-i\int d^4y\,\mathscr{L}_{\text{int}}\right]J^\nu(x)\exp\left[i\int d^4y\,\mathscr{L}_{\text{int}}\right] = J^\nu(x) - \mathscr{L}_{\text{int}}'\left(\frac{1}{i}\frac{\partial}{\partial J_\mu(x)}\right), \tag{7.223}$$

which can be rewritten as follows, by multiplying on the left both sides of the above equation by $\exp[i\int d^4y\,\mathscr{L}_{\text{int}}]$

$$J^\nu(x)\exp\left[i\int d^4y\,\mathscr{L}_{\text{int}}\right] = \exp\left[i\int d^4y\,\mathscr{L}_{\text{int}}\right]\left[J^\nu(x) - \mathscr{L}_{\text{int}}'\left(\frac{1}{i}\frac{\partial}{\partial J_\mu(x)}\right)\right]. \tag{7.224}$$

We can now use this identity to prove that equation (7.220) is the solution of equation (7.219) by calculating $J^\nu(x)Z[J]$ as follows

$$\begin{aligned}
J^\nu(x)Z[J] &= \mathscr{N}J^\nu(x)\exp\left[i\int d^4x\,\mathscr{L}_{\text{int}}\right]Z_0[J] \\
&= \mathscr{N}\exp\left[i\int d^4y\,\mathscr{L}_{\text{int}}\right](J^\nu(x) - \mathscr{L}_{\text{int}}')Z_0[J] \\
&= \mathscr{N}\exp\left[i\int d^4y\,\mathscr{L}_{\text{int}}\right]\hat{H}^{\nu\mu}\left(\frac{1}{i}\frac{\partial Z_0[J]}{\partial J^\mu}\right) - \mathscr{L}_{\text{int}}'\exp\left[i\int d^4y\,\mathscr{L}_{\text{int}}\right]Z_0[J] \\
&= \hat{H}^{\nu\mu}\left(\frac{1}{i}\frac{\partial Z[J]}{\partial J^\mu}\right) - \mathscr{L}_{\text{int}}'Z[J],
\end{aligned} \tag{7.225}$$

which is exactly equation (7.219). Notice, that to go from the second to the third equality, we have used equation (7.212) to rewrite the first term, proportional to JZ_0, and we have used the fact that $\mathscr{L}_{\text{int}}'$ and $\exp(-i\int d^4y\,\mathscr{L}_{\text{int}})$ commute and their order can be therefore interchanged. We then made use of equation (7.220) to make $Z[J]$ explicitly appear. This then proves that equation (7.220), and therefore equation (7.142) is a solution of the differential equation (7.219).

References

[1] Feynman R P and Hibbs A R 2010 *Quantum Mechanics and path Integrals* amended edn (New York: Dover)

[2] Gòmez-Reino C and Linares J 1987 Optical path integrals in gradient-index media *J. Opt. Soc. Am.* A **4** 1337

[3] Robson C W, Tamashevich Y, Rantala T T and Ornigotti M 2021 Path integrals: from quantum mechanics to photonics *APL Photon.* **6** 071103

[4] Hillery M and Zubairy S 1982 Path-integral approach to problems in quantum optics *Phys. Rev.* A **26** 451

[5] Hillery M and Zubairy S 1984 Path-integral approach to the quantum theory of the degenerate amplifier *Phys. Rev.* A **29** 1275

[6] Bechler A 1999 Quantum electrodynamics of the dispersive dielectric medium–a path integral approach *J. Mod. Opt.* **46** 901

[7] Difallah M, Szameit A and Ornigotti M 2019 Path-integral description of quantum nonlinear optics in arbitrary media *Phys. Rev.* A **100** 053845

[8] Longhi S 2009 Quantum optical analogies using photonic structures *Laser Photon. Rev.* **3** 243

[9] Born M and Wolf E 2020 *Principles of Optics* (Cambridge: Cambridge University Press)

[10] Jackson J D 1998 *Classical Electrodynamics* (New York: Wiley)

[11] Khandekar D C and Lawande S V 1975 Exact propagator for a time dependent harmonic oscillator with and without a singular perturbation *J. Math. Phys.* **16** 384

[12] Olver F W J, Lozier D W, Boisvert R F and Clarck C W (ed) 2010 *NIST Handbook of Mathematical Functions* (Cambridge: Cambridge University Press)

[13] Svelto O 1998 *Principles of Lasers* 4th edn (Dordrecht: Kluwer Academic)

[14] Brown L S 1992 *Quantum Field Theory* (Cambridge: Cambridge University Press)

[15] Boyd R W 2008 *Nonlinear Optics* 3rd edn (Amsterdam: Elsevier)

[16] Glauber R J 1963 Coherent and incoherent states of the radiation field *Phys. Rev.* **131** 2766

[17] Sudarshan E C G 1963 Equivalence of semiclassical and quantum mechanical descriptions of statistical light beams *Phys. Rev. Lett.* **10** 277

[18] Loudon R 2000 *The Quantum Theory of Light* (Oxford: Oxford University Press)

[19] Schleich W P 2001 *Quantum Optics in Phase Space* (New York: Wiley)

[20] Mandel L and Wolf E 1995 *Optical Coherence and Quantum Optics* (Cambridge: Cambridge University Press)

[21] Butcher P N and Cotter D 1990 *The Elements of Nonlinear Optics* (Cambridge: Cambridge University Press)

[22] Byron F W and Fuller R W 1992 *Mathematics of Classical and Quantum Physics* (Mineola, NY: Dover)

[23] Gerry C and Knight P L 2004 *Introductory Quantum Optics* (Cambridge: Cambridge University Press)

[24] Hillery M and Drummond P D 2014 *The Quantum Theory of Nonlinear Optics* (Cambridge: Cambridge University Press)

[25] Landau L D and Lifshitz E M 1980 *The Classical Theory of Fields* 4th edn (Oxford: Butterworth-Heinemann)

[26] Dutra S M and Furuya K 1998 The permittivity in the Huttner-Barnett theory of QED in dielectrics *Europhys. Lett.* **43** 13

[27] Vogel W and Welsch D K 2006 *Quantum Optics* (New York: Wiley)

[28] Gupta S N 1950 Theory of longitudinal photons in quantum electrodynamics *Proc. Phys. Soc.* A **63** 681

[29] Bleuer K 1950 Eine neue methode zur behandlung der longitudinalen und skalaren photonen *Helv. Phys. Acta* **23** 567

[30] Itzykson C and Zuber J B 1980 *Quantum Field Theory* (New York: Dover)

[31] Coleman S 2019 *Quantum Field Theory* (Singapore: World Scientific)

[32] Bernard C W and Weinberg E J 1977 Interpretation of pseudoparticles in physical gauges *Phys. Rev.* D **15** 3656

[33] Landau L D and Lifshitz E M 1980 *Electrodynamics of Continuous Media* 4th edn (Oxford: Butterworth-Heinemann)

[34] Crosse J A and Scheel S 2010 Effective nonlinear Hamiltonians in dielectric media *Phys. Rev.* A **81** 033815

[35] Scheel S and Welsch D G 2006 Causal nonlinear quantum optics *J. Phys. B: At. Mol. Opt. Phys.* **39** S711

[36] Scheel S and Buhmann S Y 2008 Macroscopic quantum electrodynamics–concepts and applications *Acta Phys. Slov.* **58** 675

[37] Reshef O, Giese E, Alam M Z, De Leon I, Upham J and Boyd R W 2017 Beyond the perturbative description of the nonlinear optical response of low-index materials *Opt. Lett.* **42** 3225

[38] Ryder L H 1996 *Quantum Field Theory* (Cambridge: Cambridge University Press)

[39] Srednicki M 2007 *Quantum Field Theory* (Cambridge: Cambridge University Press)

[40] Waxer L J, Bromage J and Kruschwitz B E 2023 *Petawatt Laser Systems* (Bellingham, WA: SPIE Spotlight)

[41] Maggiore M 2005 *A Modern Introduction to Quantum Field Theory* (Oxford: Oxford University Press)

[42] Peskin M E and Schroeder D V 2019 *An Introduction to Quantum Field Theory* (Boca Raton, FL: CRC Press)

[43] Barnett S M and Radmore P M 2005 *Methods in Theoretical Quantum Optics* (Oxford: Oxford Science Publications)

IOP Publishing

A Field Theory Approach to Photonics

Marco Ornigotti

Chapter 8

Light–matter interaction in 2D materials

This chapter is focused on a particular application of field theory to a problem that is closely related to photonics, i.e., the study of light–matter interaction in 2D materials.

2D materials are generally defined as crystalline materials that consist of a single atomic layer. The prototype of such materials is undoubtedly graphene, discovered by Novoselov and Geim in 2004[1][1]. Since then, many different classes and species of 2D materials have been investigated, such as transition metal dichalcogenides (TMDs) [2], black phosphorous [3], boron nitride [4], and Weyl materials [5, 6] to name a few. In photonics, these materials have been gathering a lot of attention because of their unique optical properties, such as ultrafast broadband responses [7–9] and large nonlinear responses [10–13], making them a promising candidate for realising hybrid integrated devices combining electronic, spintronic, and optical properties in a single chip [14].

From a theoretical point of view, 2D materials constitute a very interesting playground, as they enable easy access to relativistic electron dynamics [15], topological phenomena like the quantum [16] and fractional [17] Hall effect, superconductivity [18], and they also offer an experimentally accessible test ground for the AdS/CFT correspondence [19, 20]. Moreover, thanks to the work of Marino and co-workers in the 90s [21–25] on quantum electrodynamics (QED) in low dimensional systems, the so-called *pseudo quantum electrodynamics* (PQED), 2D materials have also become an interesting playground where to explore the application of techniques and methods of QED for solving problems in photonics.

This chapter, therefore, focusses on discussing one simple example of how the formalism of PQED and path integrals can be combined to calculate the optical

---

[1] A discovery that was worthy of a Nobel prize in Physics in 2010 (https://www.nobelprize.org/prizes/physics/2010/press-release/), highlighting the importance and technological impact 2D materials were thought to possess at that time.

doi:10.1088/978-0-7503-5789-0ch8   8-1

conductivity of graphene. This example will serve multiple purposes: it will allow showcasing the techniques and methods of QED and PQED applied to a concrete problem in photonics, and will also serve as a possible starting point for future developments in the field. To do so, section 8.1 first presents some key properties of graphene and discusses how the dynamics of electrons in it can be described from a field theory perspective through a massless Dirac equation. A detailed derivation of the various physical properties of graphene is given, for example, in the book by Katsnelson [15]. Section 8.3 then defines the field theory problem of PQED coupled graphene and presents a detailed derivation of the conductivity. The results presented in this section follow those of reference [26], but their derivation is based on the effective action method highlighted in appendix B of chapter 7 and is presented in a more didactical manner, to emphasise different aspects of the calculation. We do not, however, present a rigorous derivation of the Lagrangian for the pseudo electromagnetic field in 2D, as this would require a long dive into topological field theory, which is outside the scope of this book. Instead, following the line of reasoning of reference [24], we present a concise derivation of the PQED Lagrangian from the QED one based on constraining the motion of electrons on a plane, i.e., assuming interaction with a 2D material. This is done in section 8.2.

Finally, in section 8.4 a brief perspective on how to use the methods and techniques illustrated in this chapter to solve other problems involving the nonlinear light–matter interaction in 2D materials, or describe materials other than graphene is given.

8.1 A primer on graphene

In this section, we summarise some results about graphene that will be useful in the rest of the chapter. Section 8.1.1 summarises the lattice properties of graphene and introduces both the real and reciprocal lattice, as well as the Dirac points. The Hamiltnoinan for graphene is then derived in section 8.1.2 starting from general assumptions about electrons in periodic materials, and the low energy approximation is discussed in section 8.1.3. Finally, the continuum limit and the connection with Dirac equation and field theory is made in section 8.1.4.

8.1.1 The crystalline structure of graphene

Graphene is a two-dimensional material consisting of carbon atoms disposed in a honeycomb lattice. This hexagonal lattice can be broken down in two displaced triangular lattices, one made only with atoms occupying site A (blue dots in figure 8.1(a)), and the other one made only with atoms occupying site B (black dots in figure 8.1(a)). For this reason, the unit cell (shaded in gray in figure 8.1(a)) contains two atoms, one belonging to the triangular lattice A, and the other one to the triangular lattice B. The Bravais vectors spanning the honeycomb lattice of graphene are defined as

(a)

(b)

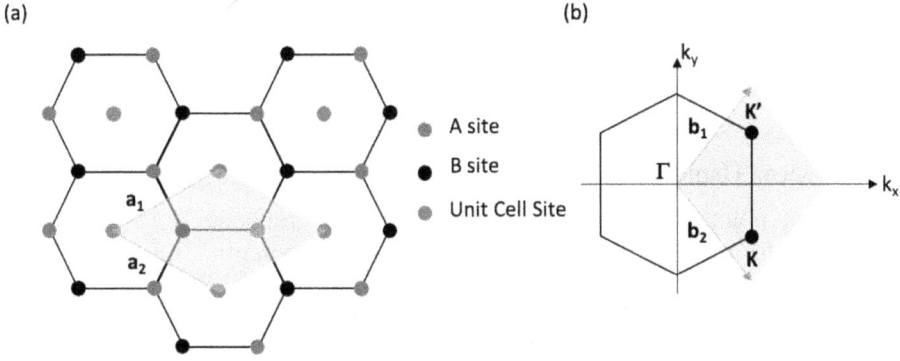

* A site
* B site
* Unit Cell Site

Figure 8.1. (a) Lattice structure of graphene. The unit cell, containing two atoms (one belonging to lattice A and the other to lattice B), is shaded in gray. The basis vectors are chosen to be $\mathbf{a}_1 = (a/2)(3, \sqrt{3})$, and $\mathbf{a}_2 = (a/2)(3, -\sqrt{3})$. (b) Reciprocal space lattice structure of graphene (the Wigner–Seitz cell is shaded in grey). The reciprocal lattice basis vectors are defined as $\mathbf{b}_1 = (2\pi/3a)(1, \sqrt{3})$ and $\mathbf{b}_2 = (2\pi/3a)(1, -\sqrt{3})$. The two inequivalent Dirac points $\mathbf{K}$ and $\mathbf{K}'$ are defined as $\mathbf{K} = (2\pi/3a)(1, -1/\sqrt{3})$, and $\mathbf{K}' = (2\pi/3a)(1, 1/\sqrt{3})$, respectively.

$$\mathbf{a}_1 = \frac{a}{2}(3\hat{\mathbf{x}} + \sqrt{3}\hat{\mathbf{y}}), \tag{8.1a}$$

$$\mathbf{a}_2 = \frac{a}{2}(3\hat{\mathbf{x}} - \sqrt{3}\hat{\mathbf{y}}), \tag{8.1b}$$

where $a \simeq 1.42$ Å is the lattice constant for graphene.

The reciprocal lattice for graphene is shown in figure 8.1(b). As can be seen, it is also a honeycomb lattice, rotated by 90 degrees with respect to the lattice in real space. The Wigner–Seitz cell, depicted as a grey shaded area in figure 8.1(b), contains two inequivalent points, conventionally labelled as $\mathbf{K}$ and $\mathbf{K}'$ and called Dirac points. The basis vectors for the reciprocal lattice of graphene, defining its Brilloiun zone, are

$$\mathbf{b}_1 = \frac{2\pi}{3a}(\hat{\mathbf{x}} + \sqrt{3}\hat{\mathbf{y}}), \tag{8.2a}$$

$$\mathbf{b}_2 = \frac{2\pi}{3a}(\hat{\mathbf{x}} - \sqrt{3}\hat{\mathbf{y}}), \tag{8.2b}$$

and the position of the Dirac points are given as

$$\mathbf{K} = \frac{2\pi}{3a}\left(1, -\frac{1}{\sqrt{3}}\right), \tag{8.3a}$$

$$\mathbf{K}' = \frac{2\pi}{3a}\left(1, \frac{1}{\sqrt{3}}\right). \tag{8.3b}$$

The two inequivalent Dirac points are, as we will see below, the points at which the valence and conduction bands of graphene touch, and they are the relevant points for defining the low energy Hamiltonian of graphene.

8.1.2 Graphene Hamiltonian

The starting point to derive the Hamiltonian for graphene is to consider the Hamiltonian for an electron in a crystalline (periodic) potential, i.e.,

$$\hat{H} = \int d^3r \; \hat{\psi}^\dagger(\mathbf{r}) \left[\frac{\mathbf{p}^2}{2m} + \phi_c(\mathbf{r}) + V_B(\mathbf{r}) \right] \hat{\psi}(\mathbf{r}) \equiv \int d^3r \; \hat{\psi}^\dagger(\mathbf{r}) \mathcal{H}_0 \hat{\psi}(\mathbf{r}), \qquad (8.4)$$

where $\phi_c(\mathbf{r})$ is the Coulomb potential, $V_B(\mathbf{r})$ is the crystalline (periodic) potential, $\hat{\psi}(\mathbf{r})$ is the electron wave function operator.

To calculate the graphene Hamiltonian, we use the following Bloch representation for the electron wave function operator

$$\hat{\psi}(\mathbf{r}) = \frac{1}{\sqrt{N}} \sum_{k} \sum_{n=0}^{N} \exp(i\mathbf{k} \cdot \mathbf{R}_n) [\phi_A(\mathbf{r} - \mathbf{R}_n)\hat{a}_k + \phi_B(\mathbf{r} - \mathbf{R}_n)\hat{b}_k], \qquad (8.5)$$

where $\phi_{A,B}(\mathbf{r})$ are $2p_z$-orbital wavefunctions at site A and B (forming an orthogonal normalised basis set), $\mathbf{R}_n = j\mathbf{a}_1 + m\mathbf{a}_2$ is a Bravais lattice vector of graphene (see figure 8.1(a)), and $\hat{a}_k$, $\hat{b}_k$ are the electron site operators, annihilating (creating) an electron on site A or B, respectively.

Let us first focus on calculating the explicit expression of the free Hamiltonian. Substituting equation (8.5) into equation (8.4) gives the following expression for the free Hamiltonian

$$\begin{aligned}\hat{H}_0 = \sum_{k,k'} \sum_{n,m} \exp[i(\mathbf{k}' \cdot \mathbf{R}_m - \mathbf{k} \cdot \mathbf{R}_n)] \Big\{ &\hat{a}_k^\dagger \hat{a}_{k'} \int \frac{d^3r}{N} \phi_A^*(\mathbf{r} - \mathbf{R}_n)\mathcal{H}_0\phi_A(\mathbf{r} - \mathbf{R}_m) \\ + &\hat{a}_k^\dagger \hat{b}_{k'} \int \frac{d^3r}{N} \phi_A^*(\mathbf{r} - \mathbf{R}_n)\mathcal{H}_0\phi_B(\mathbf{r} - \mathbf{R}_m) \\ + &\hat{b}_k^\dagger \hat{a}_{k'} \int \frac{d^3r}{N} \phi_B^*(\mathbf{r} - \mathbf{R}_n)\mathcal{H}_0\phi_A(\mathbf{r} - \mathbf{R}_m) \\ + &\hat{b}_k^\dagger \hat{b}_{k'} \int \frac{d^3r}{N} \phi_B^*(\mathbf{r} - \mathbf{R}_n)\mathcal{H}_0\phi_B(\mathbf{r} - \mathbf{R}_m) \Big\}.\end{aligned} \qquad (8.6)$$

To calculate the integrals appearing above, we make the change of variables $\mathbf{x} = \mathbf{r} - \mathbf{R}_n$, so that, for example,

$$\int \frac{d^3r}{N} \phi_B^*(\mathbf{r} - \mathbf{R}_n)\mathcal{H}_0\phi_B(\mathbf{r} - \mathbf{R}_m) = \int \frac{d^3x}{N} \phi_B^*(\mathbf{x})\mathcal{H}_0\phi_B(\mathbf{x} - (\mathbf{R}_m - \mathbf{R}_n)), \qquad (8.7)$$

and similarly for the other integrals above. Now, because the lattice is periodic, $\mathbf{R}_m - \mathbf{R}_n \equiv \mathbf{R}_\ell$ is itself a lattice vector. This gives us the possibility to calculate the summation over n and m of the exponential terms appearing above as follows

$$\sum_{n,m} \exp[i(\mathbf{k}' \cdot \mathbf{R}_m - \mathbf{k} \cdot \mathbf{R}_n)] = \sum_n \exp[i(\mathbf{k}' - \mathbf{k}) \cdot \mathbf{R}_n] \sum_\ell \exp(i\mathbf{k}' \cdot \mathbf{R}_\ell) = \delta_{k,k'} \exp(i\mathbf{k}' \cdot \mathbf{R}_\ell). \quad (8.8)$$

After this simplification, the integrals appearing in equation (8.6) are all of the form

$$I_{\alpha\beta} = \sum_\ell \exp(i\mathbf{k} \cdot \mathbf{R}_\ell) \int \frac{d^3x}{N} \phi_\alpha^*(\mathbf{x}) \mathcal{H}_0 \phi_\beta(\mathbf{x} - \mathbf{R}_\ell), \quad (8.9)$$

with $\alpha, \beta \in \{A, B\}$.

Next, we assume nearest-neighbour coupling only[2], i.e., we allow each unit cell to only interact with its direct neighbours. As a result of this approximation, we get the constraint $|\ell| \leqslant 1$, with $\ell = 0$ terms corresponding to the on-site energies $E_{A,B}$ of the two lattices in the unit cell, and the terms with $\ell = \pm 1$ corresponding to the coupling between neighbouring cells. Using figure 8.1(a) as visual guidance, the nearest-neighbours of the B sites (black dots) in the unit cell are the A sites (blue dots), and they can be reached by the lattice vector $\mathbf{R}_{\ell-1} = \mathbf{R}_\ell - \mathbf{a}_2$ and $\mathbf{R}_{\ell+1} = \mathbf{R}_\ell - \mathbf{a}_1$. Substituting this into the equation above and using Bloch theorem [27] to write $\phi(\mathbf{r} - \mathbf{R}_n) = \exp(i\mathbf{k} \cdot \mathbf{R}_n)\phi(\mathbf{r})$ we get

$$I_{AA} = f(\mathbf{k})E_A, \quad (8.10a)$$

$$I_{BB} = f(\mathbf{k})E_B, \quad (8.10b)$$

$$I_{AB} = I_{BA}^* = \gamma, \quad (8.10c)$$

where

$$E_i = \int \frac{d^3x}{N} \phi_i^*(\mathbf{x}) \, \mathcal{H}_0 \, \phi_i(\mathbf{x}), \quad (8.11)$$

is the on-site energy for site i,

$$\gamma = \int \frac{d^3x}{N} \phi_A^*(\mathbf{x}) \, \mathcal{H}_0 \, \phi_B(\mathbf{x}), \quad (8.12)$$

is the nearest-neighbour hopping constant (notice that $\gamma \in \mathbb{R}$), and

$$f(\mathbf{k}) = 1 + \exp(i\mathbf{k} \cdot \mathbf{a}_1) + \exp(i\mathbf{k} \cdot \mathbf{a}_2). \quad (8.13)$$

With this result, we can write the free graphene Hamiltonian as

$$\begin{aligned}
\hat{H}_0 &= \gamma \sum_k \left[f(\mathbf{k}) \hat{a}_k^\dagger \hat{b}_k + \text{h.c.} \right] \\
&= \sum_k \left(\hat{a}_k^\dagger \ \ \hat{b}_k^\dagger \right) \begin{pmatrix} 0 & \gamma f(\mathbf{k}) \\ \gamma f^*(\mathbf{k}) & 0 \end{pmatrix} \begin{pmatrix} \hat{a}_k \\ \hat{b}_k \end{pmatrix} \\
&\equiv \sum_k \left(\hat{a}_k^\dagger \ \ \hat{b}_k^\dagger \right) H(\mathbf{k}) \begin{pmatrix} \hat{a}_k \\ \hat{b}_k \end{pmatrix},
\end{aligned} \quad (8.14)$$

---

[2] What is known in solid-state physics as the tight-binding approximation.

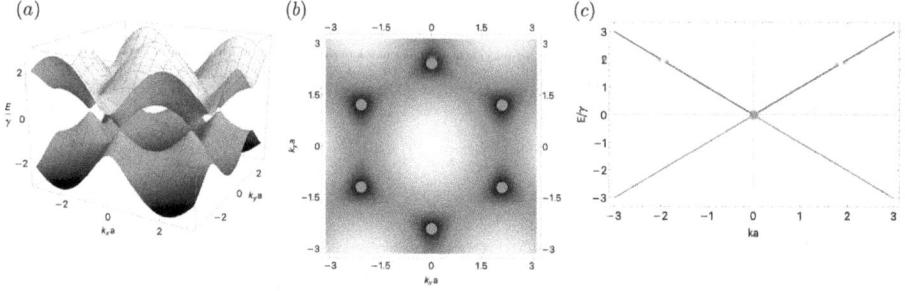

Figure 8.2. (a) Band structure of graphene, as given by equation (8.15a). The lower, dark, band is the valence band, while the upper, bright, band is the conduction band. The two bands touch into six points, corresponding to the edges of the hexagonal Brilloiun zone (see figure 8.1(b)). (b) Section, along the plane $E(\mathbf{k}) = 0$ of the band structure in panel (a). The green dots highlight the position of the six Dirac points. (c) Low energy approximation of the band structure of graphene, according to equation (8.16) and corresponding to the dispersion relation in panel (a) close to one of the Dirac points (in green). The red curve is the valence band, while the blue curve is the conduction band. All the figures in the panels have been plotted as a function of the normalised momentum $a\mathbf{k}$ and normalised energy $E(\mathbf{k})/\gamma$.

where $H(\mathbf{k})$ is the (reciprocal space) matrix representation of the free Hamiltonian. Diagonalising $H(\mathbf{k})$ gives then access to the band structure of graphene (i.e., the eigenvalues of $H(\mathbf{k})$), shown in figure 8.2, and the wave function of graphene electrons (i.e., the eigenvectors of $H(\mathbf{k})$). Notice, that since graphene possesses two atoms in the unit cell, the wave function for graphene electrons is a two-component spinor, i.e., $\psi = (\psi_A, \psi_B)^T$, where each component represents the probability amplitude of finding the electron in the correspondent site inside the unit cell [15].

The eigenvalues and eigenvectors of the free graphene Hamiltonian $H(\mathbf{k})$ are then given by

$$E(\mathbf{k}) = \pm\gamma |f(\mathbf{k})|, \tag{8.15a}$$

$$u_{\pm}(\mathbf{k}) = \frac{1}{\sqrt{2}} \begin{pmatrix} \exp\left[i\frac{\varphi(\mathbf{k})}{2}\right] \\ \pm \exp\left[-i\frac{\varphi(\mathbf{k})}{2}\right] \end{pmatrix}, \tag{8.15b}$$

where $\varphi(\mathbf{k}) = \arg[f(\mathbf{k})]$. A plot of the energy eigenvalues, i.e., valence and conduction bands, of graphene is shown in figure 8.2.

8.1.3 Low energy Hamiltonian for graphene

Now that we have the general expression for the free and light–matter interaction Hamiltonian of graphene, we can derive the low energy approximation to it. To do so, we notice, by looking at figure 8.2, that the valence and conduction bands of graphene touch at six different points. In correspondence of these points, then, $E(\mathbf{k}) = 0$, which corresponds to the six Dirac points in the reciprocal lattice of figure 8.1(b), where by symmetry, three of them will be equivalent to $\mathbf{K}$, and the other three to $\mathbf{K'}$. We can then obtain the low energy approximation to graphene Hamiltonian by Taylor expanding the band structure $E(\mathbf{k})$ around these points. In general, because $\mathbf{K}$ and $\mathbf{K'}$ are inequivalent, we would need to make this expansion for both points separately. This would result in the electron wave function to be described by a four-component spinor $\psi = (\psi_A^K, \psi_B^K, \psi_A^{K'}, \psi_B^{K'})^T$ and, since there is no interaction between electrons near the two inequivalent Dirac points, in the Hamiltonian to be represented by a block-diagonal matrix. We can, however, take account of this by redefining the Dirac points as $\mathbf{D} = (2\pi/3a)(1, \xi/\sqrt{3})$, so that by choosing $\xi = \pm 1$, we can switch between the two Dirac points. The index ξ introduced here is frequently referred to in literature as the *valley index*, since the two regions in reciprocal space around the Dirac points are known as *valleys*. In doing so, we can keep describing the electronic properties of graphene with a 2×2 Hamiltonian and a two-component spinor, plus a valley index.

Then, let us Taylor expand the band structure of graphene around the Dirac point $\mathbf{D}$, to obtain

$$E(\mathbf{k}) \simeq E(\mathbf{D}) + \nabla_k \ E(\mathbf{D}) \cdot \mathbf{q}$$
$$= \pm \left| \frac{3a\gamma}{2}(q_x + i\xi q_y) \right| = \frac{3a\gamma}{2}|\mathbf{q}|, \tag{8.16}$$

where $\mathbf{q} = \mathbf{k} - \mathbf{D}$. If we introduce the Fermi velocity as the velocity of electrons in the vicinity of a Dirac point as $3a\gamma/2 \equiv \hbar v_f$ [15], we can then write the low energy approximation to the free Hamiltonian as

$$H_0(\mathbf{q}) = \hbar v_f(\sigma_x q_x + \xi \sigma_y q_y), \tag{8.17}$$

where $\sigma_{x,y}$ are the Pauli matrices, defined as [28]

$$\sigma_x = \begin{pmatrix} 0 & 1 \\ 1 & 0 \end{pmatrix}, \ \sigma_y = \begin{pmatrix} 0 & -i \\ i & 0 \end{pmatrix}, \ \sigma_z = \begin{pmatrix} 1 & 0 \\ 0 & -1 \end{pmatrix}, \ \sigma_0 = \begin{pmatrix} 1 & 0 \\ 0 & 1 \end{pmatrix}. \tag{8.18}$$

By using minimal coupling (see chapter 2), we can introduce the interaction of graphene electrons with the electromagnetic field by substituting the electron momentum $\mathbf{q}$ with the canonical momentum $\pi = \mathbf{q} - e\mathbf{A}$, where $\mathbf{A}$ is the vector potential of the electromagnetic field. By doing so we get the following light–matter Hamiltonian for graphene electrons

$$H(\mathbf{q}) = \hbar v_f[\sigma_x(q_x - eA_x) + \xi\sigma_y(q_y - eA_y)] = H_0(\mathbf{q}) - e\hbar v_f(\sigma_x A_x + \xi\sigma_y A_y). \tag{8.19}$$

8.1.4 Dirac equation from graphene Hamiltonian

Equation (8.19) describes the light–matter Hamiltonian of graphene in reciprocal space. To make a connection with Dirac equation and be able to write a graphene Lagrangian to use in the next sections, we need to transform equation (8.19) back into real space. We can do this by simply using the position representation of momentum, i.e., $\mathbf{q} \to -i\nabla$, obtaining

$$\hat{H}_0 = \hbar v_f[\sigma_x(-i\partial_x - eA_x) + \xi\sigma_y(-i\partial_y - eA_y)]. \tag{8.20}$$

Second, we need to transform the eigenvalue equation

$$\hat{H}_0\psi = E\psi, \tag{8.21}$$

into an evolution equation. To do so we use the time-energy relation $E \to i\hbar\partial/\partial t$ and notice that in the case of graphene the energy can be written in matrix form as follows

$$E(\mathbf{q}) = \pm\gamma|f(\mathbf{q})| = \begin{pmatrix} \gamma|f(\mathbf{q})| & 0 \\ 0 & -\gamma|f(\mathbf{q})| \end{pmatrix} = \gamma|f(\mathbf{q})|\sigma_z. \tag{8.22}$$

Substituting this result and equation (8.20) into equation (8.21) we finally get the equation of motion for graphene electrons in real space, which reads

$$i\hbar\sigma_z\frac{\partial\psi}{\partial t} = \hbar v_f[\sigma_x(-i\hbar\partial_x - eA_x) + \xi\sigma_y(-i\hbar\partial_y - eA_y)]\psi. \tag{8.23}$$

This is already the equation we are looking for. However, we now rewrite it in a more field-theory-friendly manner, by introducing gamma matrices and adjoint spinors. In 2D, the gamma matrices can be written in terms of Pauli matrices, as they obey the same Clifford algebra $\{\gamma^\mu, \gamma^\nu\} = \gamma^\mu\gamma^\nu + \gamma^\nu\gamma^\mu = 2\eta^{\mu\nu} = \{\sigma^\mu, \sigma^\nu\}$, as follows

$$\gamma^0 = \sigma_z, \quad \gamma^1 = i\sigma_y, \quad \gamma^2 = -i\sigma_x. \tag{8.24}$$

We then introduce the normalised time as $\tau = v_f t$ (in analogy with $\tau = ct$ in special relativity). The results we are going to get will be the same for the other valley, but this will allow us to keep a two-spinor description and write a two-spinor version of Dirac equation in a simple manner. This allows us to write equation (8.23) as a relativistic Dirac equation for the two-spinor ψ, i.e.,

$$i\hbar\gamma^\mu\left(\partial_\mu + \frac{ie}{\hbar}A_\mu\right)\psi = 0, \tag{8.25}$$

where $\mu = \{0, 1, 2\} = \{\tau, x, y\}$ and $\xi = 1$ has been assumed, for simplicity. To write the Lagrangian from which the equation above is derived as equation of motion, and to correctly take into account the valley index ξ, we need a couple more ingredients. First, we introduce the Feynman slash notation

$$i\slashed{\partial} = i\gamma^\mu\partial_\mu, \tag{8.26}$$

then we introduce the so-called *flavour index* [23], which accounts for the spinor components $\psi_{A,B}$ and the valley to which they belong. In the case of graphene, since we have two lattice sites and two valleys, the flavour index $a \in \{1, 2, 3, 4\}$. This can be seen as an internal index for the wave function, i.e., $\psi \to \psi_a$. Moreover, to write a Lagrangian, we also need the Dirac adjoint $\bar{\psi}_a$ to the spinor ψ_a, defined as follows

$$\bar{\psi}_a = \psi_a^\dagger\gamma^0, \tag{8.27}$$

or, in matrix form

$$\bar{\psi}_a = \begin{pmatrix} \psi_A^* & \psi_B^* \end{pmatrix} \begin{pmatrix} 1 & 0 \\ 0 & -1 \end{pmatrix} = \begin{pmatrix} \psi_A^* & -\psi_B^* \end{pmatrix}. \tag{8.28}$$

To write the graphene Lagrangian correspondent to the Dirac equation (8.25), we can now use the usual Dirac Lagrangian for electrons [29], adapted to the case of graphene (meaning, we use the spinor ψ_a and its adjoint $\bar{\psi}_a$, together with the gamma matrices $\{\gamma^0, \gamma^1, \gamma^2\}$) as follows

$$\mathcal{L}_g = i\bar{\psi}_a \slashed{\partial} \psi_a + j^\mu A_\mu, \tag{8.29}$$

where we have defined the electron current as

$$j^\mu = \bar{\psi}_a \gamma^\mu \psi_a. \tag{8.30}$$

Notice, that as in QED [29], electrons couple with the electromagnetic field through their current j^μ. This is the Lagrangian of a two-dimensional, four-flavoured Dirac field, representing electrons in graphene in the low energy limit, i.e., near the Dirac points.

8.2 Light–matter interaction on a plane

Now that we have derived the Dirac Lagrangian for graphene electrons interacting with the electromagnetic field, we would like to write the partition function for the light–matter interaction problem of graphene interacting with the electromagnetic field in a manner similar to the QED partition function, i.e.,

$$Z[J] = \mathcal{N} \int \mathcal{D}A_\mu \mathcal{D}\bar{\psi}\mathcal{D}\psi \exp\left[-i \int d^4x \, (\mathcal{L}_{em} + \mathcal{L}_{matter} + \text{source terms})\right], \tag{8.31}$$

where $\mathcal{L}_{matter} = i\bar{\psi}\slashed{\partial}\psi + J^\mu A_\mu$ is the Dirac Lagrangian density for electrons [29], $\mathcal{L}_{em} = -F_{\mu\nu}F^{\mu\nu}/4$ is the standard electromagnetic Lagrangian density, and 'source terms' indicate the source currents for the electromagnetic field A_μ, as well as for the spinors ψ and $\bar{\psi}$ [29]. Remember, that as discussed in chapter 6, taking the integral with respect to A_μ also requires a suitable gauge-fixing condition, which we implicitly assume holds, for simplicity.

Notice, that the integral in the exponent of equation (8.31) is a four-dimensional integral, since we are considering QED. However, if we want to use this form of the partition function to look at QED effects in graphene, we need to reduce the dimensionality of the problem, from 4D to (2 + 1)D, since graphene is only defined on a 2D plane. What we then need to do is to find a way to define a *2D electromagnetic field*, compatible with the partition function above, and that is capable of describing the interaction of light with graphene. The proper, i.e., formal, way to do this has been extensively discussed by Marino in reference [23] and requires the use of methods from topological field theory, which are beyond the scope of this book. Instead of going through the rigorous derivation, here we present a simpler argument, also used by Marino in his book [24], which is based on the following observation: since the electrons in graphene are constrained to live on a plane, a way of deriving a suitable partition function for *pseudo QED*, i.e., for a 2D electromagnetic field interacting with graphene electrons, is to assume that the current generated by graphene electrons only has components on the plane, and no components out of plane. This allows us to first say that $J^3(x) = 0$, and also to give the following explicit expression for the current $J^\mu(x)$ appearing in the matter Lagrangian above[3]

$$J^\mu(x) = j^\mu(\mathbf{R}, t)\delta(z), \qquad (8.32)$$

where now $\mu \in \{0, 1, 2\}$, and $\mathbf{R} = x^1 \hat{\mathbf{x}} + x^2 \hat{\mathbf{y}}$ emphasises the fact, that the graphene current $j^\mu(\mathbf{R}, t)$ only has in-plane components and depends only on the in-plane coordinates $\{x^1, x^2\}$. Moreover, we assume that the temporal coordinate has been Wick rotated to imaginary time, i.e., that $x^0 = it$. This has the effect of transforming the Minkowski metric into the standard Euclidean metric, i.e., $\eta_{\mu\nu} = \delta_{\mu\nu}$ and to transform the exponential appearing in the partition function $Z[J]$ from oscillating, i.e., $\exp(iS)$, to decaying, i.e., $\exp(-S)$. This transformation has some deep implication in the derivation of PQED (see reference [23]), but here we take it as a convenient transformation, that will allow us to do calculations more easily.

To adapt equation (8.31) to the 2D case of graphene, let us first concentrate on the electromagnetic part and integrate over the vector potential A_μ. Using equation (6.101) and choosing a proper gauge-fixing procedure as discussed in chapter 6, we get

$$\mathscr{L}[J] = \exp\left[-\frac{e^2}{2}\int d^4x\, d^4y J^\mu(x)G^{\mu\nu}(x - y)J_\nu(y)\right] \qquad (8.33)$$

If we now use equation (8.32) for the current and use the momentum representation (6.102) for the propagator, we get, after integrating over the z coordinate and dividing the momentum integral in equation (6.102) into its transverse and longitudinal parts,

$$\mathscr{L}[j] = \exp\left[\frac{e^2}{2}\int d^3x\, d^3y\, j^\mu(x)\tilde{G}_{\mu\nu}(x - y)j^\nu(y)\right], \qquad (8.34)$$

where now the propagator is given by[4]

$$\tilde{G}_{\mu\nu}(x - y) = \int d\omega\, \frac{d^2k}{(2\pi)^4}\, dk_z\, \frac{\exp\{i[\mathbf{k}\cdot(\mathbf{x} - \mathbf{y}) - \omega(t - t')]\}}{\omega^2 + k_\perp^2 + k_z^2}, \qquad (8.35)$$

where $k_\perp^2 = k_x^2 + k_y^2$ is the transverse momentum and k_z is the longitudinal one. Notice, moreover, that $\mathbf{k}\cdot(\mathbf{x} - \mathbf{y})$ has no component in the z-direction. This allows us to calculate directly the integral in the longitudinal momentum using the integral [30]

---

[3] Notice, that we are using J^μ to indicate the 4D current, while we reserve the symbol j^μ to indicate the 2D current on the plane, as done in equation (8.29).

[4] Remember that in the Euclidean metric, the signature is always positive.

$$\int \frac{dx}{a^2 + b^2 + x^2} = \frac{\pi}{\sqrt{a^2 + b^2}}, \tag{8.36}$$

provided that $a^2 + b^2 \in \mathbb{C}$. Using this result, we can rewrite the propagator as

$$\tilde{G}_{\mu\nu}(x - y) = \frac{1}{2} \int d\omega \frac{d^2k}{(2\pi)^3} \frac{\exp\{i[\mathbf{k} \cdot (\mathbf{x} - \mathbf{y}) - \omega(t - t')]\}}{\sqrt{\omega^2 + k_\perp^2}}. \tag{8.37}$$

This is the propagator for a fully $(2 + 1)$-dimensional electromagnetic field constrained to propagate on a plane. To derive the associated Lagrangian density, and then be able to write the full PQED problem, we can use the following argument.

We know that the propagator G of a field is formally the inverse of a certain differential operator $\hat{L}$, such that $\hat{L}G = \delta(x - y)$. For the free electromagnetic field, the propagator in momentum space is proportional to $1/(\omega^2 - k^2)$, which corresponds, once Fourier transformed back into position space, to the quantity $1/(\partial_t^2 + \nabla^2) = 1/(\partial^\mu \partial_\mu)$, i.e., to the inverse of the electromagnetic Kernel $\hat{R}(x - y)$. If we use a similar line of reasoning for the propagator in equation (8.37), we can formally write that

$$\tilde{G} \propto \frac{1}{2\sqrt{\omega^2 + k_\perp^2}} \rightarrow \frac{2}{\sqrt{\partial_t^2 + \nabla_\perp^2}} = \frac{2}{\sqrt{-\partial^\mu \partial_\mu}}, \tag{8.38}$$

where now $\mu = \{0, 1, 2\}$ only spans the $(2 + 1)$-dimensional space. Hence, the differential operator for the $(2 + 1)$-dimensional electromagnetic field is $2/\sqrt{\partial^\mu \partial_\mu}$, and the Lagrangian density for such an electromagnetic field can be readily written as [23, 24]

$$\mathcal{L}_{pem} = -\frac{1}{4} F_{\mu\nu} \left[\frac{2}{\sqrt{-\partial^\alpha \partial_\alpha}} \right] F^{\mu\nu}. \tag{8.39}$$

The full PQED problem of light–matter interaction for graphene can be then written using the above expression as the electromagnetic Lagrangian density and equation (8.29) for the graphene Lagrangian density, so that the (PQED) graphene partition function can be written as

$$Z[B] = \mathcal{N} \int \mathcal{D}A_\mu \, \mathcal{D}\bar{\psi}_a \, \mathcal{D}\psi_a \, \exp \left[-\int d^3x \, (\mathcal{L}_{pem} + \mathcal{L}_g + ej^\mu B_\mu) \right], \tag{8.40}$$

where $B^\mu(x)$ is the source term associated to the electronic current in graphene. This partition function describes all the electronic and optical properties of graphene, and we will use it in the next section to derive the explicit expression of the optical conductivity of graphene.

8.3 Optical conductivity of graphene

In this section, we derive the expression of the DC conductivity of graphene, using the partition function in equation (8.40). The defintion of the conductivity can be cast using the so-called Kubo formula [31]

$$\sigma^{\mu\nu} = \lim_{\omega,\mathbf{k}\to 0}\left[\frac{i}{\omega}\langle j^{\mu}(\omega,\mathbf{k})j^{\nu}(\omega,\mathbf{k})\rangle\right],\qquad(8.41)$$

which allows one to calculate the conductivity from the current–current correlation function. Our strategy is to calculate the current–current correlation function from the partition function in equation (8.40) using the effective action method discussed in appendix B of chapter 7, and also making the explicit path integration with respect to electronic and photonic degrees of freedom. Then, once we have access to this quantity, we will take the limit and derive the DC conductivity for graphene and compare it with existing results. This section follows the results presented in reference [26].

8.3.1 Current correlator from effective action

To start using the effective action method, let us rewrite the partition function in equation (8.40) as

$$Z[B] = \exp\{-S[B]\},\qquad(8.42)$$

and introduce the effective action as

$$\Gamma[\phi^{\mu}] = \int d^{3}x\, B_{\mu}(x)\phi^{\mu}(x) - S[B],\qquad(8.43)$$

where, according to the definition given in appendix B, $\phi^{\mu}(x) = \partial S/\partial B_{\mu}(x)$. Recalling that the second derivative of the effective action with respect to the field $\phi^{\mu}(x)$ is the inverse of the propagator, we can then write

$$\frac{\partial^{2}\Gamma}{\partial\phi^{\mu}(x)\partial\phi^{\nu}(y)} = -\frac{\partial B_{\nu}(y)}{\partial\phi^{\mu}(x)} \equiv \frac{1}{\Pi^{\mu\nu}(x-y)}.\qquad(8.44)$$

The physical meaning of $\Pi^{\mu\nu}(x-y)$ will be discussed below. To connect this result with the current–current correlator, let us observe that

$$\frac{\partial^{2}S[B]}{\partial B_{\mu}(x)\partial B_{\nu}(y)} = \frac{\partial}{\partial B_{\nu}(y)}\left[i\frac{\partial}{\partial B_{\mu}(x)}\log Z[B]\right]$$

$$= \frac{1}{Z^{2}}\left[\frac{\partial Z[B]}{\partial B_{\mu}(x)}\frac{\partial Z[B]}{\partial B_{\nu}(y)} - Z\frac{\partial^{2}Z[B]}{\partial B_{\mu}(x)\partial B_{\nu}(y)}\right]\qquad(8.45)$$

$$= e^{2}\left[\langle j_{a}^{\mu}(x)j_{a}^{\nu}(y)\rangle - \langle j_{a}^{\nu}(y)\rangle\langle j_{a}^{\mu}(x)\rangle\right],$$

where the expectation value is interpreted in path integral sense as

$$\langle j_{a}^{\nu}(x)\rangle = \frac{1}{Z[B]}\int \mathcal{D}A_{\mu}\,\mathcal{D}\bar{\psi}_{a}\,\mathcal{D}\psi_{a}\, j_{a}^{\mu}(x)\exp(-S_{pqed}),\qquad(8.46)$$

where we have defined, for convenience, $S_{pqed} = \int d^{3}x\,(\mathscr{L}_{pem} + \mathscr{L}_{g} + j_{a}^{\mu}B_{\mu})$. Notice, moreover, that the second term in the last line of equation (8.45) vanishes by virtue of Wick's theorem (see chapter 5). Therefore, we get that

$$\left\langle j_a^{\mu}(x)\, j_a^{\nu}(y) \right\rangle = \frac{1}{e^2} \frac{\partial^2 S[B]}{\partial B_{\mu}(x)\partial B_{\nu}(y)}. \tag{8.47}$$

We can now use equation (7.204) to link equations (8.47) and (8.44) through the relation

$$\int d^3z \left[\frac{1}{e^2} \frac{\partial^2 S[B]}{\partial B_{\mu}(x)\partial B_{\nu}(z)} \right]\left[\frac{\partial^2 \Gamma}{\partial \phi^{\nu}(z)\partial \phi^{\sigma}(y)} \right] = \delta_{\mu\sigma}\delta(x-y), \tag{8.48}$$

from which, following the same formal discussion below equation (7.204), we can conclude that

$$\left\langle j_a^{\mu}(x)\, j_a^{\nu}(y) \right\rangle = \left[\frac{1}{e^2} \frac{\partial^2 \Gamma}{\partial \phi^{\mu}(x)\partial \phi^{\nu}(y)} \right]^{-1} = \Pi^{\mu\nu}(x-y). \tag{8.49}$$

To understand the origin and physical meaning of the tensor $\Pi^{\mu\nu}(x-y)$, let us calculate the current–current correlator by explicitly solving the path integral in equation (8.40).

8.3.2 Current correlator from direct path integration

To start with, let us calculate the fermionic path integral in equation (8.40). Substituting then equation (8.31) into equation (8.40) and introducing the vector field $E_{\mu} = A_{\mu} + B_{\mu}$, for convenience of notation, the fermionic path integral in equation (8.40) gives the following result[5]

$$I_f = \int \mathcal{D}\bar{\psi}_a\, \mathcal{D}\psi_a\, e^{-\int d^3x\, \bar{\psi}_a[-i\slashed{\partial} - e\slashed{E}]\psi_a} = \det\left(-i\slashed{\partial} - e\slashed{E}\right). \tag{8.50}$$

To calculate the determinant above, let us define quantity $M(x, y) = [(i\slashed{\partial} - e\slashed{E})]\delta(x-y)$ and think of it as a matrix. Then, we rewrite $M(x, y)$ as

$$M(x, z) = \int d^3y\, M_0(x, y)\tilde{M}(y, z), \tag{8.51}$$

where

$$M_0(x, y) = (-i\slashed{\partial})\delta(x-y), \tag{8.52a}$$

$$\tilde{M}(y, z) = \delta(y-z) - eS_F(y-z)\slashed{E}, \tag{8.52b}$$

where $S(y-z)$ is the fermionic propagator (for graphene electrons)

$$S_F(y-z) = \int \frac{d^3k}{(2\pi)^3}\, \frac{i\slashed{k}}{k^2 - i\varepsilon}\, e^{ik(y-z)}. \tag{8.53}$$

---

[5] To solve this integral, it is useful to think to the spinor ψ_a as a Grassmann variable, and apply the results of appendix B of chapter 6. See also reference [29] for a primer on fermionic path integrals.

We now have the matrix $M(x, y)$ written as the product of two matrices. Then, its determinant will be $\det M = \det M_0 \det \tilde{M}$, but since M_0 is independent on the background field E_μ, $\det M_0$ will only amount to a constant and can be therefore absorbed into the global normalisation factor $\mathcal{N}$ of the partition function. This leaves us then with the following result

$$\det M = \det \tilde{M} = \det(\mathbb{I} - eS_F(y - z)\mathbf{E}). \tag{8.54}$$

Using the identity

$$\det M = \exp\{\log\det M\} = \exp\{\mathrm{Tr}\log M\}, \tag{8.55}$$

and expanding $\tilde{M}$ in a Taylor series with respect to e, we finally get

$$I_f = \exp\left\{\mathrm{Tr}\sum_{n=1}^{\infty}\left(\frac{1}{n}\right)[eS_F(x - y)\mathbf{E}]^n\right\}, \tag{8.56}$$

where the trace operator Tr stands for trace in spacetime (i.e., integration $\int d^3x$) and also Dirac trace (i.e., summation over spinor indices). It is instructive to write down the explicit expression for the first few terms of the expansion above:

$$I_f = \exp\left\{e\int d^3y\ \mathrm{tr}[S_F(x - y)\mathbf{E}(y)] \right.$$
$$\left. + e^2\int d^3y d^3z\ \mathrm{tr}[S_F(y - z)\mathbf{E}(z)S_F(z - y)\mathbf{E}(y)]+\text{higher-order terms}\right\}, \tag{8.57}$$

where now $\mathrm{tr}[x]$ indicates only the Dirac trace [29]. The first term evaluates to zero when put back into the partition function and integrated with respect to the electromagnetic field, thanks to Furry's theorem [32]. The second term, on the other hand, is known in QED as the *polarisation tensor*, and represents the photon self-energy [29]. As we will see below, it is precisely the quantity $\Pi^{\mu\nu}(x - y)$ defined in equation (8.49).

Before substituting the expression of I_f back into equation (8.40) and proceeding with the integration with respect to the electromagnetic field, let us give an explicit expression to the polarisation tensor, by rewriting I_f in momentum space by introducing the Fourier transform of the field $E_\mu(x)$ as

$$E_\mu(x) = \int \frac{d^3k}{(2\pi)^3}\ E_\mu(k)e^{ik\cdot x}, \tag{8.58}$$

where $d^3k = d^2k\ d\Omega$, with Ω being the frequency associated to the imaginary time, and $\Omega = i\omega$ holds, with ω the real frequency.

After some little algebra, involving integration with respect to spatial coordinates, we get the following expression for I_f

$$I_f = \exp\left[\int \frac{d^3q}{(2\pi)^3}E_\mu^*(q)\Pi^{\mu\nu}(q)E_\nu(q)\right], \tag{8.59}$$

where the polarisation tensor is defined as

$$\Pi^{\mu\nu}(q) = e^2 \int \frac{d^3k}{(2\pi)^3} \, \text{tr}\left\{ \frac{\not{k}}{k^2 - i\varepsilon} \gamma^\mu \frac{i(\not{k} + \not{q})}{(k + q)^2 - i\varepsilon} \gamma^\nu \right\}. \tag{8.60}$$

The explicit expression for the polarisation tensor for reduced QED theories, like PQED, has been calculated at the one- and two-loop level in reference [33], and reads[6]

$$\Pi^{\mu\nu}(q) = -\frac{e^2 \sqrt{q^2}}{16}\left[1 + \left(\frac{92 - 9\pi^2}{18\pi}\right)\frac{\alpha c}{v_f}\right]\left(\delta^{\mu\nu} - \frac{q^\mu q^\nu}{q^2}\right) + \frac{e^2}{2\pi}\left(n + \frac{1}{2}\right)\varepsilon^{\mu\nu\alpha}q_\alpha, \tag{8.61}$$

where $q^2 = q_\perp^2 + \Omega^2$ and $n \in \mathbb{N}$. The second term, in particular, is associated with the topological properties of graphene, i.e., the quantum Hall effect, and $\alpha \simeq 1/137$ is the fine structure constant [29].

We can now substitute the expression above for I_f into equation (8.40) and perform the integration with respect to the electromagnetic field to see explicitly that $\Pi^{\mu\nu}(q)$ is, indeed, the current–current correlation function. Doing so and representing the PQED action in momentum space gives the following expression for the partition function

$$Z[B] = \mathcal{N} \int \mathscr{D}A_\mu \exp\left\{-\int \frac{d^3q}{(2\pi)^3}\left[A_\mu^*(q)\tilde{G}^{\mu\nu}A_\nu(q) - E_\mu^*(q)\Pi^{\mu\nu}(q)E_\nu(q)\right]\right\}. \tag{8.62}$$

Once again, the integral above is Gaussian in the vector potential A_μ and we can then perform the integration analytically. In doing so, we need first to insert $E_\mu(q) = A_\mu(q) + B_\mu(q)$ back in the exponent above and then introduce an *effective electromagnetic source* term $\phi^\mu(q) = \Pi^{\mu\nu}(q)B_\nu(q)$, and then perform the change of variables

$$\chi_\mu(q) = A_\mu(q) - \frac{\phi^\nu(q)}{\tilde{G}^{\mu\nu}(q)} = A_\mu(q) - \Pi_{\mu\nu}(q)\phi^\nu(q), \tag{8.63}$$

where the last equality follows from equation (7.204), i.e., by the fact that $\tilde{G}^{\alpha\beta}\Pi_{\mu\nu} = \delta_\mu^\alpha \delta_\nu^\beta$. Implementing this change of variables into equation (8.62) gives

$$Z[B] = \mathcal{N} \int \mathscr{D}\chi_\mu \exp\left\{-\int \frac{d^3q}{(2\pi)^3}\left[\chi_\mu^*(q)\tilde{G}^{\mu\nu}(q)\chi_\nu(q) - \phi^\mu(q)^*\Pi_{\mu\nu}(q)\phi^\nu(q)\right.\right.$$
$$\left.\left. - B_\mu(q)^*\Pi^{\mu\nu}(q)B_\nu(q)\right]\right\}. \tag{8.64}$$

Notice how only the first term in the exponential depends on $\chi_\mu(q)$, and when we take the integral with respect to χ_μ the resulting term will amount to an inessential constant, that can be absorbed into the normalisation factor $\mathcal{N}$. If we then define

---

[6] We actually use the expression given in reference [26], rather than the one given in reference [33], since the former is more elegant and compact.

$$S[B] = -\int \frac{d^3q}{(2\pi)^3} \, [\phi^\mu(q)^*\Pi_{\mu\nu}(q)\phi^\nu(q) + B_\mu(q)^*\Pi^{\mu\nu}(q)B_\nu(q)], \qquad (8.65)$$

we can then write the partition function as

$$Z[B] = \exp\{-S[B]\}, \qquad (8.66)$$

which is in the same form as equation (8.42). If we use the definition of ϕ^μ in terms of the source term B_μ, i.e., $\phi^\mu = \Pi^{\mu\nu}B_\nu$, we see that the integrand in equation (8.65) can be written as $B(\Pi\,\Pi\,\Pi + \Pi)B$, i.e., the first term is higher-order in the polarisation tensor Π and can therefore be neglected[7], leaving

$$S[B] = -\int \frac{d^3q}{(2\pi)^3} \, [B_\mu(q)^*\Pi^{\mu\nu}(q)B_\nu(q)], \qquad (8.67)$$

from which we can get the final result

$$\frac{\partial^2 Z[B]}{\partial B_\mu(q)\partial B_\nu(q)} \Big|_{B=0} = \Pi^{\mu\nu}(q) = \big\langle \, j_a^\mu(q)j_a^\nu(q)\big\rangle, \qquad (8.68)$$

which is equivalent to equation (8.49). Hence, we have justified, by using both the method of effective action and direct computation of the fermionic and electromagnetic path integrals, that the current–current correlation function is, essentially, the polarisation tensor $\Pi^{\mu\nu}(q)$, i.e., the self-energy correction to the PQED photon propagator due to the presence of graphene electrons.

We are now ready to calculate the DC conductivity of graphene.

8.3.3 DC conductivity of graphene

To calculate the zero frequency (i.e., DC) conductivity of graphene, we use the Kubo formula (8.41), where we represent the current correlator with the polarisation tensor (8.61), since, as we have proven above, these two quantities are identified with each other. Let's look at the $\mathbf{k} \to 0$ limit first. The structure of equation (8.61) suggests that as the momentum goes to zero, the components Π^{00} and Π^{0j} of the polarisation tensor vanish, as they are proportional to either k^2 or k^j. Hence, only the components $\Pi^{j\ell}$ of the polarisation tensor remain nonzero in the limit of vanishing momentum, and these terms are then those defining the current–current correlation function, whose explicit expression is then given by[8]

$$\big\langle \, j_a^\ell(q)j_a^m(q)\big\rangle = \frac{N_f e^2}{16}\left[1 + \left(\frac{92 - 9\pi^2}{18\pi}\right)\frac{\alpha c}{v_f}\right]\left\{\frac{\delta^{\ell m}(\Omega^2 + k^2) - k^\ell k^m}{\sqrt{\Omega^2 + k^2}}\right.$$
$$\left. + \frac{N_f e^2}{2\pi}\left(n + \frac{1}{2}\right)\varepsilon^{\ell m0}\Omega\right\}, \qquad (8.69)$$

---

[7] This would correspond to higher order loop corrections to the photon self-energy.
[8] Notice that this expression, contrary to that in reference [26], is missing a term v_f multiplying the momentum $\mathbf{k}$. This is due to the fact, that here we are working with a scaled time coordinate $\tau = v_f t$.

where the term N_f comes from the summation over the flavour indices. Before continuing any further, it is useful to look at this term more closely. The number of flavours in graphene is given as the sum of two quantities: the electron spin (up or down), and the valley index, i.e., whether the electron finds itself in the vicinity of the Dirac point $\mathbf{K}$ or $\mathbf{K}'$. We can name these quantities N_{spin} and N_{valley}, respectively, so that $N_f = N_{\text{spin}} + N_{\text{valley}}$. Although one would at this point be tempted in saying that, according to this discussion, $N_f = 4$, this is not always the case, because of the peculiar properties of graphene valleys. Notice, however, that the two electron spins give the same contribution to the calculation of $\Pi^{\mu\nu}(q)$, and this $N_{\text{spin}} = 2$ always. For the valley contribution, things are more delicate. In normal situation, i.e., for pristine, undoped graphene, the two valleys are connected by time reversal symmetry, as can be seen by applying the time reversal operator $\hat{T}$[9] to the graphene Hamiltonian (8.19), obtaining

$$\hat{T}H_0(\mathbf{q}) = \hbar v_f(\sigma_x q_x + \xi \sigma_y^* q_y) = \hbar v_f[\sigma_x q_x + (-\xi)\sigma_y q y], \qquad (8.70)$$

where we have used the fact that $\sigma_y^* = -\sigma_y$. Applying time reversal operator to the graphene Hamiltonian in the valley ξ leads to the same Hamiltonian, but for the valley $-\xi$. Thus, time reversal symmetry is conserved, and the contribution of the two valleys is the same, setting $N_{\text{valley}} = 2$. When time reversal symmetry is conserved, we then have that $N_f = N_{\text{spin}} + N_{\text{valley}} = 4$.

As can be seen from equation (8.69), the current–current correlator contains a term proportional to Ω^2 (the first term in equation (8.69)) and a term proportional to Ω (the second term in equation (8.69)). If we want to test the symmetry of the correlator under time reversal, we first need to transform it back to real time. This is done by setting $\Omega = i\omega$ in equation (8.69). As a result of this, the term proportional to Ω^2 is manifestly invariant under time reversal symmetry, as $\hat{T}(i\omega)^2 = (-i\omega)^2 = (i\omega)^2$. The term proportional to Ω, on the other hand, is not, since $\hat{T}(i\omega) = -i\omega \neq i\omega$. If time reversal symmetry is conserved, as in the case of pristine graphene, the two valleys contribute in the same manner to the conductivity, and the second term will cancel out, since it will have opposite sign in the two valleys due to time reversal symmetry. In this case, then, the DC conductivity of graphene is given as

$$\sigma^{j\ell} = \lim_{\omega,\mathbf{k}\to 0} \frac{i}{\omega}(\langle j^\ell(\omega, \mathbf{k})j^m(\omega, \mathbf{k})\rangle\big|_{\text{valley } \mathbf{K}} + \langle j^\ell(\omega, \mathbf{k})j^m(\omega, \mathbf{k})\rangle\big|_{\text{valley } \mathbf{K}'})$$
$$= \frac{e^2}{4}\left[1 + \left(\frac{92 - 9\pi^2}{18\pi}\right)\frac{\alpha c}{v_f}\right]\delta^{\ell m}, \qquad (8.71)$$

which agrees with the results in reference [26] and with the standard literature on graphene conductivity, once a factor of $1/\hbar$[10] is restored.

---

[9] The time reversal operator essentially amounts to complex conjugation.

[10] That has been set equal to one, for convenience, in the expression of the partition function, i.e., $Z = \exp(-S/\hbar)$.

As discussed in reference [26], the non-time-reversal-symmetric part of the current contributes to the inter-valley current, ultimately generating a valley quantum Hall effect. It is worth pointing out, that the origin of this term (and, therefore, of the correspondent conductivity) is purely topological and reflects the topological nature of quantum Hall effect in graphene. We can derive the explicit expression of the DC valley conductivity as

$$
\begin{aligned}
\sigma^{j\ell} &= \lim_{\omega, \mathbf{k} \to 0} \frac{i}{\omega} \left(\langle \, j^{\ell}(\omega, \mathbf{k}) j^{m}(\omega, \mathbf{k}) \rangle \, \big|_{\text{valley } \mathbf{K}} - \langle \, j^{\ell}(\omega, \mathbf{k}) j^{m}(\omega, \mathbf{k}) \rangle \, \big|_{\text{valley } \mathbf{K}'} \right) \\
&= \frac{2e^2}{\pi} \left(n + \frac{1}{2} \right) \varepsilon^{\ell m}.
\end{aligned}
\tag{8.72}
$$

8.4 Extension to other materials and nonlinear properties

We conclude this chapter with a brief perspective on how the methods and techniques illustrated above can be useful to study different materials from graphene, and even more complicated light–matter interaction scenarios.

PQED is a very useful framework, not only to describe graphene, but in general for looking at the electronic and optical properties of any 2D material. The only constraint to using PQED in a more general case is that the material under investigation admits a Dirac-like description, i.e., a Lagrangian similar to that of equation (8.29). For example, adding a mass term to equation (8.29) would allow a description of gapped 2D materials, or considering four-component spinors ψ_a instead of two-component spinors would allow for the description of four-band systems, instead of two-band ones. These simple modifications already allow for a broader class of materials, including transition metal dichalcogenides and topological insulators, to be described within this framework.

From the gauge field side, adding topological terms, such as Chern–Simons terms for example, would allow for the inclusion of topological light–matter interaction. We have seen in the example above, that a Hall-like term already naturally emerges from the polarisation tensor due to the contribution of the vacuum gauge (i.e., electromagnetic) field fluctuations. Adding explicitly topological terms would also allow accounting for purely topological properties of the gauge field itself, thus opening the framework to encompass, for example, also structured light–matter interaction.

Finally, the calculations presented above are limited to the linear response, i.e., they assume linear interaction between light and matter. This can be seen in the PQED Lagrangian by the absence of self-interaction terms of the gauge field (i.e., optical nonlinearities), or terms containing higher order powers of the current (i.e., electron–electron nonlinearities). In addition to that, the expansion of the fermionic determinant we have used in the previous section to derive the polarisation tensor and relate it to the current–current correlation function only contains terms that are quadratic (and therefore linear) in the source terms. Higher order terms in the expansion of equation (8.56) might instead give access to higher-order correction to the conductivity and, eventually, to the nonlinear conductivity, which is normally a difficult quantity to calculate within the context of the Kubo formula. Finally, adding explicit electromagnetic nonlinearities, combined with the techniques described in chapter 6, might also lead to novel ways of describing nonlinear optics in 2D materials.

It is worth noticing that the topic presented in this chapter, and in general this way of looking at materials from a field theory perspective, is typical of condensed matter physics, as the books by Marino [24], Zaneen [19], Hartnoll [20] (and many others) clearly hint at. In this context, however, the properties of matter are very well discussed, but the interaction with light is typically always only relegated to the linear regime. Combining the framework introduced in this chapter with more photonics content, like for example bringing in optical nonlinearities, structured light, or topological properties of optical beams, could create a much richer framework, with high potential for discovering new things, or, at the very least, providing a unified description of light and matter.

References

[1] Novoselov K S, Geim A K, Morozov S, Jiang Y Z D, Dubonos S V, Grigorieva I V and Firsov A A 2004 Electric field effect in atomically thin carbon films *Science* **306** 666

[2] Autere A, Jussila H, Dai Y, Wang Y, Lipsanen H and Sun Z 2018 Nonlinear optics with 2D layered materials *Adv. Mater.* **30** 1705963

[3] Geim A K and Novoselov K S 2007 The rise of graphene *Nat. Mater.* **6** 183

[4] Castro Neto A H, Guinea F, Peres N M R, Novoselov K S and Geim A K 2009 The electronic properties of graphene *Rev. Mod. Phys.* **81** 109–62

[5] Tchoumakov S, Civelli M and Goerbig M O 2016 Magnetic-field-induced relativistic properties in type-I and type-II Weyl semimetals *Phys. Rev. Lett.* **117** 086402

[6] Soluyanov A A, Gresch D, Wang Z, Wu Q and Troyer M 2015 Type-II Weyl semimetals *Nature* **527** 495

[7] Sun Z, Hasan T, Torrisi F, Popa D, Privitera G, Wang F, Bonaccorso F, Basko D M and Ferrari A C 2010 Graphene mode-locked ultrafast laser *ACS Nano* **4** 803–10

[8] Sun Z, Hasan T, Torrisi F, Popa D, Privitera G, Wang F, Bonaccorso F, Basko D M and Ferrari A C 2010 Graphene mode-locked ultrafast laser *ACS Nano* **4** 803–10

[9] Hasan T, Sun Z, Wang F, Bonaccorso F, Tan P H, Rozhin A G and Ferrari A C 2009 Nanotube-polymer composites for ultrafast photonics *Adv. Mater.* **21** 3874–99

[10] Liu M, Yin X, Ulin-Avila E, Geng B, Zentgraf T, Ju L, Wang F and Zhang X 2011 A graphene-based broadband optical modulator *Nature* **474** 64

[11] Sun Z, Hasan T and Ferrari A C 2012 Ultrafast lasers mode-locked by nanotubes and graphene *Physica E: Low-Dimens. Syst. Nanostruct.* **44** 1082–91

[12] Sun Z, Martinez A and Wang F 2016 Optical modulators with 2D layered materials *Nat. Photon.* **10** 227

[13] Liu X, Guo Q and Qiu J 2017 Emerging low-dimensional materials for nonlinear optics and ultrafast photonics *Adv. Mater.* **29** 1605886

[14] Yu S, Wu X, Wang Y, Guo X and Tong L 2017 2D materials for optical modulation: challenges and opportunities *Adv. Mater.* **29** 1606128

[15] Katsnelson M I 2012 *Graphene–Carbon in Two Dimensions* (Cambridge: Cambridge University Press)

[16] Zhang Y W, ad Tan Y, Stormer H L and Kim P 2005 Experimental observation of the quantum Hall effect and Berry's phase in graphene *Nature* **438** 201

[17] Bolotin K I, Ghahari F, Shulman M D, Stormer H L and Kim P 2009 Observation of the fractional quantum Hall effect in graphene *Nature* **462** 196

[18] Cao Y, Fatemi V, Fang S, Watanabe K, Taniguchi T, Kaxiras E and Jarillo-Herrero P 2018 Unconventional superconductivity in magic-angle graphene superlattices *Nature* **556** 43

[19] Hartnoll S A, Lucas S and Sachdev S 2018 *Holographic Quantum Matter* (Cambridge, MA: MIT Press)

[20] Zaanen J, Liu Y and Sun Y W 2015 *Holographic Duality in Condensed Matter Physics* (Cambridge: Cambridge University Press)

[21] do Amaral R L P G and Marino E C 1992 Canonical quantisation of theories containing fractional powers of the daalembertian operator *J. Phys. A: Math. Gen.* **25** 5183

[22] Marino E C 1991 Complete bosonization of the Dirac fermion field in 2+1 dimensions *Phys. Lett.* B **263** 63

[23] Marino E C 1993 Quantum electrodynamics of particles on a plane and the Chern-Simons theory *Nucl. Phys.* B **408** 551

[24] Matrino E C 2017 *Quantum Field Theory Approach to Condensed Matter Physics* (Cambridge: Cambridge University Press)

[25] Teber S 2012 Electromagnetic current correlations in reduced quantum electrodynamics *Phys. Rev.* D **86** 025005

[26] Marino E C, Nascimento L O, Alves V S and Morias Smith C 2015 Interaction induced quantum valley Hall effect in graphene *Phys. Rev.* X **5** 011040

[27] Haug H and Koch S W 1994 *Quantum Theory of the Optical and Electronic Properties of Semiconductors* (Singapore: World Scientific)

[28] Messiah A 2014 *Quantum Mechanics* (Dover: New York)

[29] Srednicki M 2007 *Quantum Field Theory* (Cambridge: Cambridge University Press)

[30] Olver F W J, Lozier D W, Boisvert R F and Clarck C W (ed) 2010 *NIST Handbook of Mathematical Functions* (Cambridge: Cambridge University Press)

[31] Kubo R 1957 Statistical-mechanical theory of irreversible processes. I. General theory and simple applications to magnetic and conduction problems *J. Phys. Soc. Jpn.* **12** 570

[32] Peskin M E and Schroeder D V 2019 *An Introduction to Quantum Field Theory* (Boca Raton, FL: CRC Press)

[33] Kotikov A V and Teber S 2014 Two-loop fermion self-energy in reduced quantum electrodynamics and application to the ultrarelativistic limit of graphene *Phys. Rev.* D **89** 065038

www.ingramcontent.com/pod-product-compliance
Lightning Source LLC
Chambersburg PA
CBHW080525220326

41599CB00032B/6203